T0181617

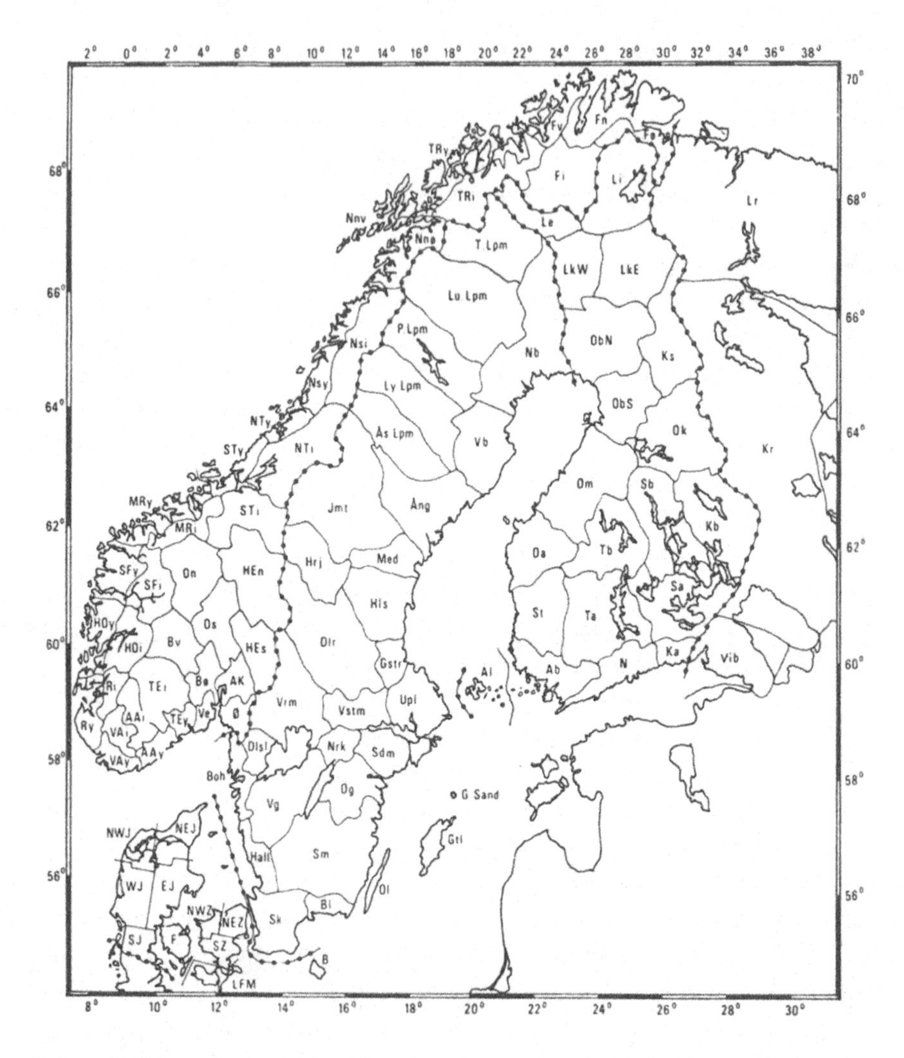

List of abbreviations for the provinces used throughout the text, on the map and in the following tables.

DENMARK

SJ	South Jutland	LFM	Lolland, Falster, Møn
EJ	East Jutland	SZ	South Zealand
WJ	West Jutland	NWZ	North West Zealand
NWJ	North West Jutland	NEZ	North East Zealand
NEJ	North East Jutland	B	Bornholm
F	Funen		

SWEDEN

Sk.	Skåne
Bl.	Blekinge
Hall.	Halland
Sm.	Småland
Öl.	Öland
Gtl.	Gotland
G. Sand.	Gotska Sandön
Ög.	Östergötland
Vg.	Västergötland
Boh.	Bohuslän
Dlsl.	Dalsland
Nrk.	Närke
Sdm.	Södermanland
Upl.	Uppland
Vstm.	Västmanland
Vrm.	Värmland
Dlr.	Dalarna
Gstr.	Gästrikland
Hls.	Hälsingland
Med.	Medelpad
Hrj.	Härjedalen
Jmt.	Jämtland
Ång.	Ångermanland
Vb.	Västerbotten
Nb.	Norrbotten
Ås. Lpm.	Åsele Lappmark
Ly. Lpm.	Lycksele Lappmark
P. Lpm.	Pite Lappmark
Lu. Lpm.	Lule Lappmark
T. Lpm.	Torne Lappmark

NORWAY

Ø	Østfold
AK	Akershus
HE	Hedmark
O	Opland
B	Buskerud
VE	Vestfold
TE	Telemark
AA	Aust-Agder
VA	Vest-Agder
R	Rogaland
HO	Hordaland
SF	Sogn og Fjordane
MR	Møre og Romsdal
ST	Sør-Trøndelag
NT	Nord-Trøndelag
Ns	southern Nordland
Nn	northern Nordland
TR	Troms
F	Finnmark

n northern s southern ø eastern v western y outer i inner

FINLAND

Al	Alandia
Ab	Regio aboensis
N	Nylandia
Ka	Karelia australis
St	Satakunta
Ta	Tavastia australis
Sa	Savonia australis
Oa	Cstrobottnia australis
Tb	Tavastia borealis
Sb	Savonia borealis
Kb	Karelia borealis
Om	Ostrobottnia media
Ok	Ostrobottnia kajanensis
ObS	Ostrobottnia borealis, S part
ObN	Ostrobottnia borealis, N part
Ks	Kuusamo
LkW	Lapponia kemensis, W part
LkE	Lapponia kemensis, E part
Li	Lapponia inarensis
Le	Lapponia enontekiensis

USSR

Vib	Regio Viburgensis
Kr	Karelia rossica
Lr	Lapponia rossica

FAUNA ENTOMOLOGICA SCANDINAVICA
Volume 12 1983

The Empidoidea (Diptera) of Fennoscandia and Denmark

II

General Part. The Families Hybotidae, Atelestidae and Microphoridae

by

M. Chvála

SCANDINAVIAN SCIENCE PRESS LTD.
Copenhagen

Fauna entomologica scandinavica
is edited by "Societas entomologica scandinavica"

World list abbreviation
Fauna ent. scand.

Printed by
Vinderup Bogtrykkeri A/S
7830 Vinderup, Denmark

ISBN 87-87491-07-9
ISSN 0106-8377

Contents

Introduction

The present volume of 'Fauna entomologica scandinavica' is a direct continuation of volume 3 of this series, which appeared in 1975 and dealt with the subfamily Tachydromiinae of the "Empididae". However, when compared with the preceding 'tachydromiine' part, the present volume will be seen to differ in two important and closely-linked aspects – the title and the rather detailed General part. One new feature of this volume is the division of the former "Empididae" into four distinct families. This has given rise to the superfamily name Empidoidea in the title and to the detailed General part which, I hope, provides adequate evidence for the proposed new classification, based on the comparative holomorphological method, and for the presumed phylogeny of the 'higher' Brachycera.

In fact, the classification suggested here is hardly new, as several Dipterists have repeatedly pointed out during the last twenty years that the "Empididae" is an unnatural paraphyletic unit. In the classification suggested here the family Empididae is restricted to the "Empidoinea-group" of subfamilies sensu Hennig (1970), and the "Ocydromioinea-group" is split into the families Hybotidae, Atelestidae and Microphoridae. The four families of the former "Empididae" and the Dolichopodidae together form the superfamily Empidoidea. For this reason, it has been necessary to change the family name "Empididae" in the title for the superfamily name Empidoidea.

The present volume covers the Scandinavian species of the former Ocydromioinea-group of subfamilies, i.e. the Hybotinae and Ocydromiinae of the Hybotidae (the Tachydromiinae were treated in volume 3 of the series), the Atelestidae, and the Microphoridae. The next volumes in this series on the former "Empididae", which are already in preparation, will deal with the true Empididae, as currently defined, that is to say the subfamilies Oreogetoninae, Empidinae, Brachystomatinae, Hemerodromiinae and Clinocerinae (cf. Fig. 141).

The material is arranged in the same way as in the 'tachydromiine' volume. Each species begins with the valid name followed by the original combination; then the synonyms are listed chronologically with their bibliographical references, followed by a short diagnosis and a description to ensure correct differentiation from related species. The distribution in Fennoscandia and Denmark, the general distribution, and the flight-period of each species are briefly outlined, whilst more detailed information on the distribution in Scandinavia and adjacent countries is given in the 'Catalogue' at the end of the volume. Notes on biology, immature stages and habitats of the species are usually summarised in the generic diagnosis. The main diagnostic features for each species are illustrated, including the male genitalia which are often the decisive character for exact identification. As these three families include forms that are morphologically very diverse, each genus begins with a habitus drawing, generally of the type-species. Species which might occur in Scandinavia, or species closely related to the known Scandinavian species, are also mentioned and briefly diagnosed. A study of all the available type-material was an essential part of the present work, and new synonymies, lectotype designations, holotype identifications and other notes on type-material are given in a separate paragraph.

The families, subfamilies and genera treated are briefly diagnosed, with notes on classification and distribution, and keys for the identification of genera and species are also given.

As already mentioned, the General part is rather detailed because of the proposed new classification of the former "Empididae". The section on adult morphology gives more detailed information on the structure of antennae and mouthparts, the wing-venation, and the structure of abdomen and external genitalia. Further sections deal with the systematic position of the Empidoidea within the Brachycera, a review of previous classifications of the former "Empididae", and the extinct forms of the Empidoidea. All these sections, and particularly the one on comparative morphology, lead on to the sections on the presumed phylogeny and the new classification of the Empidoidea. The life history, phenology, feeding, swarming and mating habits of the adults are briefly summarised. The section on zoogeography only deals with the groups covered in the present volume (Hybotidae, Atelestidae and Microphoridae), and the presentation is the same as in the 'tachydromiine' volume. Other sections of the General part contain information on all the subfamilies of the former "Empididae", and therefore include the family Empididae of the new classification. The Dolichopodidae are not included as I am not at all qualified to offer comparative data on the morphology, phylogeny and biology of that family.

Most of the data given in the General part were very briefly mentioned in my paper 'Classification and phylogeny of Empididae, with a presumed origin of Dolichopodidae' (Chvála, 1981a), which may be regarded as a first step leading to the new classification as presented here. However, the attentive reader will find some minor discrepancies in details of morphology between that paper and the present discussion, as for instance in the male genitalia of the Empidinae and Hemerodromiinae or in the mouthparts of the Clinocerinae, where my original analysis was based mainly on published records. A number of errors contained in the papers of previous authors were the result of incorrect illustrations, inaccurate observations, or even misidentification of the species examined. For example, Bährmann (1960) based his observations on the supposed relationship of the Clinocerinae and Microphorinae mainly on *Heleodromia curtipes* (Beck.) (Clinocerinae), which is in fact a species of the genus *Microphorella* (Microphoridae). All the data on morphology given in the present volume are based exclusively on my own studies of the material involved, and all published records have been verified by new dissections. Since any new classification must be based on the world fauna, I have also studied a number of non-Palaearctic elements, including some groups of the former "Empididae" which have been very little known until now, such as the Australian and South American genera *Ceratomerus* (Ceratomerinae) and *Homalocnemis* (Brachystomatinae).

In the text that follows, the area of Fennoscandia and Denmark, which includes Norway, Sweden, Finland and Denmark, is called Scandinavia for brevity, and the species inhabiting this area are called Scandinavian species. A distribution in Norway and Sweden is referred to as a distribution in the Scandinavian Peninsula, and Fennoscandia, as used in the paragraphs on distribution, covers the Scandinavian Peninsula, Finland and the adjacent parts of the European USSR.

Acknowledgements

The author wishes to thank to the following colleagues for the loan or donation of material and/or literature, and for very valuable information on collections or localities: Dr. P. H. Arnaud, Jr. (San Francisco), Dr. P. Grootaert (Brussels), the late Prof. W. Hennig (Ludwigsburg), Dr. L. V. Knutson (Washington), Dr. V. G. Kovalev (Moscow), Dr. A. Lillehammer (Oslo), Dr. A. Løken (Bergen), Dr. T. Saigusa (Fukuoka), Mr. S. P. Schembri (Birkirkara), Mr. K. G. V. Smith (London), Dr. B. Stuckenberg (Pietermaritzburg), Dr. W. J. Turner (Pullman), Dr. H. Ulrich (Bonn), Dr. J. R. Vockeroth (Ottawa), and Dr. D. D. Wilder (San Francisco, now Washington). I am also very indebted to many colleagues in Czechoslovakia for the loan or donation of material for study: Dr. M. Barták (Praha), Dr. J. Čepelák (Nitra), Dr. J. Ježek (National Museum, Praha), Dr. P. Lauterer and Dr. J. Stehlík (Moravian Museum, Brno), the late Mr. J. Macek (Praha), Dr. J. Olejníček (České Budějovice), Dr. R. Rozkošný(Brno), Ing. K. Spitzer (České Budějovice). Dr. V. Straka (Martin), Dr. O. Syrovátka (Praha), Dr. J. Vaňhara (Brno), and many others.

Dr. H. Schumann (Berlin) kindly enabled me to study the important collections of T. Becker, H. Loew and O. Duda in the Zoologisches Museum, Berlin, and Prof. G. Morge (Eberswalde) kindly enabled me to study the material of the former Deutsches Entomologisches Institut, Eberswalde, and arranged for the loan of type-material from the Gabriel Strobl collection at Admont, Austria.

I am especially grateful to Dr. Leif Lyneborg (Copenhagen), Dr. Hugo Andersson (Lund) and Prof. Walter Hackman (Helsinki) for their very valuable help with the material under their care, as well as for their kind hospitality during my visits to Scandinavia. Nor must I overlook the very important support given to my studies by Prof. S. L. Tuxen (Copenhagen) and the late Prof. C. H. Lindroth (Lund).

I am very indebted to Dr. G. C. D. Griffiths (Edmonton) for many valuable suggestions during the preparation of the parts on phylogeny and classification and for his critical reading of some parts of the manuscript. Mr. Adrian C. Pont of the British Museum (Nat. Hist.), London, has kindly checked the English of the manuscript and offered many valuable suggestions. Finally, I am most indebted to Leif Lyneborg, the chief editor of this volume, for his enthusiastic support and for many suggestions offered during the preparations of the manuscript for publication.

The Swedish National Science Research Council offered me a grant for my visits to Lund in 1977 and Helsinki in 1980, as well as for my studies of other Scandinavian material. The Czechoslovak Ministry of Education arranged a one-month stay in Berlin in 1981, within the framework of the cultural agreement between the Charles University, Praha, and the Humboldt University, Berlin.

Material and methods

Some 7,000 specimens from Scandinavia have been examined, and at least the same number from other parts of the world, mainly from Europe. The main Scandinavian collections studied were those of the Universitetets Zoologiske Museum, Copenhagen (coll. Lundbeck, Staeger), the Zoologiska Institutionen, Lund (coll. Zetterstedt, Roth, Wallengren, Ringdahl, Andersson) and the University Museum, Helsinki (coll. Frey, Finnish Collection). The rather sparse material from Norway is scattered throughout various Scandinavian collections, as well as Berlin, London and Ottawa; in addition to the small but well-preserved Siebke collection in Oslo, there is a large alcohol collection at Bergen which has been built up during the last ten years by the Hardangervidda IBP project. However, Norway still remains the most poorly investigated country within Scandinavia, as it may be seen from the 'Catalogue'. All Scandinavian collections, except for the Hardangervidda material, consist of dry specimens.

Before dissection, specimens were first relaxed for a short time in a humidified container. When dissecting the genitalia, the tip of abdomen (including two or three pregenital segments) was removed with a special optical surgical scissors and boiled for 5 to 15 minutes (according to size and the degree of sclerotisation) in a 10% solution of potassium hydroxide. The macerated genitalia were studied in a drop of glycerol on a slide without coverslip, and after study were preserved in glass or plastic microvials with glycerol, which were attached to the pinned specimens. Other parts (antennae, mouthparts, legs, wings) were mounted in Swan's medium on permanent slides.

Drawings were prepared by means of a built-in grid in the binocular microscope 'Cytoplast SMX' (Carl Zeiss Jena) or, in the case of permanent slides, the drawings were copied on to a tracing paper from the screen of the microscope 'Visopan' (Reichert, Austria). Mouthparts were first put in a drop of glycerol on slides without coverslip for general appearance drawings, and only after dissection were mounted on permanent slides for detailed drawings with the screen microscope. The length of specimens was measured from the tip of the head (without antennae) to the tip of the terminalia.

Abbreviations used in the descriptions:

ad	anterodorsal	acr	acrostichal bristles
av	anteroventral	dc	dorsocentral bristles
pd	posterodorsal	ant.s.	antennal segment
pv	posteroventral		

Abbreviations used in figures:

(genitalia)

AE	aedeagus
A.AP	aedeagal apodeme
CE	cercus
EP	epandrium
GCX	gonocoxite
GNP	gonopods
GST	gonostyle
HY	hypandrium
H.BR	hypandrial bridge
PE	periandrium
P.FD	periandrial fold
PGN	postgonite
1 - 10	terga
I - X	sterna

(mouthparts)

Cib	cibarium
Cl	clypeus
Hyp	hypopharynx
Lac	lacinia
Lb	labium
Lbl	labella
Lbr	labrum
Mx	maxilla
Mx.p	maxillary palp
Pf	palpifer
St	stipes
Tor	tormae

GENERAL PART

Adult morphology of the Empidoidea

Head

Head moderately large, generally rounded, or deeper than either long or wide, rarely with lower facial part remarkably produced (some Clinocerinae and Tachydromiinae). Tentorium not developed; the only inner sclerites of the head capsule are the strongly developed and heavily sclerotised proximal sclerites of the mouthparts, the tormae (attached to clypeus) and cibarium. The connection with the neck is usually firm, but rather weak particularly in the Ocydromiinae; rarely the head is closely set upon the thorax with the neck connection indistinct, as in some Tachydromiinae (Drapetini) or Microphoridae *(Microphorella)*. Occiput is usually flat, sometimes slightly convex or even concave, often below. The neck is placed at about middle of head in profile, rather high up in some Clinocerinae and Microphoridae (Parathalassiini), very conspicuously so in *Dolichocephala* (Clinocerinae), where it is almost level with vertex.

Head mainly occupied by the very large compound eyes which are generally bare, microscopically pubescent in the Drapetini (Tachydromiinae), Parathalassiini (Microphoridae) and in most of the Clinocerinae. It is suggested that the plesiomorphous condition within Empidoidea is represented by holoptic eyes in males and dichoptic eyes in females. This condition is best preserved in the more primitive Empididae, in the Oreogetoninae and Empidinae (although there are many modifications, as in *Hilara* or some subgenera of *Empis* and *Rhamphomyia*), and in the

Atelestidae. The holoptic condition in males is well-preserved in those groups where aerial swarming and mating are common habits. Such sexual dimorphism may also be found in other groups, although not so distinctly. On the other hand, in the predaceous groups of the Empididae (Hemerodromiinae, Clinocerinae) and Hybotidae (Hybotinae, Tachydromiinae, or *Bicellaria* of the Ocydromiinae) the arrangement of the eyes is the same in both sexes, either holoptic or narrowly dichoptic. In highly predaceous forms, the eyes also meet on the facial part below antennae.

Holopticism is usually combined with the enlargement of some eye-facets. The large facets (usually on the upper part of the eyes) facilitate the recognition of moving objects, either the prey or the females in aerial mating swarms. If both sexes are predaceous, there is no sexual dimorphism and the specialised large facets in both sexes are either on the upper half of the eyes (most of the species capturing their prey in flight – Hybotinae, *Bicellaria*), or anteriorly (some Hemerodromiinae), or even below the antennae – when prey is captured on a solid substrate, such as vegetation or the water-surface, as in some Hemerodromiinae, Ocydromiinae (*Oropezella*) and Tachydromiinae (*Symballophthalmus, Tachyempis, Stilpon*). Large facets are present only on some areas of holoptic eyes; if the predaceous species have dichoptic eyes (Clinocerinae, some Ocydromiinae and Tachydromiinae, for instance *Platypalpus*!), then the facets are always all uniformly small, of the 'apposition' type.

It may be concluded that in recent Empidoidea the holoptic or dichoptic condition has no particular value for phylogeny as it represents a relatively recent adaptation in the mating, swarming and feeding habits of particular groups of species, and it is more important for classification at the generic or subgeneric level. In other words, holopticism and dichopticism (to various degrees) in the Empidoidea may well be due to differences in function that have only slight evolutionary significance.

Three ocelli are always present, in dichoptic eyes on the upper part of frons, in holoptic eyes on a distinct ocellar tubercle at vertex. Two pairs of ocellar bristles more or less developed, the anterior pair just behind anterior ocellus, the posterior pair between posterior ocelli. Vertex and occiput either with numerous undifferentiated hairs (a plesiomorphous condition), or with 1 to 2 pairs of distinct vertical bristles and distinct occipital bristles arranged in rows (an apomorphous condition).

Antennae

Antennae (Figs. 1–19) always 3-segmented, inserted close together, usually in a slight concavity on lower edge of frons (or frontal triangle in holoptic eyes) and between the more or less distinct triangular antennal eye-excisions; antennal sockets not developed. Antennae usually inserted at about middle of head in profile or slightly lower (especially in males with holoptic eyes), rarely above middle of head, as in some Clinocerinae and Microphoridae (tribe Parathalassiini – a condition leading to the Dolichopodidae), or much above middle of head in profile, as in the genera *Oropezella* (Ocydromiinae) or *Afroempis* (Empidinae).

First antennal segment distinctly bristled in the Empididae, the bristles rarely reduced (some Clinocerinae and Hemerodromiinae), very exceptionally (*Rhagas* of the

Oreogetoninae) 1st segment is nearly bare. This segment is long in the ground-plan of the Empididae, generally much longer than 2nd segment, although rather shorter in some Oreogetoninae, Hemerodromiinae, Clinocerinae and Brachystomatinae. In other families of the former "Empididae", in the Hybotidae, Atelestidae and Microphoridae,

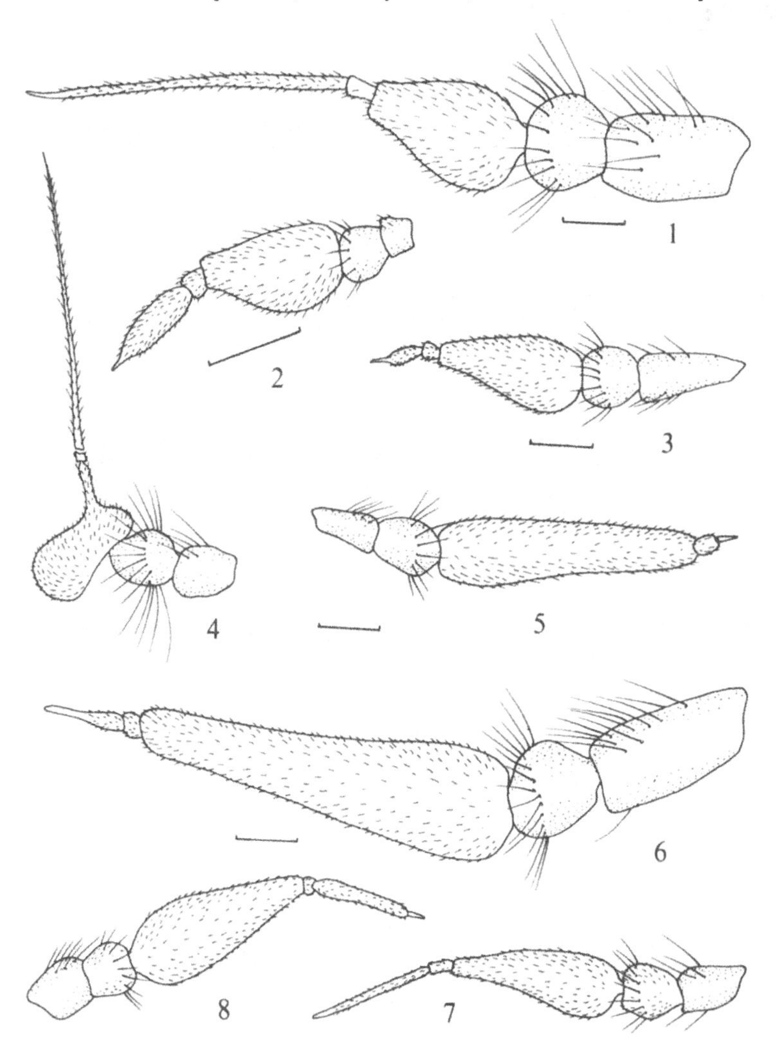

Figs. 1–8. Antennae of Empididae: Oreogetoninae and Empidinae. – 1: *Oreogeton basalis* (Loew); 2: *Rhagas unica* Walk.; 3: *Hesperempis mabelae* (Mel.); 4: *Gloma fuscipennis* Meig.; 5: *Anthepiscopus ribesii* Beck.; 6: *Rhamphomyia sulcata* (Meig.); 7: *Empis pennipes* L.; 8: *Hilara maura* (Fabr.). Scale: 0.1 mm.

1st segment is always very small, devoid of distinct bristles, covered at most with microscopic pile. In these three non-empidid families, 1st segment is usually rather poorly differentiated from the 2nd segment (as if both segments are partly fused), which led Melander (1928) to describe the antennae in *Leptopeza, Meghyperus, Microphorella* and others as "two-jointed".

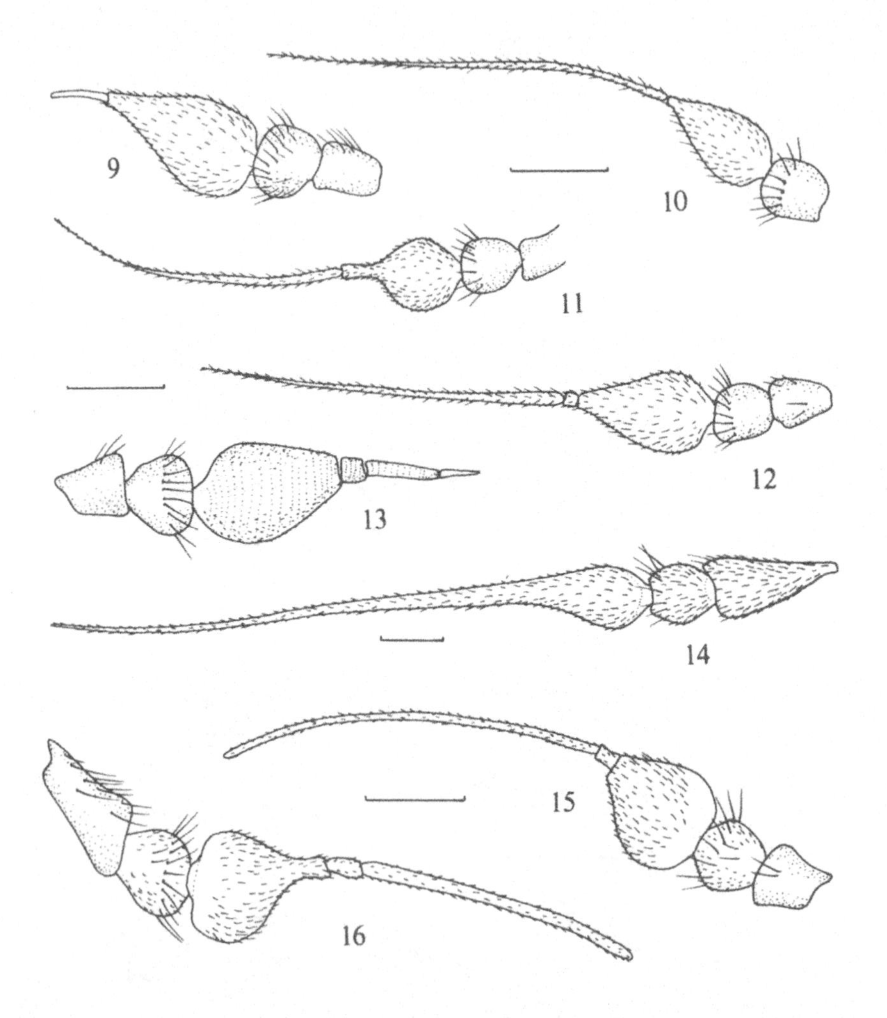

Figs. 9–16. Antennae of Empididae: Hemerodromiinae and Clinocerinae. – 9: *Chelifera precatoria* (Fall.); 10: *Chelipoda vocatoria* (Fall.); 11: *Heleodromia immaculata* Hal.; 12: *Phyllodromia melanocephala* (Fabr.); 13: *Dryodromia testacea* Rond.; 14: *Trichopeza longicornis* (Meig.); 15: *Clinocera* sp.; 16: *Wiedemannia bistigma* (Curtis). Scale: 0.1 mm.

Second antennal segment is always short and bristled, in the Hybotidae, Atelestidae and Microphoridae rather globular and armed with a circlet of distinct preapical bristles.

Third antennal segment varies considerably in shape and length, and its structure is a useful diagnostic character at the generic and specific levels. The ground-plan condition of the Empidoidea is a moderately long conical segment with a terminal style that is at most equally long (Figs. 6–8), as in some Oreogetoninae *(Rhagas, Hesperempis)*, most Empidinae, some Hemerodromiinae *(Hemerodromia, Chelifera)* and Clinocerinae *(Dryodromia)*, Brachystomatinae, some Ocydromiinae (Trichinini), in the primitive forms of the Tachydromiinae *(Symballophthalmus*, some *Platypalpus)*, and in the Microphoridae. This type of antenna is thought to represent a plesiomorphous state from which the antennae evolved in two different directions: by the lengthening of 3rd segment with parallel shortening of the style (Fig. 5), and by shortening of the segment with the style becoming much longer, arista-like (Fig. 10). It is suggested that the first evolutionary direction (lengthening of 3rd segment) was a non-progressive one, as it has developed in only a few isolated groups of the Empididae and Hybotidae: in the Oreogetoninae *(Anthepiscopus, Iteaphila)*, Empidinae *(Atrichopleura, Afroempis*, in *Empis* subgenera *Xanthempis, Lissempis, Anacrostichus*, and in *Rhamphomyia* subgenus *Lundstroemiella)*, Clinocerinae *(Niphogenia, Ceratempis)* and in the Ceratomerinae; in the Hybotidae only in the tribe Oedaleini of the Ocydromiinae and in a few *Platypalpus* species (Tachydromiinae).

On the other hand, a very short 3rd antennal segment with a conspicuously long filiform (arista-like) style seems to be a much commoner apomorphous state within the

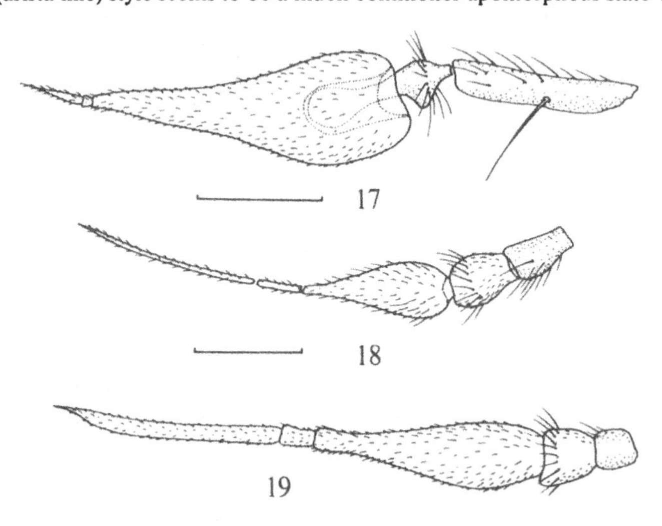

Figs. 17–19. Antennae of Empididae: Ceratomerinae and Brachystomatinae. – 17: *Ceratomerus paradoxus* Phil. ♂; 18: *Brachystoma pleurale* Frey ♀; 19: *Homalocnemis perspicuus* (Hutton) ♂. Scale: 0.2 mm.

Empidoidea. This type of antenna is rather rare in the primitive forms of the Empididae, only in *Oreogeton* and *Gloma* (Oreogetoninae), and the curious *Empis* subgenus *Rhadinempis* (Empidinae). However, this condition is common in the Hemerodromiinae and Clinocerinae (in the specialised forms of the Empididae), practically in all Hybotidae (tribe Ocydromiini of the Ocydromiinae, all Hybotinae and most of Tachydromiinae), in the Atelestidae and some Parathalassiini (Microphoridae). Abnormally-shaped antennae, deviating from the usual type, have developed in *Gloma* (Fig. 4), where 3rd segment is somewhat kidney-shaped with dorsal style, and in *Ceratomerus* (Fig. 17), where 3rd segment is long, flattened and somewhat lanceolate, somewhat resembling the curious antennae of grasshoppers of the genus *Acrida*.

The antennal style is generally terminal, slightly supra-terminal in some Tachydromiinae (*Chersodromia, Stilpon*), rarely sub-dorsal or almost dorsal, as in *Gloma* (Empididae) and *Ocydromia* (Hybotidae). The length of the style is generally the inverse of the length of 3rd antennal segment: the shorter the segment, the longer and rather arista-like the style. Rarely, the style is conspicuously stout, as stout as 3rd segment at

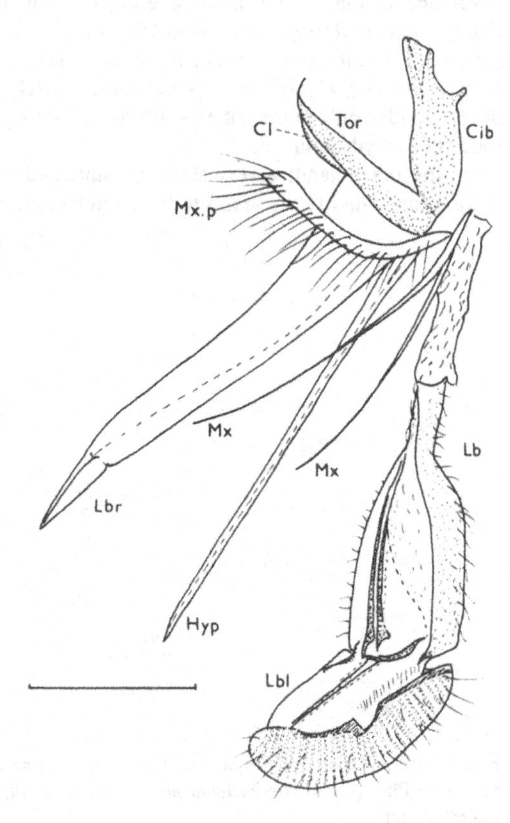

Fig. 20. Mouthparts of *Rhamphomyia sulcata* (Meig.) ♂, (Empidinae). Scale: 0.5 mm. For abbreviations see p. 13.

tip, as in *Rhagas* (Fig. 2) (Oreogetoninae), some *Trichina* (Fig. 233) and *Oedalea* (Figs. 386–391) (Ocydromiinae), and a few *Platypalpus* species (Tachydromiinae).

The style may even be completely absent, as in a few species of the tribe Oedaleini (Ocydromiinae), in the genus *Allanthalia* (Figs. 478, 479) and apparently some *Euthyneura* species. In *Trichopeza* (Clinocerinae), the 3rd antennal segment is conspicuously prolonged into a long style-like terminal part but no real style is differentiated (Fig. 14). The related Nearctic genera *Niphogenia* and *Ceratempis*, which I did not study, may also be of this type.

The antennal style is characteristically articulated, a feature used for instance by Hennig (1972) for the classification of the Brachycera. (Instead of 'segmentation' of the style, the term 'articulation' is consistently used in the following text, to prevent any confusion of antennal segments with articles of the style.) The ground-plan condition of the Empidoidea is a 2-articulated style, with a small basal article (Fig. 1). The only exception is the Brachystomatinae *(Brachystoma, Homalocnemis)* where the basal article is rather long (Figs. 18, 19). A 1-articulated style (the small basal article absent) is present, so far as I am aware, in *Hemerodromia, Chelifera* and *Chelipoda* (Hemerodromiinae), in *Heleodromia* (Clinocerinae), *Ocydromia* (Ocydromiinae) and the tribe Parathalassiini of the Microphoridae. The 1-articulated style of *Anthalia* (Fig. 477) and of some

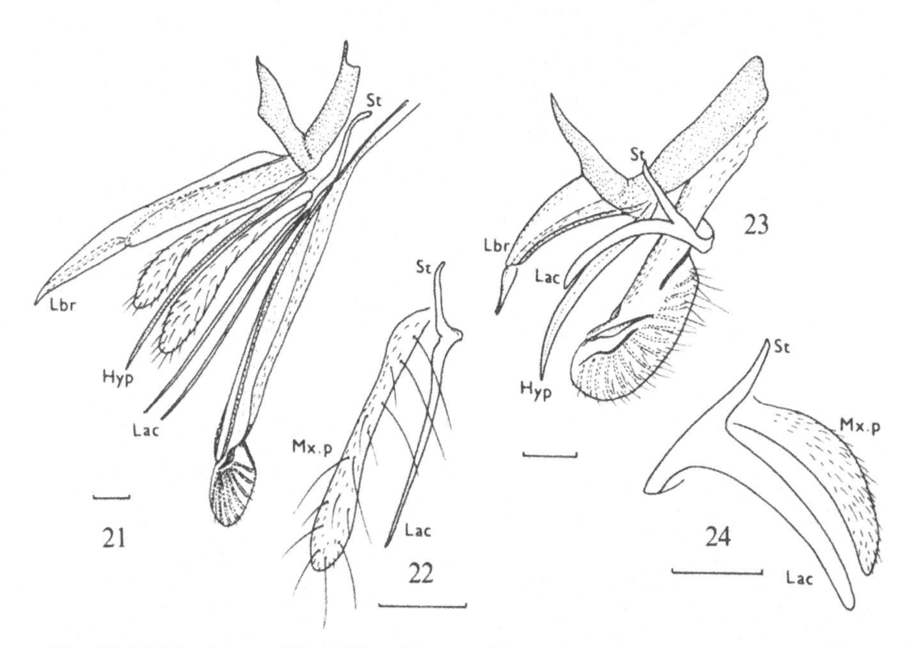

Figs. 21–24. Mouthparts of Empididae: Oreogetoniinae and Brachystomatinae. – 21: *Anthepiscopus ribesii* Beck. ♀; 22: *Gloma fuscipennis* Meig. ♂; 23–24: *Brachystoma pleurale* Frey ♀. Scale: 0.1 mm. For abbreviations see p. 13.

Euthyneura (Fig. 456) (both Ocydromiinae) is undoubtedly of a different origin: it represents an intermediate step, leading from a very short style to its entire loss as in *Allanthalia* (Fig. 478).

The 3-articulated condition, with two small basal articles as in all Cyclorrhapha (except Opetiidae), is present only in the Atelestidae, in the Nearctic *Meghyperus* species (Fig. 560) so far as I am aware. The 3-articulated styles of *Hesperempis* (Fig. 3) and *Rhagas* (Fig. 2) of the Oreogetoninae, *Hilara* (Fig. 8) of the Empidinae and *Dryodromia* (Fig. 13) of the Clinocerinae (all Empididae) differ substantially from the cyclorrhaphous condition, as the 3rd article is formed by a bare apical part of the style, and only one small basal article is present as in the 2-articulated state.

Mouthparts

Proboscis pointing downwards in primitive forms and of varying lengths, in some Empidinae conspicuously long, often several times longer than head is deep. In the highly

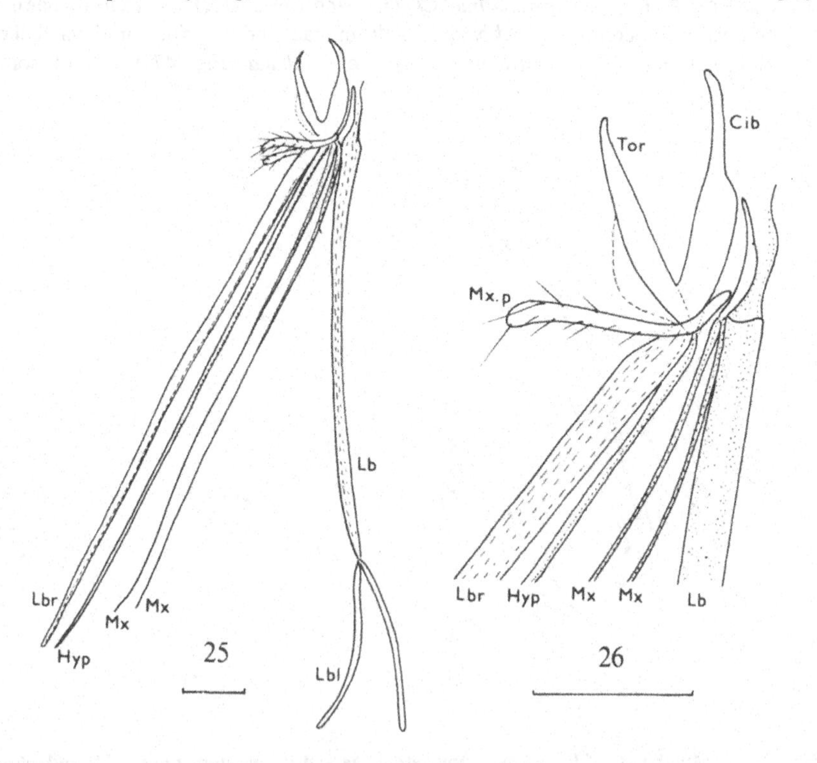

Figs. 25, 26. Mouthparts of *Empis pennipes* L. ♂, (Empidinae). Scale: 0.3 mm. For abbreviations see p. 13.

predaceous species the proboscis is generally shorter and stronger, sometimes directed backwards (Hemerodromiinae and Drapetini of the Tachydromiinae). There is a tendency for the proboscis to be directed obliquely forwards in the Oreogetoninae, as well as *Homalocnemis* (Brachystomatinae) and in the Oedaleini (Ocydromiinae). In the very specialised forms that hunt other insects in flight (Hybotinae), the proboscis points forwards, rather as in the Atelestidae and in *Microphorus* (Microphoridae).

The primitive structure of the mouthparts in the Empidoidea is shown in Fig. 20. The labrum (Lbr) is the most sclerotised part of the mouthparts, its length indicating the length of proboscis. The hypopharynx (Hyp) is stylet-like, always heavily sclerotised, attached close to labrum and generally as long as labrum. The maxillae (Mx) are in the form of thin blades or stylets, and consist of short stipites (St) and longer laciniae (Lac) in primitive species, closely attached to labium proximally. In most of the Oreogetoninae (Fig. 21) and Empidinae (Fig. 25) the laciniae are long stylet-like blades subequal to labrum and hypopharynx, shorter in *Rhamphomyia* (Fig. 20), but there is a tendency for them to be reduced in other groups. The maxillary palps (Mx.p) are always 1-segmented, ovate to cylindrical and more or less bristled, exceedingly long and slender in *Iteaphila* and *Anthepiscopus* (Oreogetoninae). In the ground-plan, the palpi are firmly attached to the maxillae, originating near tip of stipites. The labium (Lb) is about as long as labrum, but if porrect (due to its membraneous flexible basal

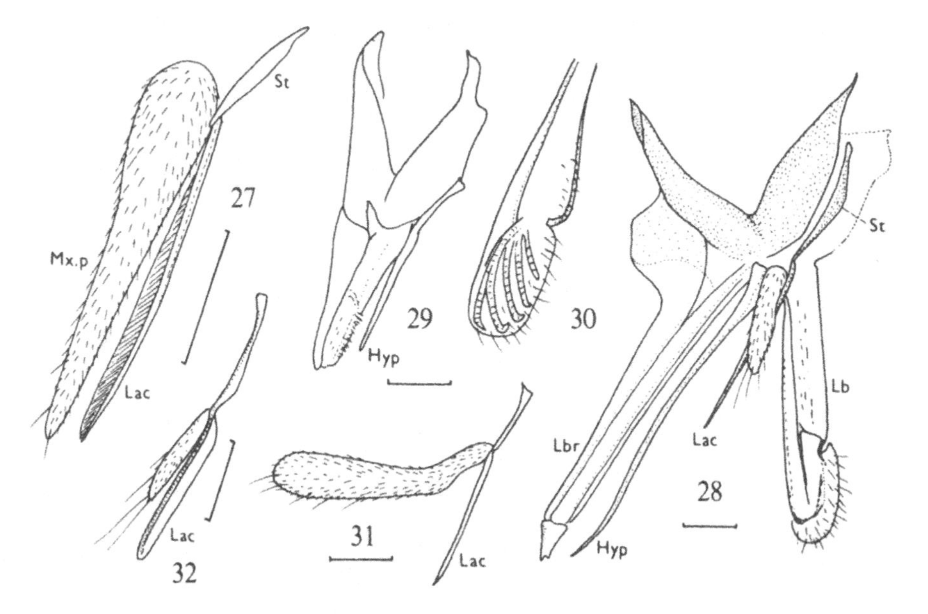

Figs. 27-32. Mouthparts of Empididae: Hemerodromiinae and Clinocerinae. - 27: *Chelifera precatoria* (Fall.) ♂; 28: *Trichopeza longicornis* (Meig.) ♀; 29-31: *Dryodromia testacea* Rond. ♂; 32: *Heleodromia immaculata* Hal. ♀. Scale: 0.1 mm. For abbreviations see p. 13.

part) it is often clearly longer than labrum. The labial palps form more or less large labellae (Lbl), usually with pseudotracheae. In *Empis* (Fig. 25) the labellae are long and slender, rather sclerotised and without pseudotracheae; a slender and remarkably sclerotised labium with scarcely differentiated labellae is very characteristic of the Hybotinae (Figs. 152, 155).

Completely developed mouthparts are present in the Empididae, Atelestidae and Microphoridae, although there are some adaptations, especially in the reduction of maxillae, in the specialised forms of the Empididae and Microphoridae. The general

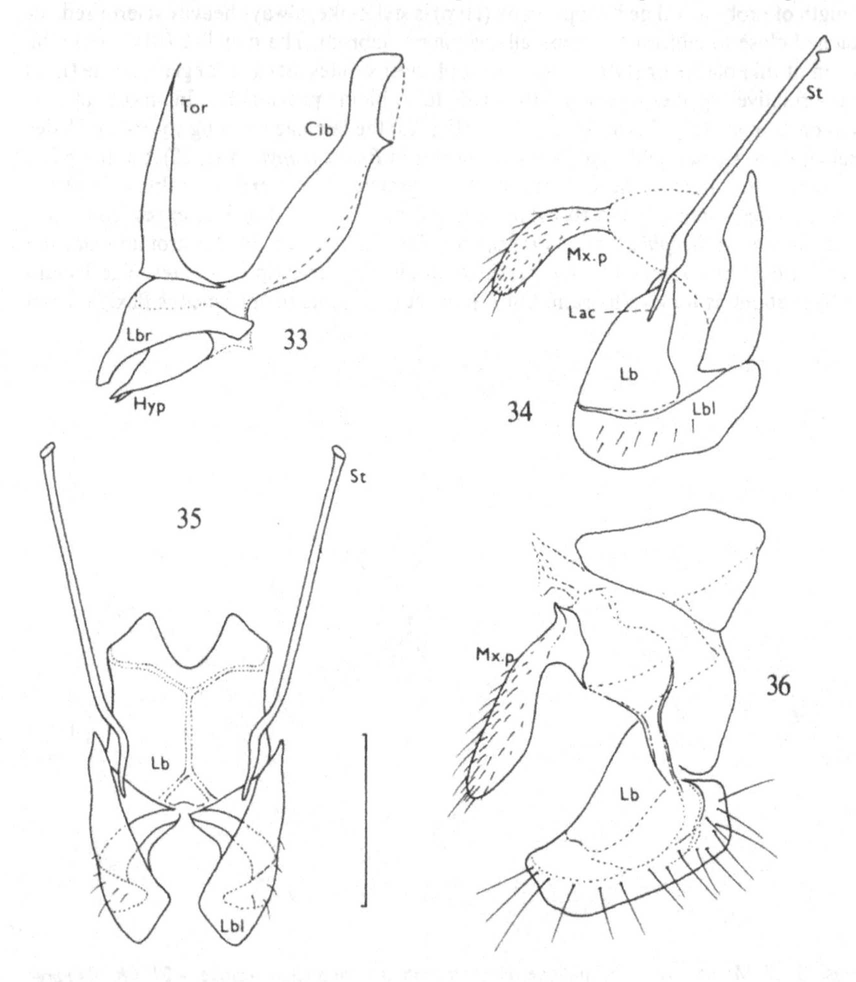

Figs. 33–36. Mouthparts of Empididae: Clinocerinae. – 33–35: *Clinocera* sp. ♂; 36: *Wiedemannia bistigma* (Curtis) ♂. Scale: 0.2 mm. For abbreviations see p. 13.

structure of the mouthparts in the Hemerodromiinae is the same as in the Empidinae, but the proboscis is adapted for special raptorial habits: the labrum is very strong and, except for *Phyllodromia,* directed backwards, and the maxillae are always completely developed (Fig. 27). A similar situation is to be found in most of the Clinocerinae, for instance in *Trichopeza* (Fig. 28), *Dryodromia* (Fig. 31) or *Heleodromia* (Fig. 32). In *Clinocera* (Fig. 34) the maxillary lacinia is very reduced but the stipites are remarkably long, and the palpus is connected with the maxilla by a membrane, not directly attached to the stipites. Krystoph (1961: 864) described the Clinocerinae - except for *Trichopeza* and *Synamphotera* (= *Dryodromia*) - as having the lacinia lost and the palpi joined to the maxillae by a "palpifer"; actually Krystoph described the same situation as in the Hybotidae. Unfortunately, I quoted this data without verification in the com-

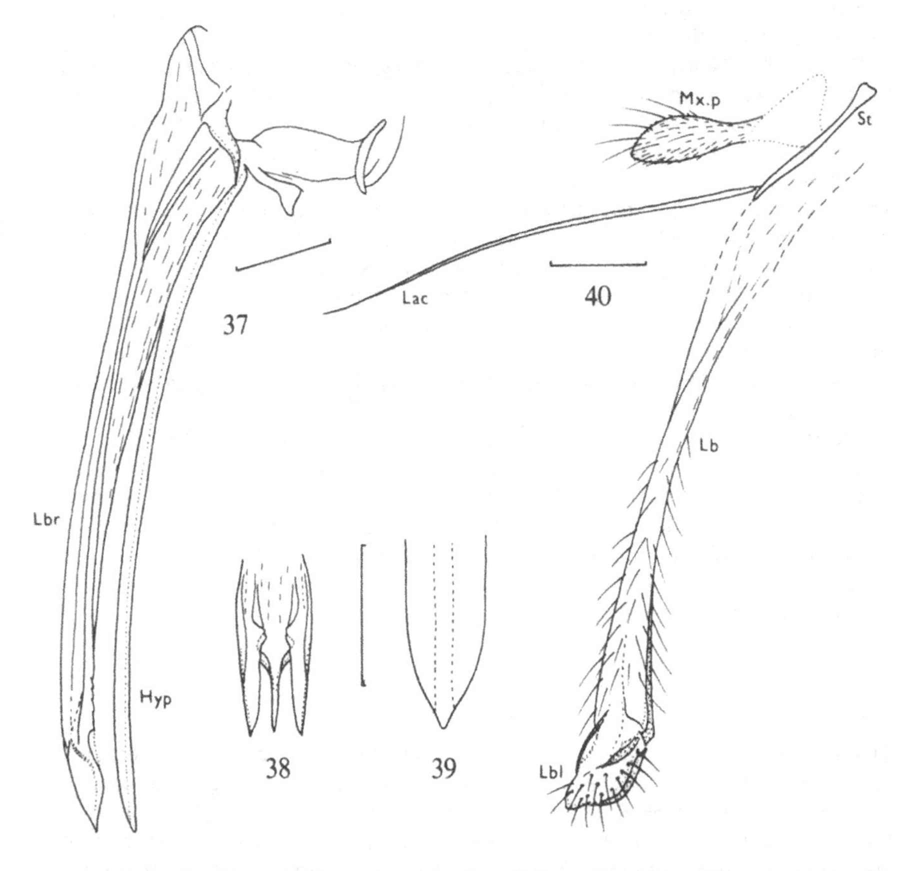

Figs. 37–40. Mouthparts of *Ceratomerus paradoxus* Phil. ♂, (Empididae: Ceratomerinae) - 38: apex of labrum; 39: apex of hypopharynx. Scale: 0.1 mm. For abbreviations see p. 13.

parative holomorphological diagram in my previous paper (Chvála, 1981a: 277). On the basis on my own recent studies, the maxillae (including lacinia, even if reduced) are always present in the Clinocerinae except for *Wiedemannia* (Fig. 36), and the palpus, if not attached directly to maxillae (as it usually is), connected with the maxillae (or: separated from the maxillae) by a fine membrane; no chitinised sclerite is present. The mouthparts in *Wiedemannia* (Fig. 36) are extremely modified, with the labrum and hypopharynx as in *Clinocera* but maxillae quite absent, very probably fused with the labellar paraphyses, and palpi attached directly to the labial wall; labellae apparently without pseudotracheae, as in *Clinocera*.

In the Brachystomatinae – only *Brachystoma* (Fig. 23) was studied as it was impossible seriously to damage by dissection the only available specimen of *Homalocnemis* in the collection at Eberswalde – the maxillae are rather broad, the stipites forwardly directed, almost at right angles to the laciniae, and the maxillae connected basally to form a ring around the labium.

The mouthparts of *Ceratomerus* (Ceratomerinae) show the ground-plan condition of the Empididae, but the proboscis (Fig. 37) is very different from what are thought to be their most closely related subfamilies. Labrum (3-pointed at tip) and hypopharynx heavily sclerotised, long and slender, at least twice as long as head is high; labium equally long, rather slender and entirely covered with distinct setae, apical labellae small and slender, without pseudotracheae but covered with sensory bristles. Maxillary palps as in *Clinocera*, of the same shape and size, and connected to the stipites by a membrane; laciniae, however, very thin (almost hair-like) and long, about two-thirds as long as labium. Mouthparts placed rather anteriorly on the head.

The mouthparts are very specialised in the Hybotidae; maxillae are reduced to only short stipites, laciniae are absent, and palpi are situated on a special sclerite, the "palpifer", which links palpus with stipites (Fig. 152). Detailed information is given in the Systematic part.

The ground-plan of the mouthparts in the Atelestidae (Fig. 566) and Microphoridae (Fig. 607) is the same as in the Empididae, with fully developed maxillae, but in the Microphoridae the palpi are usually separated from the maxillae by a fine membrane (Fig. 608) and in the Parathalassiini the laciniae are absent (at least extremely reduced).

The structure of the tormae (Tor) and cibarium (Cib) (Fig. 20) also provide good morphological characters in shape and structure, but it is not a purpose of this discussion to elaborate on the details.

Thorax

Thorax consists mainly of the large mesothorax, and the prothorax and metathorax are very reduced. Mesonotum (dorsum of thorax, thoracic shield) is more or less convex, sometimes strongly arched above and even somewhat humped (Ocydromiinae, Hybotinae). It is suggested that the plesiomorphous condition of thorax is represented by uniform pubescence of non-differentiated hairs, and in more specialised groups the pubescence is bristle-like and the bristles are strongly differentiated and/or reduced.

For details on the thoracic sclerites and their chaetotaxy see the first "empidid" volume (Chvála, 1975, Fig. 22).

One very important taxonomic feature is the structure of the prothorax (Fig. 41), especially the degree of sclerotisation of the prothoracic sternal parts, which reflects the stage of development of the thoracic muscles. The prothorax consists of the pronotum (prothoracic collar) which is always formed by a discrete and often characteristically bristled sclerite in front of the mesonotum. Laterad of the pronotum are the humeri (postpronotum), generally well-developed and convex, reduced in Drapetini (Tachydromiinae). Laterally above fore coxae there are two more or less discrete sclerites of the propleuron - episternum and epimeron. Ventrally (between the fore coxae) lies the prosternum, which in the plesiomorphous state is divided into a larger ventral basisternum and a vestigial sclerite above it, the presternum. This condition is thought to be a plesiomorphous condition within the Empidoidea, as was suggested for the "Empididae" by Collin (1961) and Tuomikoski (1966).

The primitive type of prothorax has a small isolated saddle-shaped prosternum (ac-

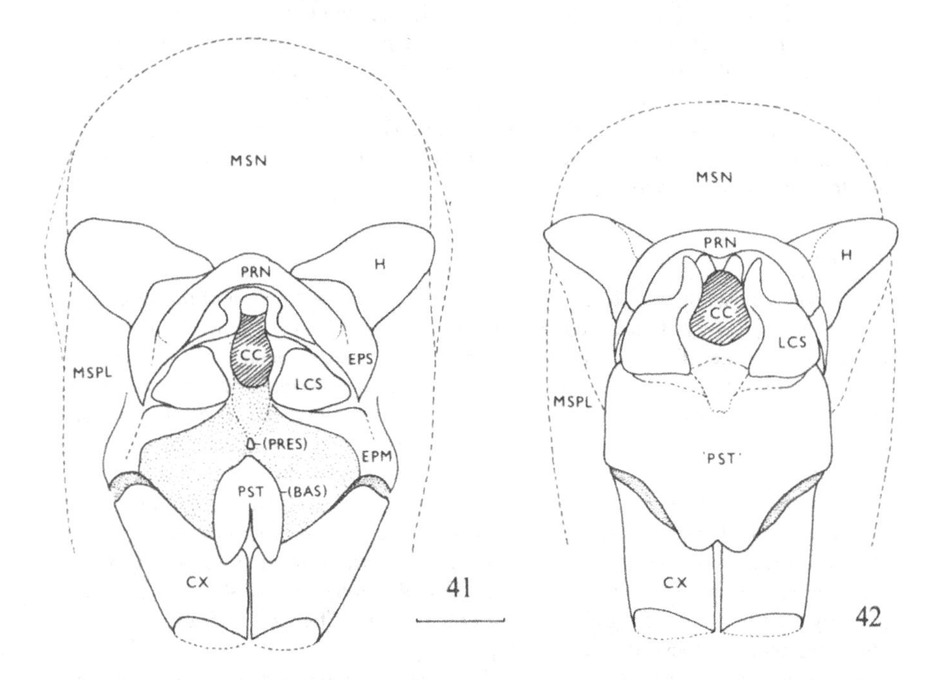

Figs. 41, 42. Thorax in anterior view (schematically, membrane dotted). - 41: *Hybos grossipes* L. 42: *Empis pennipes* L. Scale: 0.2 mm. ACS - anterior cervical sclerite; BAS - basisternum; CC - cervical cavity; CX - coxa; EPM - epimeron; EPS - episternum; H - humerus; LCS - lateral cervical sclerite; MSN - mesonotum; MSPL - mesopleura; PRES - presternum; PRN - pronotum; PST - prosternum.

tually basisternum) between the front coxae, separated by a membrane from other sclerites of the propleuron, and is generally called the "*Hybos*-type" prosternum (Fig. 41). It is present in the primitive groups of the Empididae, in the Oreogetoninae and, so far as I am aware, only in *Trichopeza* (Clinocerinae) and *Homalocnemis* (Brachystomatinae) apart from the Oreogetoninae. However, it is also the ground-plan condition of the Hybotidae, Atelestidae and Microphoridae, and is quite well-preserved in recent forms. Nevertheless, there is a strong tendency in some specialised forms, for instance in some Tachydromiinae (Hybotidae), for the prosternum to be enlarged and to be partially fused with the propleural sclerites.

The second type of prosternum, present in most Empididae (except for the Oreogetoninae and the two genera mentioned above), is thought to be the apomorphous state; the prosternum is in the form of a large sclerite, occupying the whole area between fore coxae and expanded laterally by fusion with the episterna and epimera (Fig. 42). This type of prothorax, called a precoxal bridge, was considered by Speight (1969) to be the apomorphous condition of the prothorax in cyclorrhaphous Acalyptratae as well. The pubescence or bristling of this large prosternum is a very valuable differential character at the subgeneric level in the Empidinae.

The thoracic pleura are very rarely pubescent in some species of *Hilarempis* and *Afroempis* (Empidinae), but the metapleura often have characteristic bristles arranged in "metapleural fan", a common character in *Oreogeton* (Oreogetoninae) and in most genera of the Empidinae (except *Hilara, Atrichopleura, Afroempis* or *Toreus*). In the families treated in the present volume, the thoracic pleura are always bare and the prosternum is of the "*Hybos*-type".

Wings

The wings are usually well-developed in the Empidoidea. Reduction of the wings only occurs in the very specialised predaceous forms of the Tachydromiinae and Hemerodromiinae. Brachyptery has developed in some species of *Tachydromia, Chersodromia, Stilpon* (Tachydromiinae) and *Drymodromia* (Hemerodromiinae), whilst entire loss of wings occurs in the tachydromiine genera *Ariasella, Pieltainia* and *Apterodromia.* Sexual dimorphism of the wings is very rare: in a few species of *Empis* and *Rhamphomyia* (Empidinae) the wings in females are larger and differently shaped from those of males. Wings are generally clear or uniformly clouded, rarely milky-white (some species of *Empis, Rhamphomyia, Hilara, Euthyneura, Chersodromia),* spotted *(Dolichocephala, Syneches),* striped *(Tachydromia),* or with some other distinctive pattern *(Tachypeza, Stilpon, Rhamphomyia, Empis, Homalocnemis).* A costal stigma is particularly well-developed and very useful as a differential feature in the Hemerodromiinae, Ocydromiinae and Hybotinae.

The wing membrane is generally entirely covered with microtrichia, except in some Hybotinae *(Syndyas, Hybos)*; macrotrichia are never present on the membrane. The veins are bare except for costal vein, and apparently the only exception is the genus *Oreogeton* (Empididae), where veins R_1 and R_{4+5} above, and R_{2+3} and M_1 beneath, also have microtrichia (Fig. 43).

As far as the recent species are concerned an alula is present only in the Atelestidae (Figs. 561, 562); it is also present in the Jurassic Protempididae (Fig. 139) and in the extinct hybotid genus *Trichinites*. The well-developed large axillary wing-lobe is undoubtedly in the ground-plan of the Empidoidea; it is present in the Oreogetoninae and most Empidinae (Empididae), in the Atelestidae, and in the primitive forms of the Hybotidae. In these forms the axillary excision is very acute or almost right-angled. The absence, or very slight development, of the axillary lobe is an apomorphous condition, as for instance in *Empis* subgenus *Lissempis, Rhamphomyia* subgenus *Lundstroemiella* (Empidinae), in all Hemerodromiinae, Clinocerinae, Brachystomatinae (except *Homalocnemis*) and Ceratomerinae (Empididae), in some Ocydromiinae *(Oropezella, Leptodromiella)* and in practically all Tachydromiinae (Hybotidae); in the latter subfamily the axillary excision, if present, is always very obtuse.

The complete wing-venation of the Empidoidea in its plesiomorphous state is shown in Fig. 43; it corresponds roughly with that of the extinct family Protempididae except for the shape of anal cell. In the Protempididae (Fig. 139) the anal cell (cell Cu) is still very long, apically pointed and short-petiolate, resembling in shape the anal cells when closed of lower Brachycera.

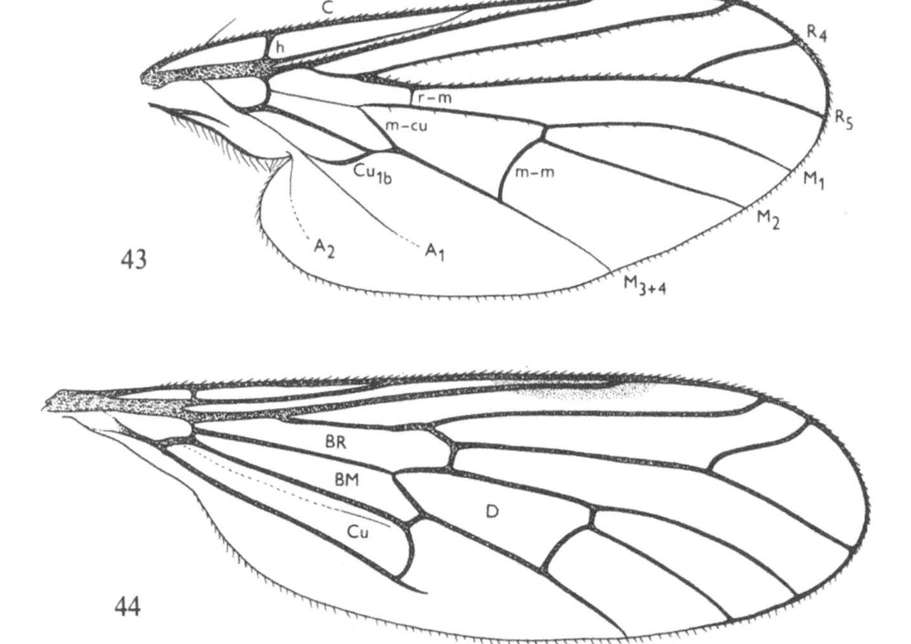

Figs. 43, 44. Wings of Empididae. - 43: *Oreogeton basalis* (Loew); 44: *Brachystoma vesiculosum* (Fabr.). For abbreviations see the text.

The primitive wing-venation may be characterised, as follows: Costa (C) running around the wing, at least as an ambient vein on posterior wing-margin. Subcosta (Sc) long, ending at costa. Radial vein (R) with four branches; R_1 long, ending at costa in apical half of wing, radial sector (rs) 3-branched, with R_{2+3} ending near the wing-tip, R_{4+5} apically forked. Media (M) 3-branched, M_1, M_2 and M_{3+4} arising independently from discal cell and running to wing-margin. Discal cell (cell D) rather blunt-tipped (not truncate) and emitting 3 veins to wing-margin. Two basal cells; both 1st basal cell (cell BR) and 2nd basal cell (cell BM) rather long. Anal cell (cell Cu) long, at least as long as 2nd basal cell or longer, the vein closing it (vein Cu_{1b}) curved to meet anal vein at an angle of 90° or more. Originally with two anal veins (A); vein A_1 closing anal cell from below and reaching posterior wing-margin, vein A_2 (retained only in some Oreogetoninae and Empidinae) present as a faint fold near axillary excision. There are four crossveins: humeral crossvein (h), anterior crossvein (r-m, ta) at about middle of

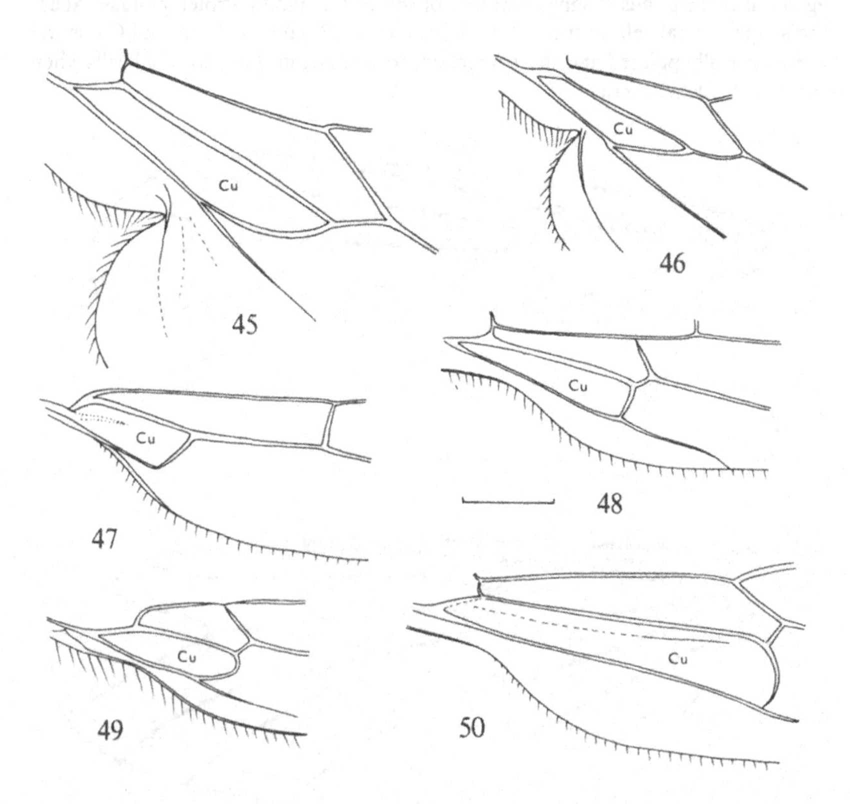

Figs. 45–50. Shape of anal cell (cell Cu) of Empididae. – 45: *Oreogeton basalis* (Loew); 46: *Empis pennipes* L.; 47: *Chelifera precatoria* (Fall.); 48: *Dryodromia testacea* Rond.; 49: *Clinocera* sp.; 50: *Brachystoma vesiculosum* (Fabr.). Scale: 0.5 mm.

discal cell, posterior crossvein (m-m, tp) closing discal cell, and basal crossvein (m-cu, tb) closing 2nd basal cell.

This primitive wing-venation may be found in the Oreogetoninae (Empididae), but the anal cell is more of the empidine-type (Fig. 45) with the vein closing it (vein Cu_{1b}) so recurrent as to join vein A_1 without making distinct angle. The primitive type of anal cell is present in the Brachystomatinae (Fig. 50), Atelestidae (Fig. 53) and Hybotinae (Fig. 51), including the extinct genus *Trichinites*. The extreme form of anal cell, a small apically rounded cell, is present in some Clinocerinae and Hemerodromiinae (Fig. 49) of the Empididae, and in all Microphoridae except *Parathalassius*. The anal cell is absent in *Hemerodromia* and *Ceratomerus* within the Empididae, and in all Tachydromiinae (Hybotidae) except for *Symballophthalmus* and *Platypalpus*.

The loss of the discal cell occurs rarely in all families except for the Microphoridae: in several genera of the Hemerodromiinae and Clinocerinae (Empididae), in *Atelestus*

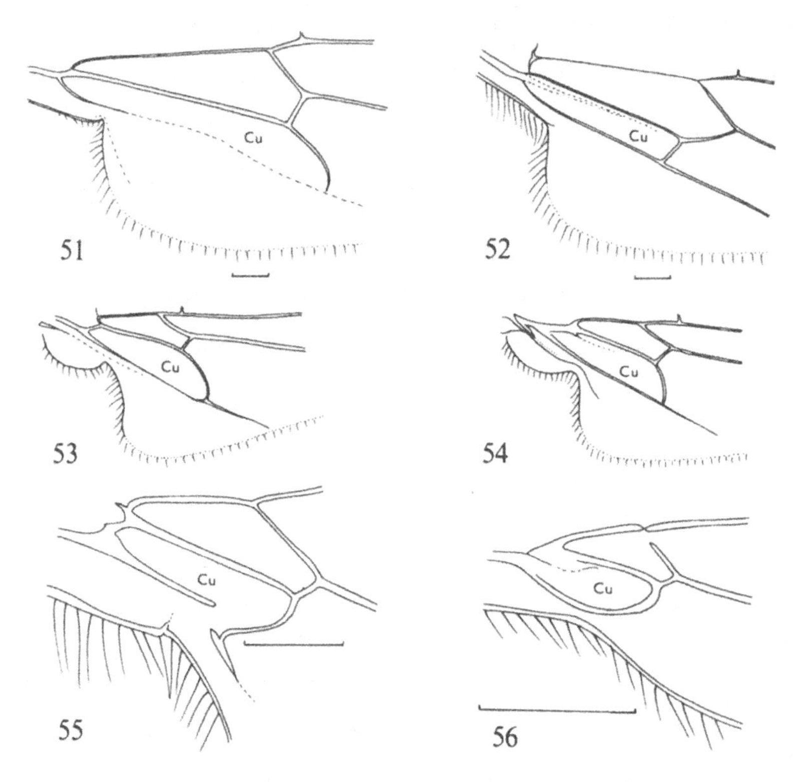

Figs. 51–56. Shape of anal cell (cell Cu) of Hybotidae, Atelestidae and Microphoridae. - 51: *Hybos grossipes* (L.); 52: *Ocydromia glabricula* (Fall.); 53: *Atelestus pulicarius* (Fall.); 54: *Meghyperus sudeticus* Loew; 55: *Microphorus holosericeus* (Meig.); 56: *Microphorella praecox* (Loew). Scale: 0.2 mm.

(Atelestidae) and some Ocydromiinae (*Bicellaria, Hoplocyrtoma*), but within the Hybotidae the absence of discal cell is part of the ground-plan of the Tachydromiinae.

The absence of the discal cell (actually the absence of the crossvein tp which closes the discal cell) is combined in the Empididae with the presence of a medial fork M_1 and M_2. However, a true *apical* fork of M_1 and M_2 (as present in the dolichopodid subfamily Chrysosomatinae, the Platypezidae or Hilarimorphidae) is clearly developed only in *Chelifera* (Hemerodromiinae), *Ceratomerus* (Ceratomerinae) and *Meghyperus* (Atelestidae).

There is a real tendency towards the shortening and/or reduction of the veins in the Empidoidea, but the wing-venation is rather constant within individual subfamilies. The Clinocerinae and Hemerodromiinae (Empididae) are an exception to this. Their wing-

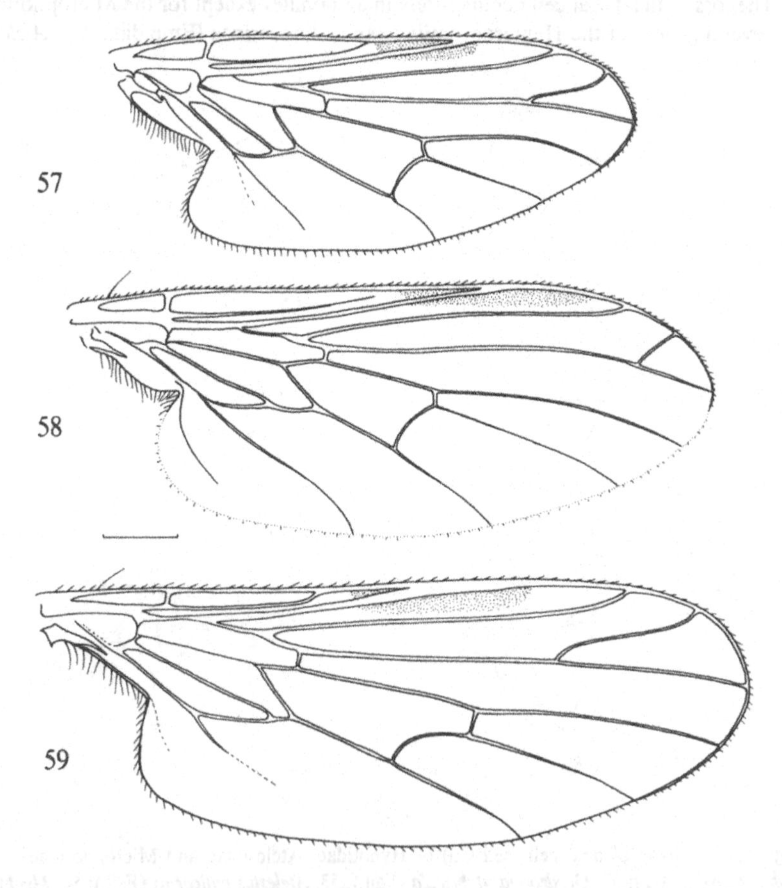

Figs. 57–59. Wings of Empididae: Oreogetoninae and Empidinae. – 57: *Iteaphila nitidula* Zett.; 58: *Empis pennipes* L.; 59: *Hilara maura* (Fabr.). Scale: 0.5 mm.

venation varies substantially from genus to genus and, moreover, there are distinct tendencies towards the multiplication of veins, as for instance in *Dryodromia* and *Dolichocephala* (Figs. 61, 64) of the Clinocerinae, with an apparent apical fork of R_{2+3} (!). The most reduced wing-venation within the Empididae is found in *Hemerodromia* (Fig. 63) and *Ceratomerus* (Fig. 62).

The Empidinae have some apomorphous characters in the wing-venation that are not known elsewhere in the family Empididae: for instance, the costa runs only to R_{4+5} (absent on posterior wing-margin), and Sc is abbreviated, not reaching the costa. This is not true of *Hilara* or *Afroempis* which obviously possess the primitive type of venation (Sc complete, C around the wing), including the rather long acute fork of R_4 and R_5 (Fig. 59). From this point of view the "open" radial fork of *Empis* (Fig. 68) and Hemerodromiinae (Fig. 67) – but not Clinocerinae (Fig. 69) – is an apomorphous state. It should be noted that a costa running around the wing (posteriorly at least as a fine

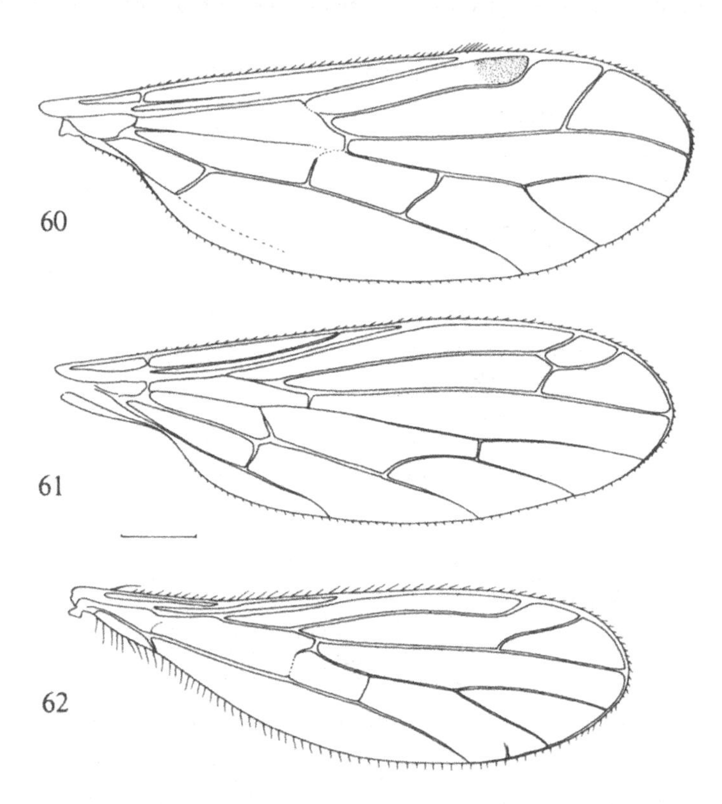

Figs. 60–62. Wings of Empididae: Hemerodromiinae, Clinocerinae and Ceratomerinae. – 60: *Chelifera precatoria* (Fall.); 61: *Dryodromia testacea* Rond.; 62: *Ceratomerus paradoxus* Phil. Scale: 0.5 mm.

ambient vein) is also found in some species outside *Hilara* (some *Rhamphomyia* species), but the incomplete Sc seems to be a constant character in the majority of empidine genera including *Atrichopleura*, *Hilarempis* or *Haplomera* and so it can be accepted as a ground-plan character of the Empidinae, as indicated in the diagram, Fig. 141.

The wing-venation of the Hybotidae, Atelestidae and Microphoridae is fully discussed in the Systematic part in the sections on phylogeny and classification.

Legs

The legs are generally long and slender, often specifically bristled or covered with a dense, sometimes long pubescence. In the Empidinae, Ceratomerinae and Ocydromiinae the hind legs are often elongate and swollen, sometimes very conspicuously so. In females of many Empidinae the legs are broadly "pennate", covered

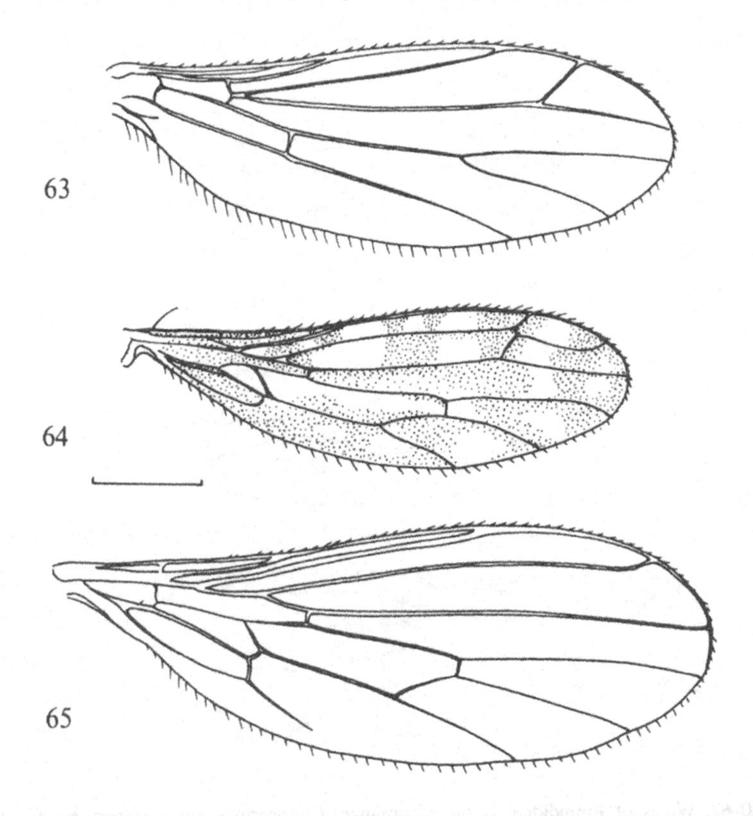

Figs. 63–65. Wings of Empididae: Hemerodromiinae and Clinocerinae. – 63: *Hemerodromia unilineata* Zett.; 64: *Dolichocephala irrorata* (Fall.); 65: *Heleodromia immaculata* Hal. Scale: 0.5 mm.

with conspicuously flattened bristles that undoubtedly assist the males in recognising flying females (which have a characteristic silhouette) in aerial mating swarms.

In the Tachydromiinae (and also some of the Hemerodromiinae and Clinocerinae) the legs are shorter and stronger, a modification for walking and climbing on vegetation, on leaves or tree trunks, on the ground, or even on sand. Conspicuously short and strong legs are present in some rapidly-running species, such as *Chersodromia, Tachydromia* or *Stilpon* (Tachydromiinae). The long slender legs of other Empidoidea are mostly for holding on to twigs or leaves, or on to grass stems, and the flies move almost exclusively by wing. In the Hemerodromiinae and Clinocerinae the front legs are very elongate, particularly the coxae, and are characteristically widely separated from the posterior two pairs; the mid coxa is in the posterior third of the thorax.

Special raptorial legs for catching and/or holding prey, functioning as clasping "mantis-like" legs with elongate and ventrally specifically bristled or spinose femora and tibiae, are developed in many groups: the front legs in the Hemerodromiinae, mid legs in some Tachydromiinae *(Platypalpus)* or Brachystomatinae *(Homalocnemis)*, and hind legs in the Hybotinae and some Ocydromiinae (Oedaleini).

These elongated and modified mid or hind legs serve for holding the prey and/or the female during copulation. In the Atelestidae and in the males of most Microphoridae, the hind tibiae and tarsi (or metatarsi) are remarkably incrassate and flattened, probably an adaptation for aerial mating. The front metatarsi in the males of most *Hilara,* and very rarely some *Empis,* are conspicuously swollen and bear a special silk gland, which plays an important role in their unusual courtship behaviour. A tubular

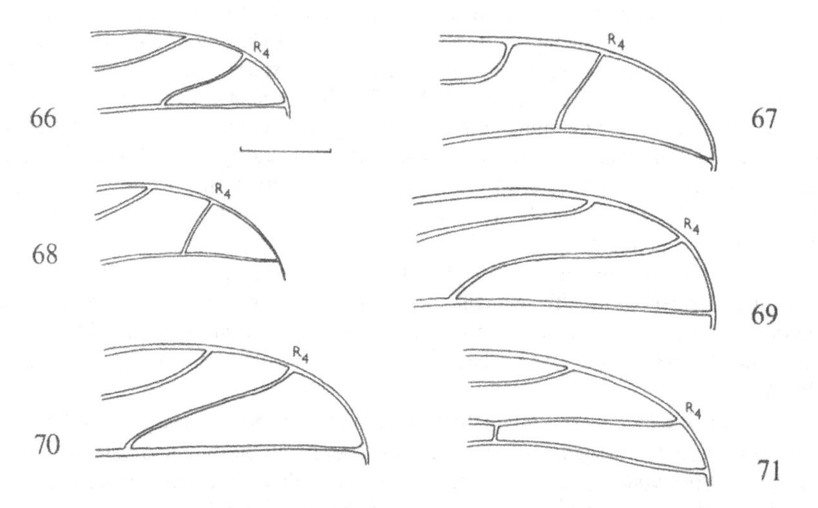

Figs. 66–71. Radial fork of R_4 and R_5 in Empididae. – 66: *Iteaphila nitidula* Zett.; 67: *Chelifera precatoria* (Fall.); 68: *Empis pennipes* L.; 69: *Trichopeza longicornis* (Meig.); 70: *Hilara maura* (Fabr.); 71: *Clinocera* sp. Scale: 0.5 mm.

gland of unclear function is present near the base of fore tibia in all Hybotidae; its opening is often characteristically pubescent or spinose, and sometimes lies on a special tubercle (Figs. 160, 179).

As far as I am aware, the tarsi are always 5-segmented in the Empidoidea, the basal segment (metatarsus) generally being the longest. The praetarsus consists of two unguiculi, more or less broad pulvilli, and a bristle-like empodium (Fig. 181). In the Clinocerinae the pulvilli and empodium are often reduced to almost vestigial *(Clinocera* subgenus *Bergenstammia)*, or the empodium is pulvilliform *(Dolichocephala, Clinocera)*.

It may be concluded that the structure and armature of the legs in the Empidoidea are very diverse, but all these adaptations, which have undoubtedly arisen through parallel development in non-related groups, reflect differences in function that clearly have an evolutionary significance. It appears that the constant presence of the tibial gland in the very heterogeneous family Hybotidae is the only character that may be of significance for phylogenetic classification of this group.

Abdomen

In the ground-plan the abdomen consists of eleven segments, the posterior three segments being more or less modified in both sexes, very conspicuously so in males. The 11th segment is represented by paired cerci (anal papillae) in the ground-plan of both sexes.

The *male abdomen* consists of eight more or less visible segments; the 9th segment (andrium, or genital segment) is highly modified, and together with the vestigial 10th segment and cerci forms the hypopygium (genitalia, or terminalia in general). The 9th segment bears genital opening and, together with the following two postgenital segments, forming the proctiger, is fully discussed in the next section. The eight segments preceding the genital segment are generally called the pregenital segments.

The anterior eight abdominal segments are almost unmodified in the primitive species of the Empididae and in some Hybotidae *(Syneches, Syndyas)* (Fig. 184), where terga and sterna of the posterior segments are all discrete and of the same shape. However, the 1st segment is always much shorter, firmly fused to the metathorax, and the 2nd and 3rd segments are generally larger than the following segments. The 8th segment, even if unmodified and identical with the preceding segments, is often more or less concealed within the 7th segment. There are 7 pairs of spiracles on the 1st to 7th abdominal segments, even if the posterior segments are remarkably modified (Fig. 128), and the 8th segment never has spiracles; this was pointed out by Bährmann (1960).

The male abdomen is commonly divided into the preabdomen (the anterior five and generally unmodified segments) and the postabdomen (the 6th and following segments). The postabdomen is usually more or less modified, influenced in various ways by the structure and size of the hypopygium. The first stage in the modification of the male postabdomen is seen in the reduction of the 8th segment in its tergal part; this state may be found in many Oreogetoninae and Empidinae, and it is quite commonly found in various degree of development in almost all groups, particularly if the hypopygium is upturned (Fig. 379). However, it can hardly be said that it is in the ground-plan of any higher taxonomic unit within the Empidoidea (as it is not constant or regularly distributed among all the primitive taxa), except for the Atelestidae. In this family the narrow vestigial 8th tergum and the large ventral sternum (Fig. 572) seem to be the ground-plan condition. Griffiths (1972) considered the reduction of the 8th tergum to a narrow band (in a ventral position, because of rotation of the postabdomen through 180°) to be an apomorphous ground-plan condition of the Cyclorrhapha. In the Empidoidea, the smaller 8th tergum is always placed dorsally and the large 8th sternum is ventral in position. It should be noted that the partial rotation of the hypopygium in the Hybotidae (lateroversion), through 90° towards the right, does not involve the preceding abdominal segment.

On the other hand, the inverted and lateral position of the hypopygium in the Microphoridae (Fig. 72) is caused by deflexion and partial rotation of the two pregenital segments; the hypopygium itself is apparently unrotated. Bährmann (1960: 525, Fig. 37) gave one erroneous example of rotation of the pregenital segments in the Empididae as currently defined, in *Heleodromia curtipes* (Beck.) of the Clinocerinae; however, examination of Becker's type-material in Berlin shows that this species belongs to the microphorid genus *Microphorella*.

The 8th tergum and sternum are always represented by discrete sclerites, even if very differently shaped. The only exception, so far as I am aware, is in *Trichopeza* of the Clinocerinae (Fig. 126), where the 8th tergum is fused with the corresponding sternum. In the Ceratomerinae (Fig. 128) the 8th segment consists of a strip-like tergum and a much larger sternum. The 7th segment is similarly shaped (even though the sternum is decidedly smaller), but it is entirely concealed within the unmodified 6th segment. In *Homalocnemis* of the Brachystomatinae (Fig. 133) the pregenital terga have a conspicuous structure, with a parallel strong sclerotisation.

Further substantial modifications of the postabdomen, even involving the 7th and

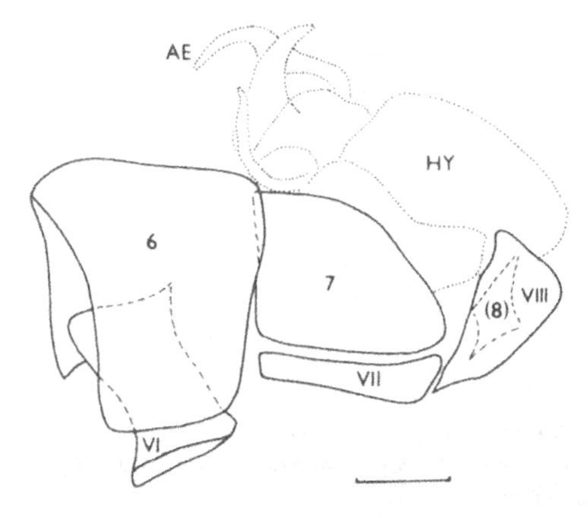

Fig. 72. Male postabdomen of *Microphorus holosericeus* (Meig.) (Microphoridae). Scale: 0.1 mm. For abbreviations see p. 13.

6th segments, are rather rare and are only of specific importance, as in some species of *Rhamphomyia* and *Empis* (Empidinae). A constant character of some phylogenetic importance is the deflexion and rotation of the pregenital segments in the Microphoridae, as mentioned above. In rare cases the 5th abdominal segment of the preabdomen is modified, as for instance in *Microphorella* (Fig. 637) of the Microphoridae, but other tergal modifications such as spines and processes on the preabdomen (as on the 4th tergum in *Empis picipes*) are specific arrangements of unknown significance and function (? courtship) and have nothing to do with the 'genital' modifications of the postabdomen.

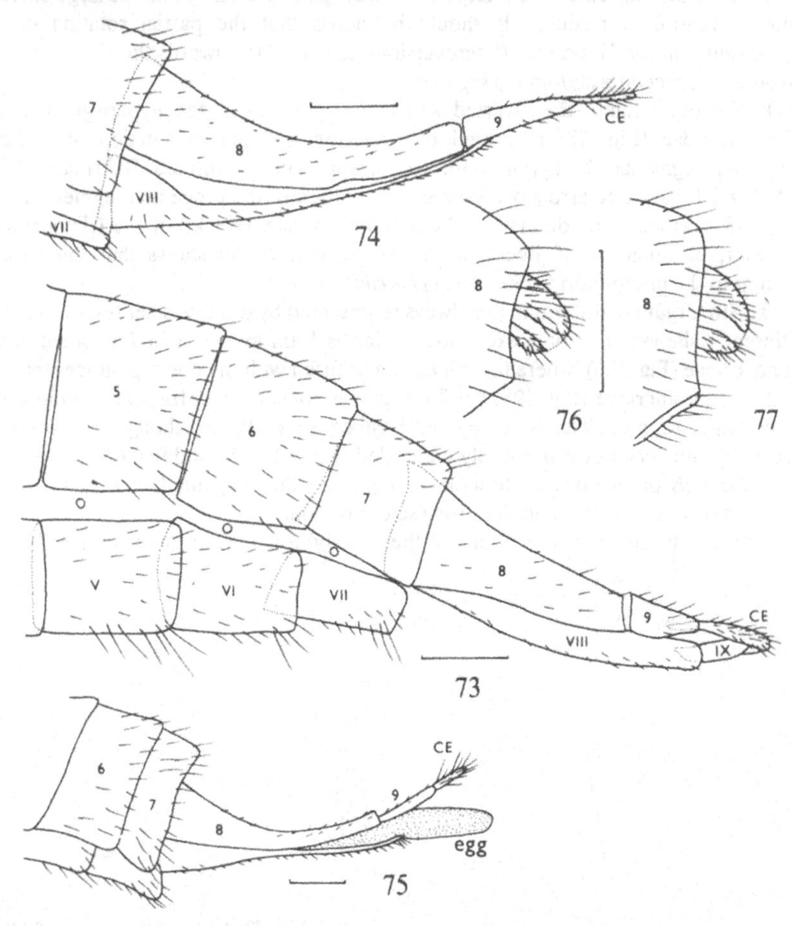

Figs. 73–77. Female abdomens of Empididae: Ocydromiinae. – 73: *Oedalea stigmatella* Zett.; 74: *Leptopeza flavipes* (Meig.); 75: the same with an egg; 76: *Ocydromia melanopleura* Loew; 77: *Ocydromia glabricula* (Fall.). Scale: 0.2 mm. For abbreviations see p. 13.

The *female abdomen* is much less modified than in males. In its primitive state, the female abdomen is somewhat telescopic, with nine visible segments, gradually narrowing towards tip; 10th segment usually vestigial and more or less membraneous, terminal cerci distinct (Fig. 185). This "telescopic" abdomen is undoubtedly a plesiomorphous condition within the Empidoidea. Apomorphous states have evolved in two different directions from this ground-plan. Firstly, by a narrowing and prolongation of the postabdomen, the "ovipositor-like" abdomen itself functions like a long ovipositor; and secondly, by a special modification of the 9th tergum which splits into two conspicuously spinose sclerites, the cerci are lost or vestigial and the preceding terga are partly modified.

The first variant of the ovipositor-like abdomen has evolved in two directions. In the Atelestidae (*Meghyperus* and *Atelestus; Acarteroptera* was not examined), the posterior five abdominal segments (segments 4–8) gradually taper, not becoming very long but remarkably narrow, the terga and sterna are reduced to small sclerites, and the narrow apical half (or more) of the abdomen is rather membraneous (Fig. 573). The cerci are long and the 9th and 10th segments very vestigial, membraneous.

In the Ocydromiinae (Hybotidae) only the tip of the abdomen is ovipositor-like. Segments 4–7 are unmodified, but the 8th segment is very narrow and long, rather like the shorter 9th segment, and all terga and sterna are heavily sclerotised (Fig. 73). The 10th segment is not distinguishable, but long slender cerci are present. In comparison with the Atelestidae, the "ovipositor" is much more sclerotised and actually consists only of the 8th segment. This condition is frequent in some genera of the Ocydromiinae, in *Oedalea* (Fig. 73), *Leptopeza* (Fig. 74), *Leptodromiella, Euthyneura* (Fig. 455), *Anthalia* and others.

The second type of female terminalia, with the cleft spinose 9th tergum (hemitergites, or acanthophorites), is only found in some Empididae and

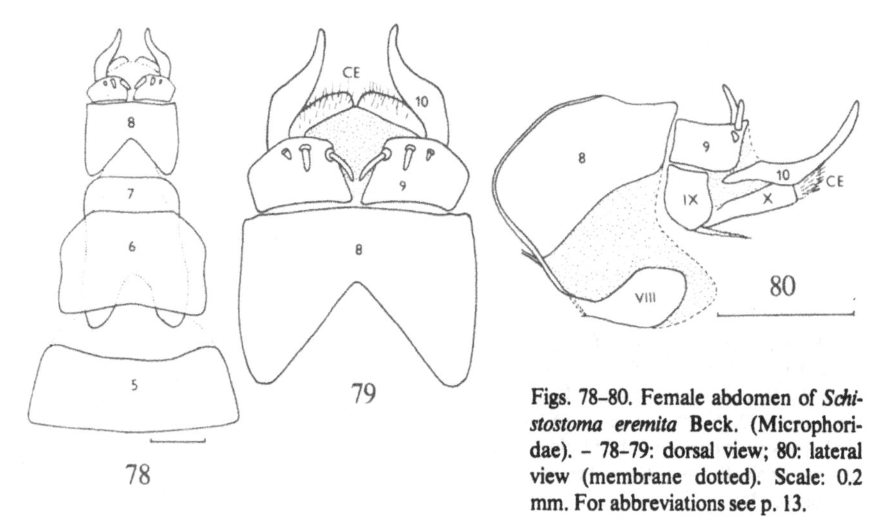

Figs. 78–80. Female abdomen of *Schistostoma eremita* Beck. (Microphoridae). – 78–79: dorsal view; 80: lateral view (membrane dotted). Scale: 0.2 mm. For abbreviations see p. 13.

Microphoridae, but it is a character that is part of the ground-plan of the Dolichopodidae (and other families of the "lower" Brachycera). In the Empididae, so far as I am aware, this type of ovipositor is developed in *Gloma* (Oreogetoninae), *Heleodromia* and *Trichopeza* (Clinocerinae), the Brachystomatinae (*Homalocnemis* ♀ was not studied) and the Ceratomerinae. Such a pattern of distribution within the Empididae clearly demonstrates that this character cannot be of any phylogenetic significance, and that it is a functional adaptation that has arisen through parallel development. In the Microphoridae this type of ovipositor is much more widely distributed: in the Palaearctic fauna it is known in all genera except *Microphorus*, but apparently it also occurs in some Nearctic species of *Microphorus* (see Hennig, 1976: 44).

In this type of ovipositor (Fig. 78) the 9th tergum is paired and distinctly spinose, and the two preceding terga are also usually modified in various ways and conspicuously bristled or spinose. The 6th tergum is unmodified, and in *Heleodromia* (Fig. 81) the 7th segment is also unchanged. The 10th segment is always strongly sclerotised, with an upturned spine-like paired tergum and in the Microphoridae with a discrete 10th sternum. The cerci, if present, are very membraneous and inconspicuous, absent in *Heleodromia* (Fig. 81) and *Trichopeza* (Fig. 83). In *Trichopeza* the 9th tergum is still in the form of an unpaired, apically bifurcate sclerite. A striking arrangement is found in *Brachystoma* (Fig. 84) where the 7th tergum is remarkably enlarged and globose, covering the very modified small apical abdominal segments and cerci.

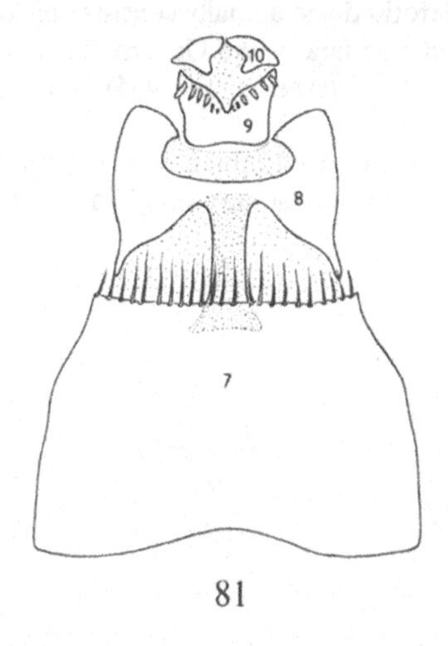

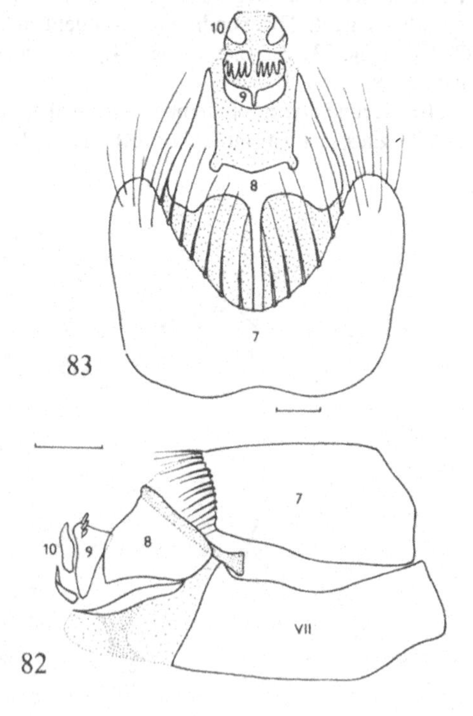

Figs. 81–83. Female abdomen of Empididae: Clinocerinae (membrane dotted). – 81: *Heleodromia immaculata* Hal., dorsal view; 82: the same, lateral view; 83: *Trichopeza longicornis* (Meig.), dorsal view. Scale: 0.1 mm. For abbreviations see p. 13.

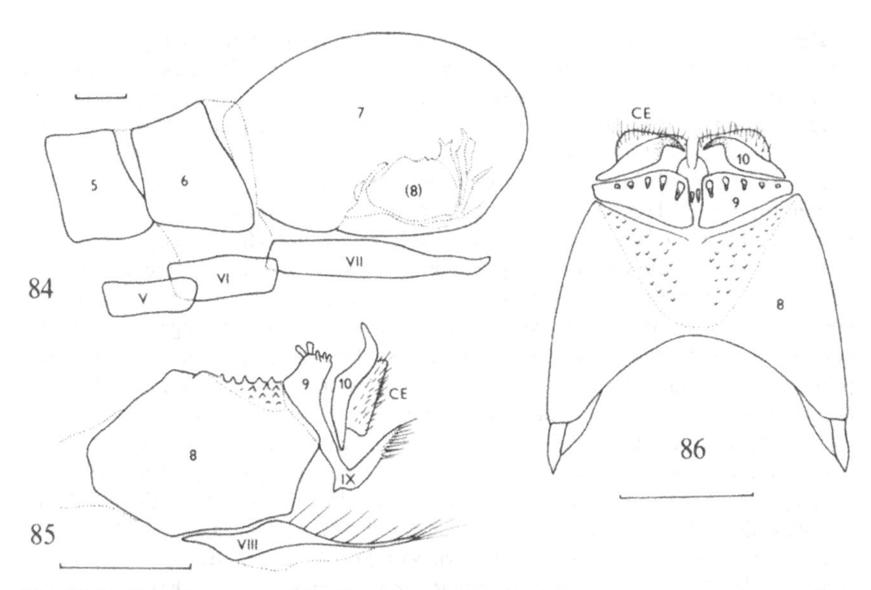

Figs. 84–86. Female abdomen of *Brachystoma vesiculosum* (Fabr.) (Empididae: Brachystomatinae). - 84, 85: lateral view; 86: dorsal view. Scale: 0.3 mm. For abbreviations see p. 13.

Male genitalia

The male genitalia consist of the highly modified 9th abdominal segment (andrium) and the proctiger, the remains of the 10th and 11th segments. The structure and homology of individual parts of the external male genitalia in the Diptera, including the "Empididae", have been intensively studied in the last twenty years, but the terminology used by various authors shows substantial differences. The first important account of the male genitalia of the "Empididae" was by Bährmann (1960). From many points-of-view this was very radical and new, as it brought many different interpretations to the customary terminology in the "family". I do not want to go into details here as it would be outside the scope of this book, nor do I feel very competent in this field of morphology.

However, a brief general survey needs to be given in order to facilitate understanding of the proposed new classification and the phylogeny of the Empidoidea, as well as to give an explanation of the terminology used in the following sections. Furthermore, the course of development of the male genitalia in the Empidoidea (Orthogenya) is vital for the correct recognition of the hypopygial structures in the whole of the Eremoneura, since Empidoidea and Cyclorrhapha are two monophyletic subgroups of the Eremoneura.

The two original sclerites of the 9th (genital) abdominal segment are highly modified and have generally been accepted by recent authors as the epandrium (9th tergum) and hypandrium (9th sternum). The lateral clasping lobes of the "lower" Diptera, which were originally bi-articulated, are a subject of some controversy as to their origin and terminology. The history of this problem has been fully discussed by Griffiths (1972) and, therefore, is only briefly summarised here.

Van Emden & Hennig (1956) suggested that the lateral lobes arose from the precoxae and styli of the Thysanura and called them gonopods; in their bi-articulated form, the basal article was the basistylus (or gonocoxite) and the distal article the dististylus (or gonostylus). This terminology was followed by Bährmann (1960), Ulrich (1972) and Hennig (1976). However, this theory was rejected by Snodgrass (1957) who believed that the lateral lobes in Diptera and all Holometabola originated from the primary lobes of the Thysanura, and called them parameres. He then called the basal articles basimeres and the distal articles telomeres (or distimeres). This theory was followed by Griffiths (1972) and Ulrich (1975), as far as the "Empididae" are concerned. It should also be noted that in Hennig's epandrial theory (see below), where the lateral clasping organs of the Eremoneura are supposed to be of tergal origin, the distal articles are called surstyli, as has also been recently followed by Afanasjeva (1980).

Quite recently Griffiths (1981), in his review of the 'Manual of Nearctic Diptera', has supported McAlpine's proposal to restore the term gonopods for the clasping organs in Diptera, and this term is also used in the present book. This means that the originally lateral clasping lobes of the Empidoidea (side-lamellae of authors) are called gonopods, and if they are (rarely) bi-articulated the distal article is called gonostylus and the basal article gonocoxite.

The second, and most important, problem in the Empidoidea is the fate of the gonopods of the Rhagionidae and other »lower« Diptera in the Empididae as recently defined. In other words, have the gonopods expanded dorsally, forming large lateral lobes, and consequently fused dorsally in the "higher" Empidoidea and Cyclorrhapha (periandrial theory); or have they shifted downwards and fused with the hypandrium, the lateral lobes then being of tergal origin (epandrial theory)? The situation in the Empididae seems to throw considerable light on the sequence of these structures in the Eremoneura.

In volume 3 of "Fauna ent. scand." (Chvála, 1975) dealing with the Tachydromiinae (a subfamily of the Hybotidae as now defined), I followed Griffiths' (1972) periandrial theory and called the large, dorsally expanded hypopygial lamellae the periandrium. I still believe that this was correct, and the situation in the true Empididae may support this view. The crux of the problem rests with the correct interpretation and homology of the lateral and dorsal sclerites of the hypopygium, the so-called side- and dorsal lamellae sensu Collin (1961), in the Oreogetoninae and Empidinae. If the periandrial theory is correct, then the "side lamellae" should be gonopods (basimeres sensu Griffiths). I feel that the most important element in the argument is the correct interpretation of the dorsal lamellae of the Empidinae. Bährmann (1960) interpreted the dorsal lamellae of the Empidinae and Hemerodromiinae as the epandrium, the cerci having been lost (except *Hilara*), and the lateral lamellae as gonopods. Ulrich (1972,

1975) suggested that the dorsal lamellae are cerci and that the lateral lamellae are the sclerites of the completely fused gonopods and epandrium. Griffiths (1972) offered an alternative explanation, that the epandrium (9th tergum) has been lost in all Eremoneura (and thus in all the Empidoidea too) and that the sclerites in question (i.e. the dorsal lamellae) are all referable to the proctiger, which represents the modified cerci and/or 10th tergum; the latter is part of the ground-plan of the Brachycera, as indicated by the situation in the Rhagionidae. A similar explanation was put forward by Hennig (1976), who suggested that the dorsal lamellae are cerci and probably the surstyli of the 10th tergum in *Empis borealis*, but, in agreement with his epandrial theory, that the side lamellae are the 9th tergum.

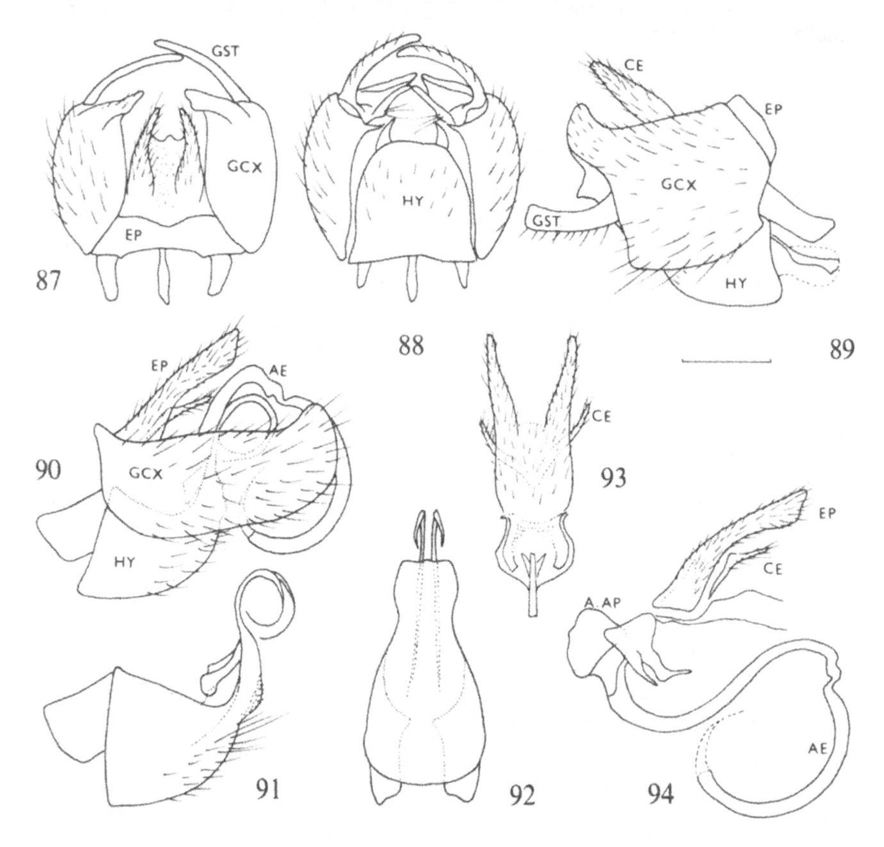

Figs. 87–94. Male genitalia of Empididae: Oreogetoninae. – 87–89: *Hormopeza copulifera* Mel. – 87: dorsal view; 88: ventral view; 89: lateral view. – 90–94: *Iteaphila nitidula* Zett. – 90: lateral view; 91: hypandrium in lateral view; 92: the same in ventral view; 93: epandrium with cerci in dorsal view; 94: aedeagal complex in lateral view. Scale: 0.2 mm. For abbreviations see p. 13.

However, it is difficult to believe that there is no continuity in the evolution of such functionally important structures as the clasping gonopods on the one hand and the vestigial proctiger on the other. If the periandrial theory is correct, then there is a logical and gradual progression in the gonopods from the lower Brachycera through the Empididae to the other families of the Empidoidea and the Cyclorrhapha. Such a sequence seems to be obvious, but even so it would seem to be illogical for the proctiger to develope in the Empididae at the expense of the 9th segment and epandrium in particular, since there is a clear tendency for a reduction of the proctiger in the Eremoneura. It is difficult to believe that in the Empididae the whole direction of hypopygial development would evolve in a quite different and complicated way. Moreover, there is no example elsewhere in the Diptera of the cerci (or other parts of proctiger) developing into such complicated sclerotised structures as are the dorsal lamellae of the Empididae.

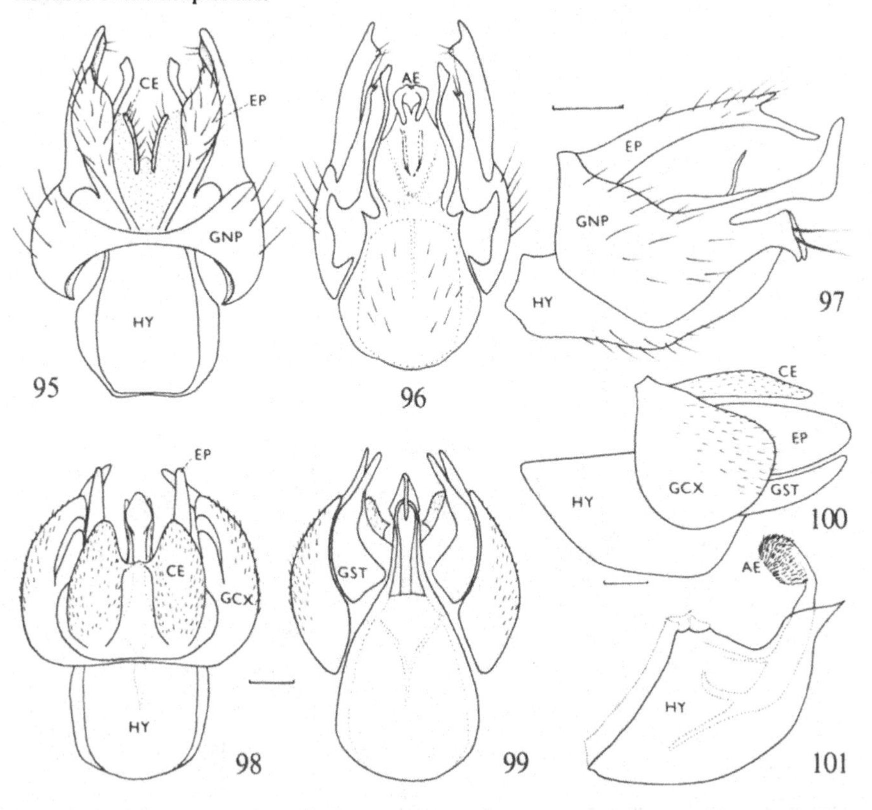

Figs. 95–101. Male genitalia of Empididae: Oreogetoninae. – 95–97: *Gloma fuscipennis* Meig. – 95: dorsal view; 96: ventral view; 97: lateral view. – 98–101: *Hesperempis mabelae* (Mel.) – 98: dorsal view; 99: ventral view; 100–101: lateral view. Scale: 0.1 mm. For abbreviations see p. 13.

For a better understanding of the general structure of the male genitalia in the Empididae, it is necessary to begin with those undoubtedly primitive forms with distinct membraneous cerci. These are the Oreogetoninae, *Hilara* of the Empidinae, or the Clinocerinae. According to Bährmann (1960), the cerci are absent in the Empidinae *(Empis, Rhamphomyia)* and Hemerodromiinae ("die gonopodentragenden Formen"). However, vestigial and very small membraneous cerci do appear to be present in these forms, as explained below.

Within the Oreogetoninae (Empididae), a very primitive state of the male hypopygium, closely resembling the ground-plan condition of the Rhagionidae, may be seen in *Hormopeza* (Figs. 87–89). There are distinct cerci above the 9th tergum (epandrium), with membraneous remnants of the 10th segment (proctiger), bi-articulated gonopods with long apical gonostyli (clasping organs), and a well-developed hypandrial sclerite beneath. The gonopods are still lateral in position, separated on the dorsum by the unmodified epandrium. In *Iteaphila* (Figs. 90–94) the gonopods are not articulated (gonostyli absent), the epandrium is already bifid apically, somewhat resembling the "dorsal lamellae" of the Empidinae, and cerci are small. The male terminalia of *Gloma* (Figs. 95–97) and *Hesperempis* (Figs. 98–101) closely resemble each other in general, with the epandrium remarkably bifurcated (dorsal lamellae), the gonopods narrowly connected dorsally (distinctly bi-articulated in *Hesperempis* but less so in *Gloma*), and

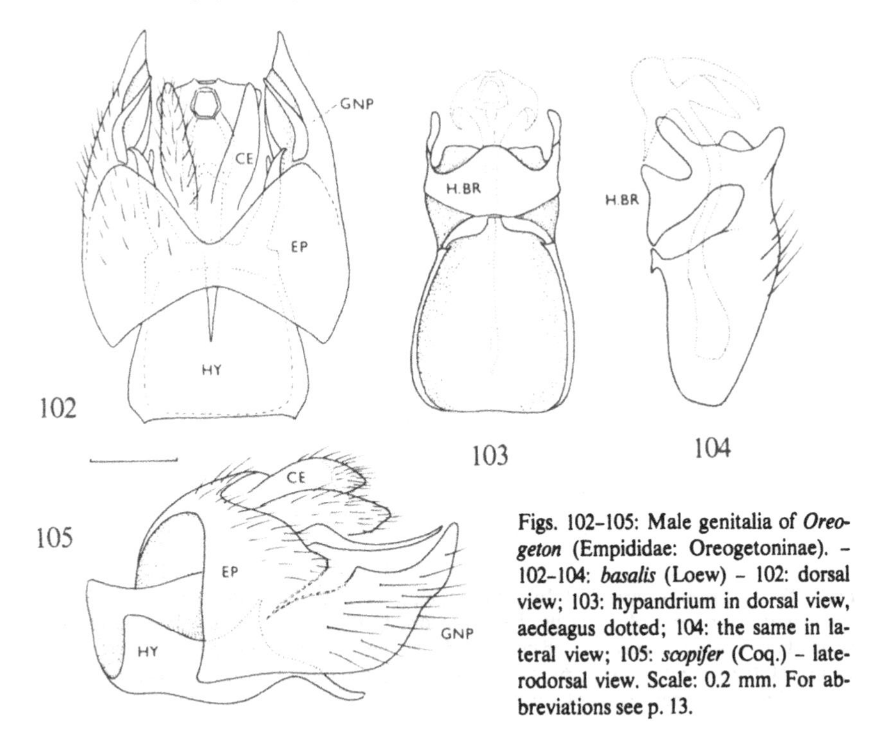

Figs. 102–105: Male genitalia of *Oreogeton* (Empididae: Oreogetoninae). – 102–104: *basalis* (Loew) – 102: dorsal view; 103: hypandrium in dorsal view, aedeagus dotted; 104: the same in lateral view; 105: *scopifer* (Coq.) – laterodorsal view. Scale: 0.2 mm. For abbreviations see p. 13.

the cerci membraneous and very conspicuously developed in *Hesperempis*. A special modification of the epandrium and gonopods (membraneous cerci always conspicuous) has arisen in *Oreogeton* (Figs. 102–105), where the bilobed epandrium (in dorsal position) appears to be partly fused anterolaterally with the gonopods, but it is possible to derive this without much difficulty from the hypopygium of *Hormopeza* (Fig. 87).

In the Palaearctic genera of the Empidinae (the non-Palaearctic genera were not studied), distinct cerci are only present in *Hilara*, as stated by Bährmann (1960). The cerci are rather small and are closely attached to the heavily sclerotised epandrial

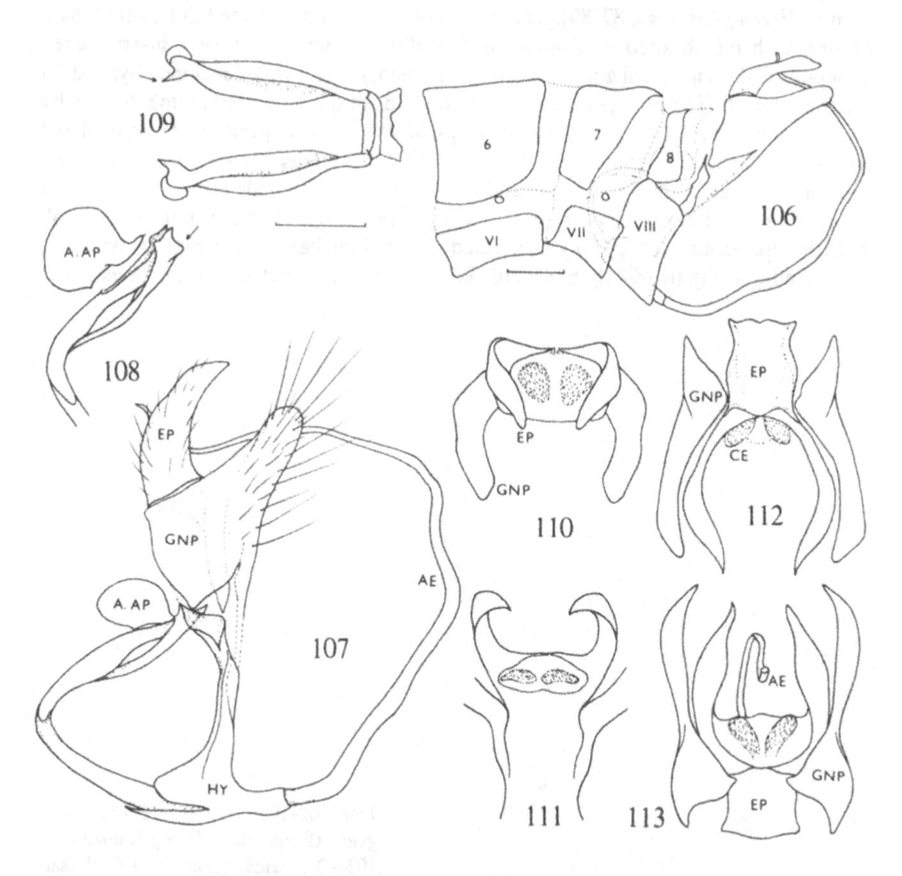

Figs. 106–113. ***Empis pennipes*** **L. ♂ (Empididae: Empidinae). – 106: postabdomen; 107: genitalia in lateral view; 108: base of aedeagus; 109: hypandrium in caudal view (arrows indicating connexion to base of aedeagus); 110: epandrium with gonopods and cerci in caudal view (schematically); 111: the same in ventral view; 112: the same in dorsal view; 113: the same in anterior view. Scale: 0.2 mm. For abbreviations see p. 13.**

(dorsal) lamellae, which are narrowly connected anteriorly (Fig. 114). The gonopods are in the form of large lateral sclerites (lateral lamellae) with vestigial gonostyli (Fig. 115). The same ground-plan condition may be found in other Empidinae, even though the individual parts are developed in various ways as regards size and shape, but the cerci are always vestigial. In the type-species of *Empis* (Figs. 110–113), very indistinct membraneous cerci are visible between the anteriorly-connected dorsal lamellae, which clearly indicates that the lamellae are of tergal origin. The same situation is found in *Rhamphomyia* (Fig. 116), where the dorsal lamellae (epandrium) are very peculiar in shape.

A similar condition is present in the Hemerodromiinae, which are obviously closely related to the Empidinae. In *Hemerodromia* (Figs. 118, 119) the gonopods are bi-articulated, with distinct gonostyli, but the cerci are absent, or at least not distinguishable in the membraneous area between the epandrial lamellae. In *Chelifera* (Figs. 120, 121) small cerci are present, and gonopods without gonostyli.

In the Clinocerinae the cerci are usually well-developed, though absent in

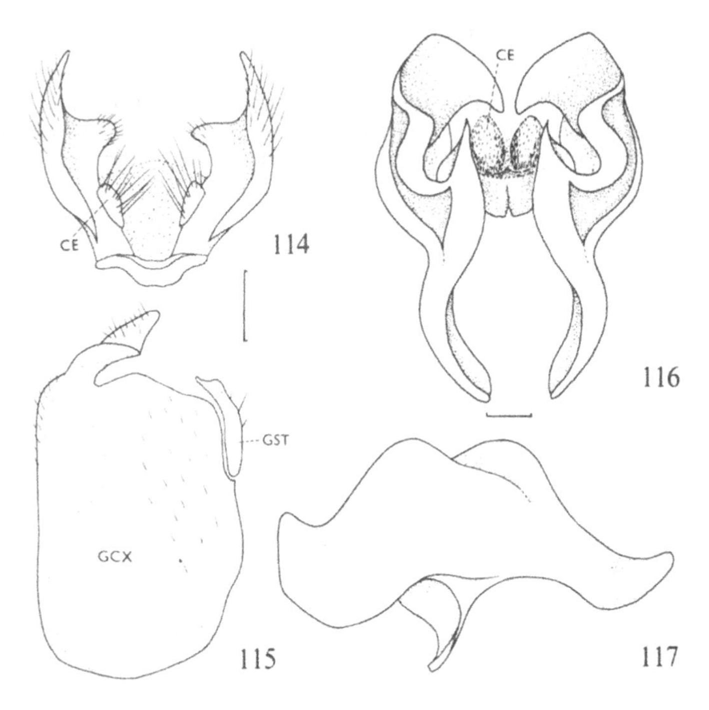

Figs. 114–117. Male genitalia of Empididae: Empidinae. – 114, 115: *Hilara maura* (Fabr.) – 114: epandrium in dorsal view; 115: gonopods in lateral view. – 116, 117: *Rhamphomyia sulcata* (Meig.) – 116: epandrium in caudal view; 117: the same in lateral view. Scale: 0.1 mm. For abbreviations see p. 13.

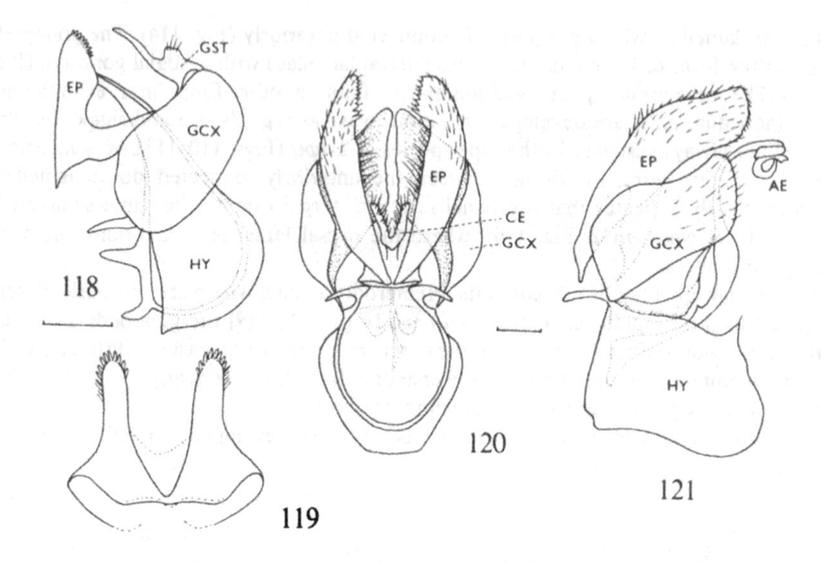

Figs. 118–121: Male genitalia of Empididae: Hemerodromiinae. – 118, 119: *Hemerodromia unilineata* Zett. – 118: lataral view; 119: epandrium in dorsal view. 120, 121: *Chelifera precatoria* (Fall.) – 120: dorsal view (aedeagus omitted); 121: lateral view. Scale: 0.1 mm. For abbreviations see p. 13.

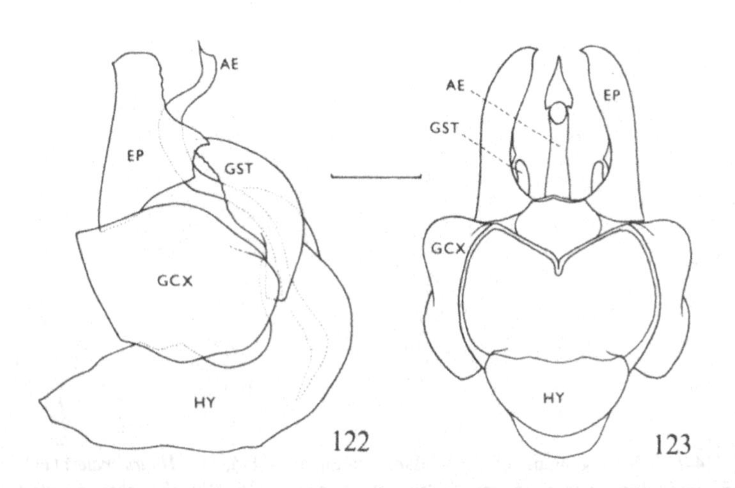

Figs. 122, 123. Male genitalia of *Dryodromia testacea* Rond. (Empididae: Clinocerinae). – 122: lateral view; 123: posterior view. Scale: 0.2 mm. For abbreviations see p. 13.

Dryodromia (Figs. 122, 123). *Clinocera* (Figs. 124, 125) has very distinctive genitalia: the gonopods (very probably with discrete inner gonostyli) have shifted posteriorly above the epandrium, and the proctiger (with small cerci) lies between the epandrial lamellae anteriorly owing to the upward direction of the whole hypopygium; in this position the epandrium becomes ventral and the gonopods dorsal. The membraneous connection between the 8th and 9th tergum, as well as the position of the proctiger, indicates that the »lateral lamellae« in this case are of tergal origin. The homology of the individual sclerites in the male genitalia of the Clinocerinae is very difficult and needs further study, especially in the unusually asymmetrical hypopygium of *Trichopeza* (Figs. 126, 127).

The Ceratomerinae are very specialised and apomorphous in many characters but the hypopygium of *Ceratomerus* (Figs. 128–132) shows the ground-plan condition of the Empididae. The discrete (although paired!) epandrium, distinct cerci and bi-articulated gonopods are plesiomorphous.

On the other hand, the Brachystomatinae, which are undoubtedly more primitive in many respects, have the male genitalia more specialised. The lateral sclerotisation of the hypandrium, the median part being mostly membraneus, is a very characteristic feature in *Brachystoma* and *Homalocnemis*. In *Homalocnemis* (Figs. 134, 135) there are moderately large cerci but only simple, long, lateral clasping organs, which are apparently the gonopods, this also supports the presence of distinct inner folds which are firmly connected to the hypandrial bridge. If this interpretation is correct, then the epandrium has been entirely lost. In *Brachystoma* (Figs. 136–138) there are large soft dorsal appendages, probably cerci (with membraneous remnants of the 10th tergum mesally), and two lateral sclerites; the apical lobe (actually inner in position) covers the hypandrium and aedeagus, and undoubtedly represents the gonopods; the basal (outer)

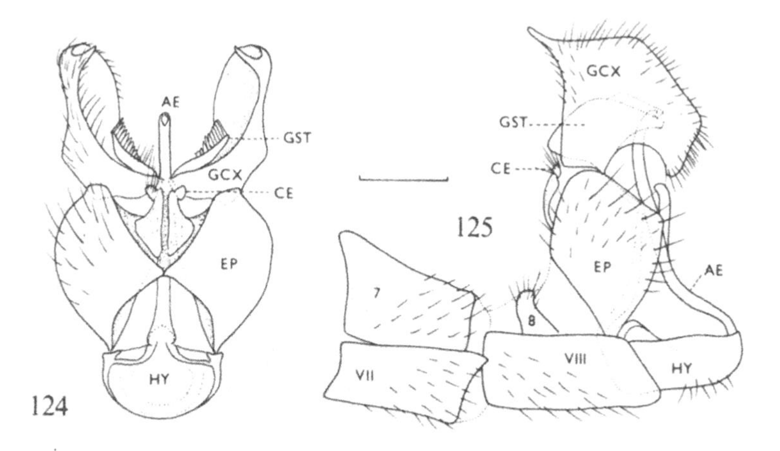

Figs. 124, 125. Male genitalia of *Clinocera* sp. (Empididae: Clinocerinae). – 124: anterior view; 125: lateral view. Scale: 0.2 mm. For abbreviations see p. 13.

lobe looks like an epandrium, but in view of its connection with the inner lobe it may well be the basal article of the gonopods (gonocoxite); if the latter suggestion is correct, then the epandrium has been lost as in *Homalocnemis*. However, the long filiform aedeagus of *Brachystoma* differs substantially from the short stout aedeagus of *Homalocnemis*.

It may be concluded that in the Empididae the dorsal lamellae of the hypopygium are the epandrium whilst the lateral lamellae are gonopods, which are still bi-articulated in several cases, with a distinct basal gonocoxite and a terminal (or rather inner) gonostyle. The distinct inner sclerites (folds), that link the gonopods with the base of aedeagus and partly form the hypandrial bridge, are identical with the periandrial folds of other Empidoidea. The cerci are always soft and more or less membraneous, and are vestigial in some subfamilies. In other families of the Empidoidea (Hybotidae, Atelestidae, Microphoridae, Dolichopodidae) the epandrium is lost and the gonopods (gonocoxites) expanded dorsally, forming a periandrium by fusion across the dorsum.

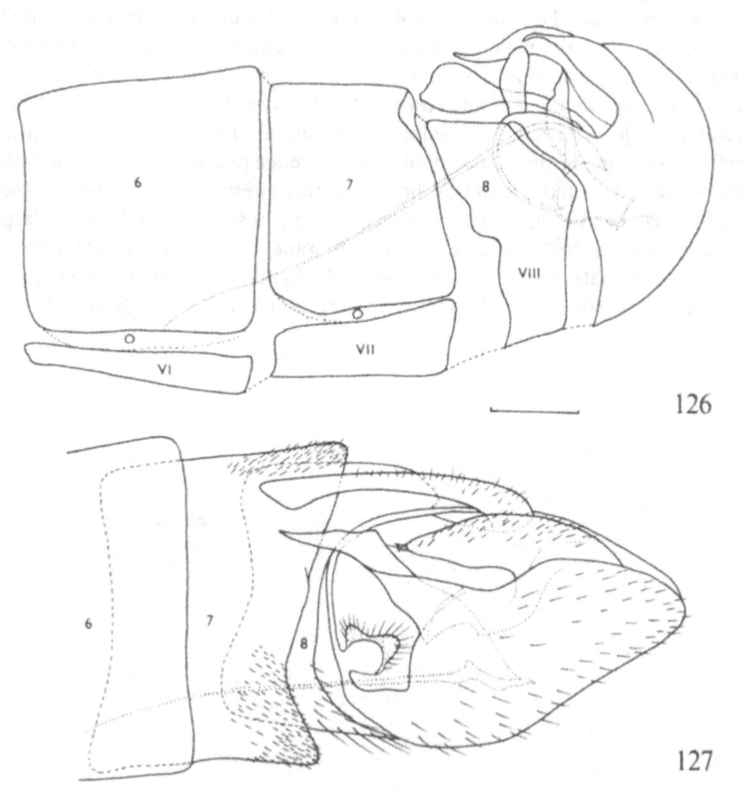

Figs. 126, 127. Male postabdomen of *Trichopeza longicornis* (Meig.) (Empididae: Clinocerinae). – 126: lateral view; 127: dorsal view. Scale: 0.2 mm. For abbreviations see p. 13.

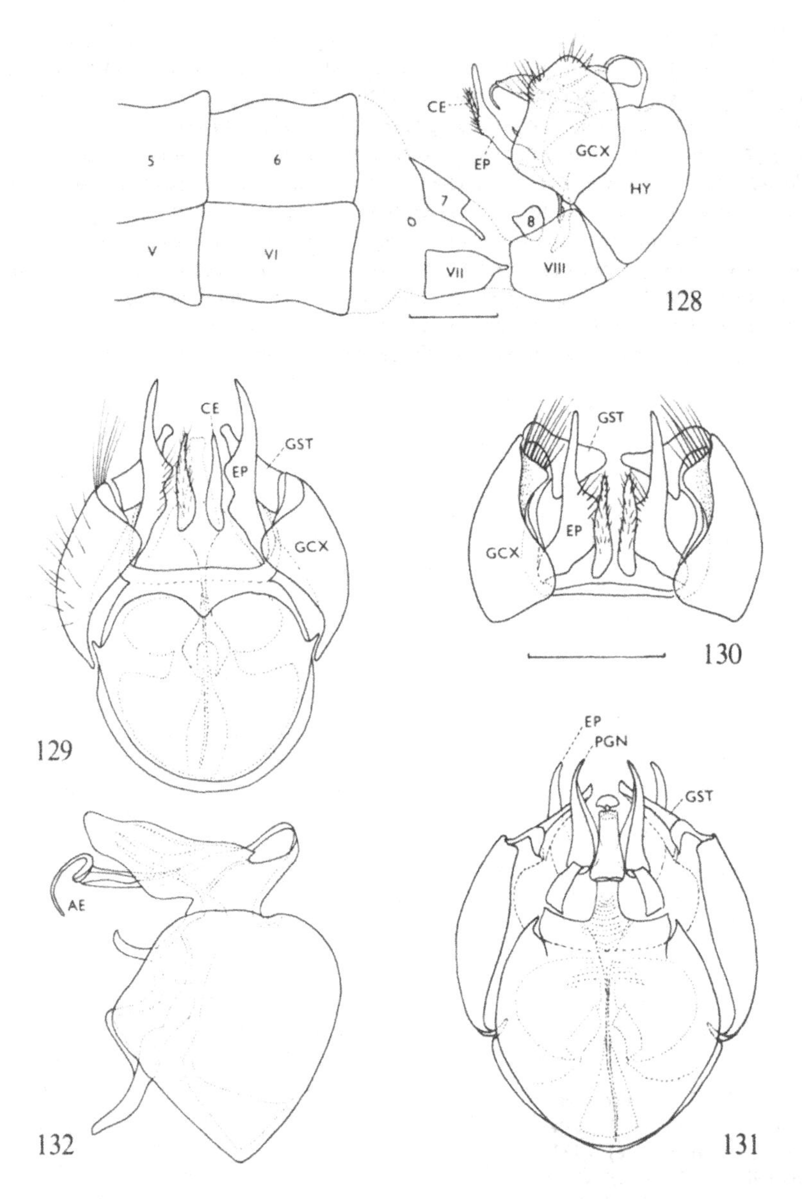

Figs. 128–132. *Ceratomerus mediocris* Coll. ♂ (Empididae: Ceratomerinae). – 128: postabdomen (macerated); 129: genitalia in anterior view; 130: gonopods with epandrium and cerci; 131: genitalia in posterior view; 132: hypandrium with aedeagus. Scale: 0.2 mm. For abbreviations see p. 13.

The hypandrium (9th sternum) is originally a large ventral sclerite covering the base of the aedeagus from below. The plesiomorphous condition in some Oreogetoninae and Empidinae is represented by a large, slightly convex ventral sheath of the aedeagus, which may gradually become turned upwards at the sides and very convex *(Hilara)*, or may gradually be reduced (Empidinae, Brachystomatinae) or completely absent (some *Rhamphomyia*). In the Hemerodromiinae and Clinocerinae the hypandrium is enlarged, and in some groups it is rather firmly connected laterally with the gonopods. In the Microphoridae the hypandrium has become dorsal in position owing to the deflexion and rotation of the pregenital segments, and forms a large and often bristled sclerite that covers the whole hypopygium from above. In the Hybotidae the hypandrium is placed on the left side (or rather above) bacause of the right-hand rotation of the hypopygium; in primitive forms it is a large flat sclerite, often prolonged posteriorly

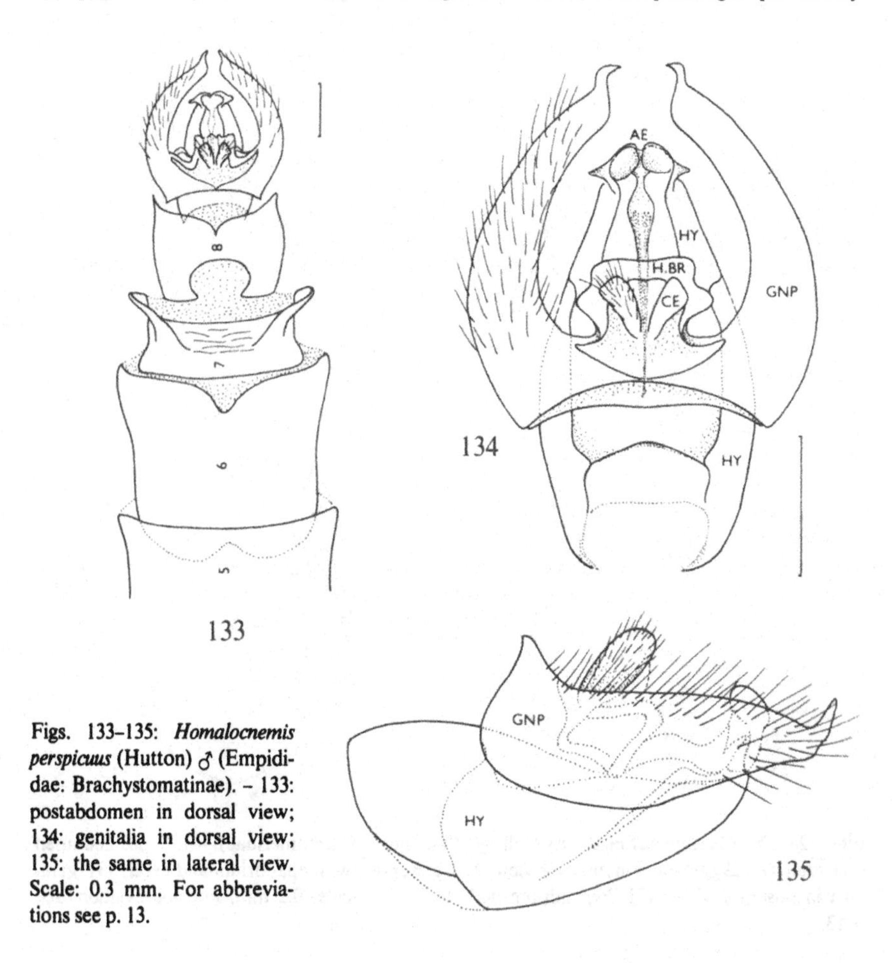

Figs. 133–135: *Homalocnemis perspicuus* (Hutton) ♂ (Empididae: Brachystomatinae). – 133: postabdomen in dorsal view; 134: genitalia in dorsal view; 135: the same in lateral view. Scale: 0.3 mm. For abbreviations see p. 13.

into hypandrial arms, or slightly convex and in the plesiomorphous state almost symmetrical. In primitive forms the hypandrium is a free sclerite, more or less loosely shielding the aedeagus from below, and with a more or less developed hypandrial ring (hypandrial bridge) around the base of the aedeagus; in some Hybotidae the aedeagus is firmly attached to the inner wall of the hypandrium.

In the ground-plan the aedeagus (penis, or phallus) is long and slender, heavily sclerotised and upcurved, and in the Empidinae often curiously bowed or wrinkled and far overlapping the whole hypopygium (Fig. 107). The structure of this simple aedeagus is a very useful and specifically important diagnostic character in the Empididae. In the Hybotidae the aedeagus is much more complicated in structure, often with paired postgonites (rarely unpaired), the sensory lobes near its base of considerable diagnostic value, or the tip of the aedeagus has peculiar long appendages (Ocydromiinae). An apomorphous state of the aedeagus in the Empidoidea is usually indicated by its shortening, the elongation of the aedeagal apodeme (or paired apodemes), weak sclerotisation, and a firmer connection with the hypandrium.

The hypopygium is symmetrical and unrotated in the Empididae and Atelestidae. So far as I am aware, the only asymmetry of the hypopygium within the Empididae is in *Trichopeza* (Clinocerinae) and very remarkably the asymmetry also involves the 8th and partly even the 7th abdominal segments (Fig. 127). Such a development is very excep-

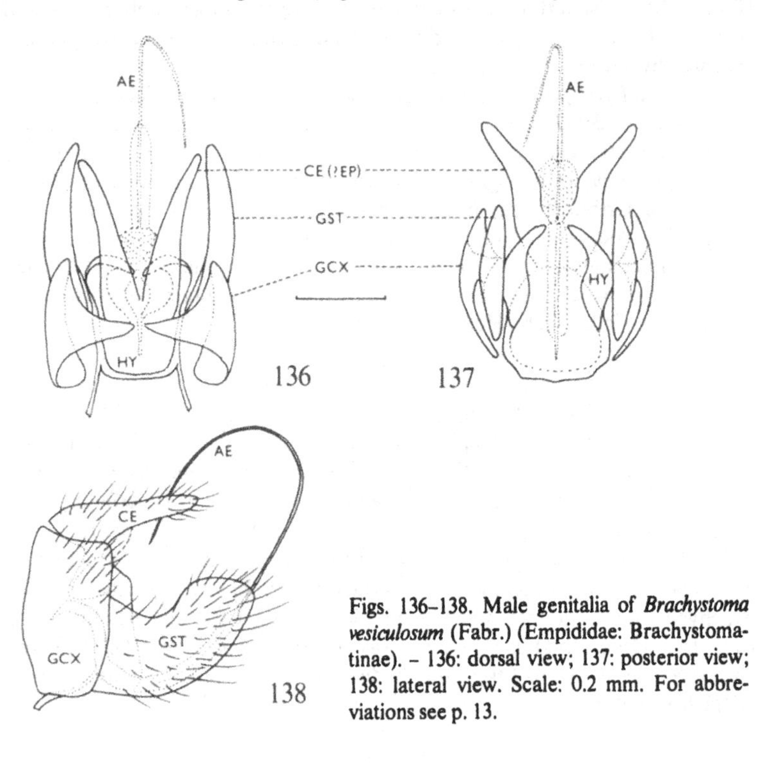

Figs. 136–138. Male genitalia of *Brachystoma vesiculosum* (Fabr.) (Empididae: Brachystomatinae). – 136: dorsal view; 137: posterior view; 138: lateral view. Scale: 0.2 mm. For abbreviations see p. 13.

tional in the Empididae and is undoubtedly an autapomorphy. Partial asymmetry is also present in some *Rhamphomyia* in the Empidinae (subgen. *Dasyrhamphomyia*, *Ctenempis*), with conspicuously enlarged epandrium (dorsal lamellae). Such asymmetry is only partial, with the sternal parts remaining intact, and in fact this arrangement makes it possible for males with abnormally developed genitalia to achieve close contact with females during mating. In the Hybotidae and Microphoridae the male hypopygium is conspicuously asymmetrical (yet almost symmetrical in some primitive forms of the Ocydromiinae and Hybotinae) and has more or less rotated: in the Hybotidae it has rotated through about 90° towards the right, whilst in the Microphoridae it has rotated by the deflexion and partial rotation of the pregenital segments.

Systematic position of Empidoidea

Brauer (1883) was apparently the first to establish that the former "Empididae" and Dolichopodidae are closely related and form a monophyletic group, on the basis of characters shown by the larvae, and he erected the group Orthogenya for these two families. This classification was followed by Verrall (1909 – superfamily Microphona), Hendel (1928 – superfamily Empidoidea: the correct superfamily name), and Hennig (1952, 1954 – family group Empidiformia), and, with some minor exceptions, has been accepted by most recent dipterists.

A more difficult problem concerns the systematic position of the Empidoidea within the suborder Brachycera, particularly because the "Empididae" appear to be somewhat intermediate between the orthorrhaphous and cyclorrhaphous Brachycera. The Empidoidea were originally considered to be the most advanced group of the Brachycera Orthorrhapha, but Lameere (1906) suggested that there was a close relationship between Brauer's group Orthogenya (= Empidoidea) and the Cyclorrhapha and placed them in a higher group Eremoneura. This classification, the correctness of which is in fact supported by my proposed new classification of the Empidoidea, was followed by Hennig (1952, 1954).

The anomalous position of the Empidoidea within the Brachycera, particularly in respect of the wing-venation and the structure of the male hypopygium, is reflected in the various classifications proposed. Enderlein (1936) recognised three suborders of Diptera, the Nematocera, Gephroneura and Brachycera. The Gephroneura consisted of the families "Empididae", Dolichopodidae and Lonchopteridae. Hendel (1937) rejected the suborder Gephroneura and reinstated the superfamily Empidoidea (originally Empididoidea), but he also included in it the Hilarimorphidae.

Aczél (1954) proposed two divisions of the order Diptera, the Orthopyga and Campylopyga. The former included the so-called lower Diptera, Nematocera and a part of Brachycera up to and including the "Empididae"; the latter included the Cyclorrhapha, with the Dolichopodidae. Aczél was in fact separating the Brachycera and Cyclorrhapha by dividing the Empidoidea down the middle, and his solution seems to be the least natural one.

Hennig (1952, 1954), in his classification of the Brachycera, adopted Lameere's (1906) original views and divided the Brachycera into two sections: (1) Tabanomorpha, with subsections Tabaniformia and Asiliformia; and (2) Muscomorpha (= Eremoneura), with subsections Empidiformia (= Orthogenya, Empidoidea) and Musciformia (= Cyclorrhapha). The systematic position of the Empidoidea, forming a monophyletic group within the Cyclorrhapha, was confirmed by Griffiths' (1972) studies on the male postabdomen in the Cyclorrhapha, and indirectly by Bährmann (1960) and Ulrich (1972). The Eremoneura is characterised by a set of apomorphous characters given in the section 'Classification' for the family Empididae. The Cyclorrhapha, which are a monophyletic subgroup of the Eremoneura, may be characterised by the following apomorphous characters: (1) larval head-capsule reduced, pupation within the larval skin (puparium); (2) male genitalia twice rotated, through 360° (hypopygium circumversum); (3) male 8th tergum in the form of a narrow band placed ventrally (the position after first rotation throuoh 180°).

It should be noted that Hennig (1972), on the basis of the articulation of the antennal style, subsequently rejected his recognition of the Eremoneura (Empidoidea and Cyclorrhapha) as a monophyletic group, and reinstated the original classification which postulates that the Empidoidea are more closely related to the lower Brachycera and form a monophyletic group with the Asiloidea (Asiliformia). He claimed that the antennal style is 7-articulated in the ground-plan of the Tabaniformia, 2-articulated in the Asiloidea and Empidiformia, but 3-articulated in the Cyclorrhapha. As mentioned above in the section on antennal morphology, the articulation of the antennal style is not as constant as Hennig supposed. The number of articles in the style varies considerably, not only in the Empidoidea but also in other brachycerous families (see Chvála, 1981a: 226). Later Hennig (1976) himself once again adopted the Eremoneura, particularly when a 2-articulated style was found in the cyclorrhaphous family Opetiidae.

If my view of the phylogeny and classification of the superfamily Empidoidea is correct, the breakdown of the former "Empididae" (with the Empididae in the new restricted sense representing a more primitive lineage and the Atelestidae and Cyclorrhapha descending from an ancestor common only to them) clearly shows that Lameere's (1906) group Eremoneura was well-founded.

This classification also shows very clearly that the suborder Brachycera cannot be divided into two further suborders, Brachycera and Cyclorrhapha, because the Empidoidea (as non-cyclorrhaphous flies) are then separated unnaturally from the Cyclorrhapha. The family Atelestidae links brachycerous and cyclorrhaphous flies and, as an undoubted intermediate between the Empidoidea and the primitive groups of the Cyclorrhapha (? Platypezidae), it clearly demonstrates the monophyly of the Empidoidea and Cyclorrhapha (= Eremoneura). Certain other classifications of Diptera based more on a 'natural evolutionary pattern', as for instance that of Oldroyd (1977), are not considered here as they seem to be rather biassed and in fact they do not directly affect the classification of the Empidoidea.

Historical review of "Empididae"

The variety and diversity of species and genera included in the former "Empididae" are clearly reflected in the various different classifications of the family. It is curious that the "Empididae" were divided into three distinct families up to the second half of last century. Although practically all the higher taxa (families and orders) of the classical entomologists have subsequently been split into two or more taxa of the same level, the "Empididae" is one of the few exceptions to this.

The genus *Empis* was erected by Linné in 1758 and included only the large species, the smaller species being described in the genus *Musca*. Further genera were not described until later, especially by Meigen and Macquart. Meigen (1804) used a generic classification (genera *Empis, Tachydromia, Hybos*), as did Fabricius (1805), and the family was formally erected by Latreille (1809), his 8th family Empides. The same classification was used by Fallén (1815), family Empidiae, but a year later, in his 'Supplementum' (Fallén, 1816), he proposed a new classification of the family Empidiae, with a division into three families which corresponded to Meigen's (1804) three genera, the Hybotinae, Empidiae and Tachydromiae. The family "Tachydromiae" included the small raptorial forms (*Tachydromia, Hemerodromia* etc.) and this step of including the Hemerodromiinae and Clinocerinae in the Tachydromiae was followed until Schiner's (1862) reclassification. Fallén's proposal to recognise three separate families (sometimes with the family suffix differently spelled) was followed for instance by Meigen (1820, 1822), Wiedemann (1828–30), Zetterstedt (1842) and Scholz (1851).

Latreille (1825) included all the orthorrhaphous Brachycera in his 2nd family Tanystoma (1st family Nemocera), and the "Empididae" were placed in his 7th tribe Hybotini and 8th tribe Empides; the Hybotini included the modern Hybotinae and Ocydromiinae (except *Bicellaria*), whilst all the other genera were included in the Empides. This tribal classification was followed by Macquart (1834), although earlier (Macquart, 1827) he had divided the "Empididae" into two families, the Empides (as defined by Latreille) and Hybotides (including a new genus *Microphorus*). Griffith (1832) accepted Latreille's (1825) family Tanystoma but only recognised two tribes, the Asilici Latr. and Hybotini Latr., the latter including all the "Empididae".

Schiner (1862) was apparently the first author to recognise a single family "Empididae" (s.lat.), arranging all the "empidide" forms in five subfamilies: Tachydromiinae, Hemerodromiinae, Empidinae, Hybotinae and Ocydromiinae. In fact, this classification has been more or less followed by all subsequent authors up to the present.

No real attempt at a natural classification of the "Empididae", based on the morphological characters of the adults, was made until the first half of this century. Melander (1928) divided the family into seven subfamilies, arranged from the most specialised to the most primitive: Tachydromiinae, Hemerodromiinae, Clinocerinae, Empidinae, Ocydromiinae, Hybotinae and Brachystomatinae. He placed the two peculiar New Zealand and South American genera *Ceratomerus* and *Homalocnemis* in the Clinocerinae and Brachystomatinae respectively. Collin (1928) placed these two genera in their own subfamilies, Ceratomerinae and Homalocneminae, but later (Collin, 1933) he himself rejected the Homalocneminae and placed *Homalocnemis* in the Hybotinae. Smith (1969) again transferred it to the Brachystomatinae, as Melander (1928) had originally done, and this classification seems to be correct.

Hennig (1970) recognised two further subfamilies, the Microphorinae and Atelestinae, which are now treated as families, and Chvála (1976) distinguished the subfamily Oreogetoninae, which now includes the primitive forms of the restricted family Empididae.

Collin's (1961) classification in his excellent revision of the "Empididae" in 'British Flies' was very simplified and certainly unnatural. It must be regarded as an unfortunate grouping of non-related forms for the quicker and simpler identification of British species, as Collin himself noted several times in the text. His broad concept of the Hybotinae, which included forms with a small saddle-shaped prosternum, was correctly criticised by Tuomikoski (1966) and has not been followed. Unfortunately, the unnatural assignment of the Hemerodromiinae and Clinocerinae to a single subfamily Hemerodromiinae has been followed by many subsequent authors.

Extinct forms

Family Empididae

The oldest known ancestor of the Empidoidea is about 160 million years old. It was described by Usachev (1968) as *Protempis antennata* in the extinct family Protempididae, from Upper Jurassic shales. The positive and negative compression fossil of the single female (holotype) was found in 1963 at the celebrated locality in the Karatau range of hills in Kazakhstan, NW of Dzhambul, Central Asia. The specimen, which is complete and lying on its right side, was fully described and illustrated by Usachev (1968: 624, Fig. 6), but Kovalev (letter of 11.1.1982) drew my attention to some errors in Usachev's original description. During a visit to Moscow in June 1982, I had an

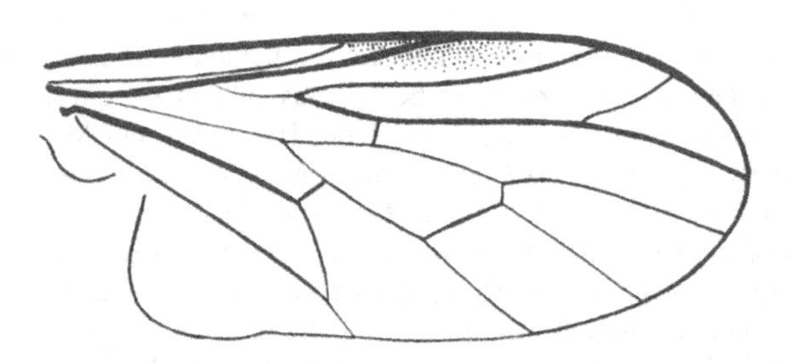

Fig. 139. Wing of *Protempis antennata* Usachev (Protempididae), holotype ♀ (V. G. Kovalev del.).

opportunity to re-examine the holotype of *Protempis antennata:* in fact, the alula is present in *Protempis* as shown in Fig. 139 (not mentioned or illustrated by Usachev) and the costa runs right around the wing, distinct even on the posterior wing-margin (Usachev mistakenly described it as ending "in ¼ distance between R_5 and M_1").

All the main characters of *Protempis antennata* agree with the ground-plan of the Empididae except for the presence of the alula and the shape of the anal cell (Fig. 139); the latter is very long and short-petiolate, closely resembling the closed anal cell of some "lower" Brachycera (Tabanoidea, Stratiomyoidea, Asiloidea or Bombylioidea). Perhaps the shape of the anal cell justifies the assignment of *Protempis* to a separate family, as this type of anal cell is unknown in recent Empididae.

The antennae of *Protempis* are of plesiomorphous structure with a rather short apical 2-articulated style (with a small basal article). The proboscis is not visible: a structure resembling a rather long proboscis projecting diagonally below the head is the right fore tibia, the tarsus being disconnected; the legs are simple. The wings have a distinct, almost rectangular axillary lobe and well-developed alula: the alula appears to have been entirely lost in the recent Empididae, although it is still retained in the extinct genus *Trichinites* of the Hybotidae and in all recent Atelestidae. Otherwise, the wing-venation of *Protempis* is plesiomorphous within the Empidoidea, with vein C running around the wing, Sc complete and ending in the costa, R_4 and R_5 forked, discal cell emitting 3 veins to the wing-margin, and rs short; the only substantial difference seems to be the unusually long-petiolate anal cell. The whole insect (body length 3.4 mm, wing 2.8 mm) has much in common with both the Oreogetoninae and some Empidinae, but unfortunately the prosternum is not visible.

All other extinct forms of the Mesozoic period are from Cretaceous amber, the fossilised resin of trees, and all belong to the families Hybotidae, Microphoridae or Dolichopodidae. According to Kovalev (in litt.) several extinct forms of Cretaceous Empididae have recently been found by the Palaeontological Institute of the USSR Academy of Sciences, including some Oreogetoninae, but descriptions are not yet available.

Family Hybotidae

The oldest known member of this family is *Trichinites cretaceus* Hennig, 1970, described from the holotype ♀ from the Lower Cretaceous of Lebanon, from the Jazzine district. As the extinct genus *Microphorites* (see below) was described from the same locality and the same horizon, the two branches of the Hybotidae and Microphoridae had evidently already separated. *Trichinites* is thought to be a common ancestor of the whole family (Ocydromiinae, Hybotinae, Tachydromiinae), having many of the synapomorphies of the Hybotidae (C running to M_1, Sc incomplete, R_{4+5} unforked) but anal cell still in a plesiomorphous state, at least as long as 2nd basal cell and the vein closing it (Cu_{1b}) rather straight, very much as in the recent *Syneches* (Fig. 150). In both *Hybos* and the Atelestidae the vein closing the anal cell is still longer and distinctly curved (Fig. 148). The antennae of *Trichinites* are plesiomorphous, as in the primitive Ocydromiinae (Trichinini), and the thorax has numerous hairs on the mesonotum. As the fore tibiae do not have a tubular gland and wing alula is still slightly developed in *Trichinites*, the Atelestidae may well have separated of shortly before this period. Nevertheless, it is difficult to decide if the Atelestidae were already distinct in the early Cretaceous, or if *Trichinites* is a common ancestor of both the Hybotidae and Atelestidae, as was suggested by Hennig (1970). I would suggest that the Atelestidae, which are thought to share a common ancestor with the Cyclorrhapha, are in fact a very ancient group of flies from which the Hybotidae had already separated before the Lower Cretaceous; this hypothesis may be supported by the fact that the Platypezidae (Kovalev, 1979a) and the Microphoridae were also in existence at that time.

Cretoplatypalpus archeus Kovalev, 1978 (♂♀, Cenomanian) and *Archiplatypalpus cretaceus* Kovalev, 1974 (♂, Coniak-Santonian) were both described from the ambers of western Taymyr, Siberia. They are extinct Hybotidae from the late Cretaceous and appear to be direct ancestors of the subfamily Tachydromiinae; this means that the three subfamilies of the Hybotidae had already differentiated by the Upper Cretaceous. Further data on these fossil genera and species, and their presumed systematic positions, were given in my earlier paper (Chvála, 1981a: 232).

Electrocyrtoma electrica Cockerell, 1917 (? sex), described from Burmese amber which is probably Upper Cretaceous, is another extinct form of the Tachydromiinae closely related to *Cretoplatypalpus* according to Kovalev (1978).

Family Microphoridae

The Lower Cretaceous *Microphorites extinctus* Hennig, 1971, was described from Lebanese amber, from the same period as *Trichinites*. The single holotype female has all the main morphological characters of the recent Microphoridae, including the very characteristic wing-venation with very small basal cells, anal cell equally small and rounded at tip, radial sector originating opposite humeral crossvein, and no alula. The axillary lobe of wing is very slightly developed, practically absent, a condition that is thought to be plesiomorphous within the Microphoridae; the large axillary lobe of recent Microphorini is considered to be a secondarily apomorphous state. Further characters (antennae inserted at middle of head in profile, style 2-articulated, eyes bare, 2nd basal cell closed, abdomen telescopic with cerci) are part of the ground-plan of the recent tribe Microphorini. The distribution of the main characters in *Microphorites*, in comparison with the Microphorini and Parathalassiini, is shown in Fig. 142. The enlarged upper facets of the dichoptic female eyes in *Microphorites* are very unusual, but the apically weaker veins Sc and R_1 (see Hennig, 1971: 17, Fig. 21) can hardly be truly abbreviated veins, as I previously believed (Chvála, 1981a: 234), and I do not think that this character is an apomorphy.

Cretomicrophorus rohdendorfi Negrobov, 1978, was described from the male holotype from the Upper Cretaceous, Coniak-Santonian, of the eastern Taymyr, Siberia, from the same locality as *Archiplatypalpus*. This species has a large number of the synapomorphies of the tribe Parathalassiini of the Microphoridae, except for the bare eyes. I now believe that the complete vein Sc reaching the costa is a ground-plan character of the Microphoridae, which differs from my earlier interpretation (Chvála,

1981a: 234). However, there are also three apomorphous characters that are only found in the Dolichopodidae: (1) C runs to M_1 (absent on posterior wing-margin); (2) 2nd basal cell joined to discal cell (crossvein tb closing 2nd basal cell completely absent); and (3) posterior four tibiae bristled. In other words, *Cretomicrophorus* possesses several features of the Dolichopodidae which do not occur in recent Microphoridae, although the discal cell still emits three veins to the wing-margin as in the Microphoridae. It

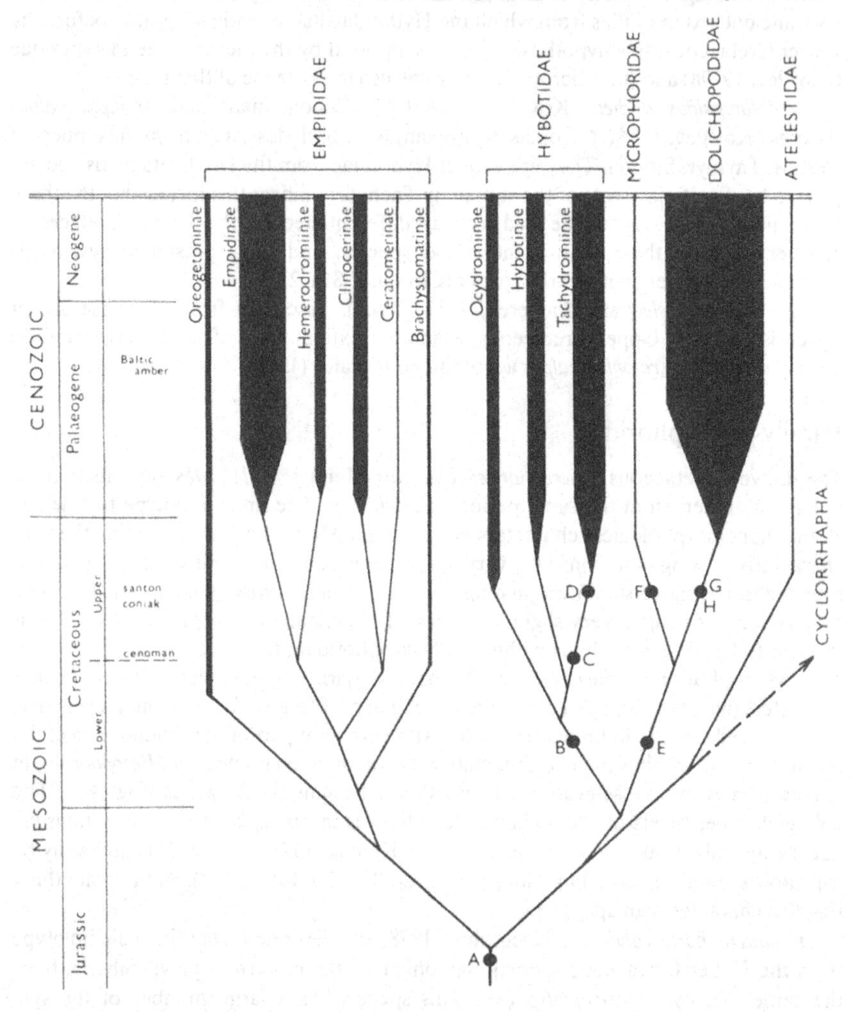

Fig. 140. The presumed phylogeny of Empidoidea. A – *Protempis* (Protempididae); B – *Trichinites;* C – *Cretoplatypalpus;* D – *Archiplatypalpus;* E – *Microphorites;* F – *Cretomicrophorus;* G – *Archichrysotus;* H – *Retinitus.*

means that *Cretomicrophorus* lies rather on the boundary between the Microphoridae and Dolichopodidae and may well indicate the period when the Dolichopodidae were differentiating from the Microphoridae.

Family Dolichopodidae

The genera *Archichrysotus* (2 species) and *Retinitus* (1 species), from the same Upper Cretaceous amber as *Cretomicrophorus*, are assigned to the Dolichopodidae. The main reason for their inclusion in this family is the 2-branched media (vein M_2 absent), a character known only in the Dolichopodidae and not at all in recent Microphoridae, where the media is always 3-branched. However, in both these genera the antennal style is 1-articulated (as in the Parathalassiini) and the eyes bare (as in the Microphorini); moreover, *Archichrysotus* still has a complete crossvein tb closing the 2nd basal cell, and consequently the discal cell is completely separated from the 2nd basal cell. In view of these three characters, the extinct genera *Archichrysotus* and *Retinitus* cannot be regarded as absolutely identical with recent Dolichopodidae; their existence supports the view that the Dolichopodidae, so far as it is known from the few scattered fossil forms that have been described, were not completely differentiated from the Microphoridae in the late Cretaceous.

The Coenozoic Empidoidea are not discussed here as they cannot contribute very much to new understanding of phylogeny at the family level. However, there are large collections in various museums (London, Copenhagen, Berlin, Amsterdam, Brussels) of Baltic amber, from the late Eocene to early Oligocene periods, and Dominican amber (Santo Domingo, Stuttgart), from the late Oligocene or early Miocene periods. This material will undoubtedly prove to be very important for our understanding of the phylogeny of the genera and species, but most of it is still undetermined and those parts that have been studied (Meunier, 1908) were usually misidentified generically. The specimens described from Burmese amber (Cockerell, 1917) appear to be Upper Cretaceous, not Tertiary, and are roughly as old as Canadian amber. A comprehensive study of Baltic amber, its distribution and its fauna, was recently published by Larsson (1978).

Phylogeny

To work out the phylogeny of the Empidoidea we need to analyse our knowledge of the few fossil forms and of the relationships between the recent taxa, which are reflected in the classification. The phylogeny of the Empidoidea shown in Fig. 140 is based mainly on an analysis of the morphological characters of recent species and on the relationships between the higher taxa. The lengths of the palaeontological periods in Fig. 140 are given in a comparative scale, and the thickness of the subfamily and family columns indicates the number of recent species. It must also be borne in mind that only a few fossil forms are known and they need not necessarily to be on a direct line of descent. I agree with Tuxen (1980) that "the phylogenetical line will not be a true line, but a line with offshoots, and those are the types we may by chance find in fossil states" and "if we find them, we cannot, except in rare cases, deduce a direct descent from them". For instance, *Trichinites* may well be a side-branch from a hybotid ancestor in which the tibial gland was not developed; or *Archichrysotus*, with its closed 2nd basal cell and 1-articulated antennal style, may well not be an ancestor of the Dolichopodidae or a primitive form of both the Parathalassiini (Microphoridae) and the Dolichopodidae, but may only be a side-branch of the Dolichopodidae which became extinct in the late Cretaceous or early Palaeogene period.

As shown in Fig. 140, the Eremoneura (Muscomorpha) are thought to have evolved along two phylogenetic lines. One line, with complete wing-venation (rs 3-branched) and symmetrical unrotated male genitalia with discrete lateral gonopods and median 9th tergum (epandrium), led to the Empididae. The second line had a reduced wing-venation (rs 2-branched) and originally symmetrical unrotated male genitalia with a developed periandrium, the epandrium being lost. The Jurassic Protempididae are tentatively considered to be the precursors of both these lines, though they may well belong to the first line.

The second phylogenetic line was obviously a more progressive one; the primitive forms with symmetrical male genitalia led to the Atelestidae and Cyclorrhapha, which are thought to form a monophyletic group. They probably differentiated in the early Cretaceous period as the first known extinct members of the Cyclorrhapha are known from the Lower Cretaceous (Hennig, 1971; Kovalev, 1979a). This suggests that at that period the Cyclorrhapha had already separated from the Atelestidae (the few recent atelestids appear to be a relict group of ancient pre-cyclorrhaphous forms) and *Trichinites* would then be a precursor of the hybotid line only.

Another phylogenetic line separated from the "atelestid" line, probably during the early Cretaceous, and it may be characterised by the apomorphous asymmetry of the male genitalia. This line had again bifurcated during the Lower Cretaceous period (see *Trichinites* and *Microphorites*), to produce the "hybotid" and "microphorid" phylogenetic lines.

The hybotid line is characterised by many apomorphous features (see Classification) but the main character is the rotation of the male genitalia through about 90° towards the right. This line has led to the recent Hybotidae, and the three recent subfamilies (Ocydromiinae, Hybotinae, Tachydromiinae) were undoubtedly distinct by the early Upper Cretaceous (*Cretoplatypalpus*).

The microphorid line developed in quite a different way from the Hybotidae, by the deflexion (or also partial rotation) of the pregenital segments (but not the hypopygium), and by the very distinctive wing-venation. This line led to two recent families, the Microphoridae and Dolichopodidae, which are supposed to be a monophyletic subgroup within the Empidoidea. In view of the existence of several somewhat "intermediate" extinct forms from the Upper Cretaceous period, and of the evolutionary radiation of the Dolichopodidae during the Palaeogene (abundant material in Baltic

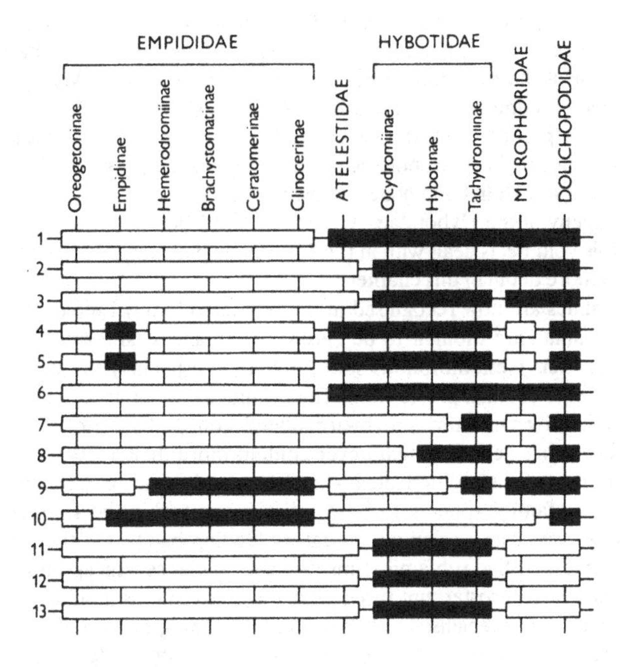

Fig. 141. Distribution of some characters among the families and subfamilies (except for the Dolichopodidae) of the Empidoidea. Where common possession of apomorphous or plesiomorphous conditions is interpreted as due to synapomorphy or symplesiomorphy, continuous corresponding rectangles are given.

Plesiomorphous conditions (white rectangles)	Apomorphous conditions (black rectangles)
1. Gonopods forming separate lateral lamellae	Gonopods fused basally, forming periandrium
2. Hypopygium symmetrical	Hypopygium asymmetrical
3. Hypopygium not rotated	Hypopygium rotated or deflexed
4. Costa running around the wing	Costa running to M_1 or R_{4+5}
5. Sc complete, reaching costa	Sc incomplete
6. rs with 3 branches, R_{4+5} forked	rs with 2 branches, R_{4+5} not forked
7. Discal cell present	Discal cell absent
8. M with 3 branches	M with 2 (or 1) branches
9. Axillary lobe developed	Axillary lobe not developed
10. Prosternum small, isolated	Prosternum large, fused with episterna
11. Fore tibiae without tubular gland	Fore tibiae with tubular gland
12. Maxillary lacinia present	Maxillary lacinia absent
13. Palpi attached to stipites	Palpi removed from stipites, connected through a sclerotized palpifer

amber), it is thought that the Dolichopodidae originated relatively late, not before the Upper Cretaceous period. The Lower Cretaceous *Microphorites* might be considered to be a precursor of the common microphorid line, and the recent Microphoridae (with about 40 species, compared with nearly 4,500 species of recent Dolichopodidae) may well be a relict group within the recent fauna, rather like the Atelestidae. The distribution of the main morphological characters among the extinct and recent Microphoridae and Dolichopodidae is shown in Fig. 142.

The phylogeny of the Hybotidae, Atelestidae and Microphoridae, the three families treated in this volume, is dealt with in the Systematic part. However, the Empididae are discussed in more detail in this chapter.

Six subfamilies are now recognised in the restricted family Empididae, and of these the Oreogetoninae are thought to be closest to the ancestral forms, with a full set of characters in their plesiomorphous state (Fig. 141), especially the wing-venation and the primitive type of small prosternum. The primitive type of male genitalia, with a discrete unmodified epandrium and bi-articulated gonopods, is present in *Hormopeza* (Fig. 87). The other genera, which never contain more than a few recent species, exhibit a wide range of structural details in the genitalia and of antennal structure. In general, the subfamily appears to be a relict group within the Empididae, and has retained what are clearly primitive predatory feeding habits. On the other hand, the large and very diversified subfamily Empidinae possesses several apomorphous characters, such as a large prosternum fused with the episterna (as in all the other empidid subfamilies), a more specialised wing-venation (C ending at wing-tip, Sc incomplete), and male genitalia usually with a conspicuously developed cleft dorsal epandrium, which forms remarkable "dorsal lamellae". The Empidinae have also modified their original predatory feeding habit, and have mostly adopted nectar feeding.

The subfamilies Hemerodromiinae and Clinocerinae have transferred their larval development into water and, as a result of their markedly predaceous habits as adults, both have developed in parallel a similar general appearance (prolongation of the front coxae, and posterior four legs shifted posteriorly) which led Collin (1961) and some other authors to combine these two groups into the single subfamily Hemerodromiinae. However, the Hemerodromiinae are obviously more closely related to the Empidinae than to the Clinocerinae, having male genitalia and mouthparts of very similar structure. The loss of the axillary lobe on the wing, a character common to the Hemerodromiinae and Clinocerinae, as well as to the Ceratomerinae and Brachystomatinae, is rarely found in the Empidinae.

The Clinocerinae have male genitalia that are more primitive in their ground-plan, with distinct cerci (except *Dryodromia*), but they are remarkably specialised (asymmetrical) in *Trichopeza*, and which resemble more closely the genitalia of the Oreogetoninae in general structure. The mouthparts are more specialised in the Clinocerinae, where the maxillae are reduced in *Clinocera* and quite absent in *Wiedemannia*.

The Ceratomerinae *(Ceratomerus)* have male genitalia resembling the Clinocerinae in their ground-plan, but the wing-venation is very reduced (basal cells confluent, anal cell absent, apical fork to M_1 and M_2 present) and the antennae are strikingly shaped.

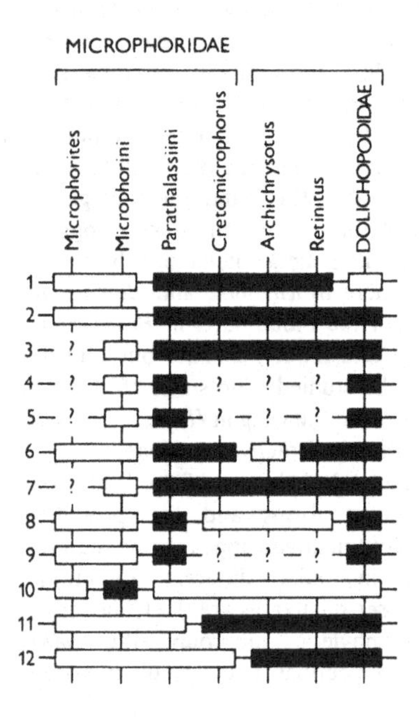

Fig. 142. Distribution of some characters among the extinct and recent Microphoridae and Dolichopodidae. For explanations see Fig. 141.

Plesiomorphous conditions (white rectangles)	Apomorphous conditions (black rectangles)
1. Antennal style 2-articulated	Antennal style l-articulated
2. Antennae placed below or at middle of head	Antennae placed above middle of head
3. Clypeus flat, separated as distinct sclerite	Clypeus convex and/or not separated from face
4. Maxillary lacinia present	Maxillary lacinia absent
5. Labial paraphyses free from maxillary stipites	Labial paraphyses attached to maxillary stipites
6. Crossvein t_b complete	Crossvein t_b incomplete (or absent)
7. Males holoptic	Males dichoptic
8. Eyes bare	Eyes pubescent
9. ♀ abdomen telescopic, 9th tergum complete	9th tergum in ♀ cleft and spinose, abdomen blunt-ended
10. Axillary lobe on wing absent or slightly developed	Axillary lobe well developed
11. Costa running around the wing	Costa running to M_1 or R_{4+5}
12. M with 3 branches, M_2 present	M with 2 branches, M_2 absent

However, the long slender and heavily sclerotised vertical proboscis (shifted forwards on the head) is very unlike other "hemerodromiine" forms and resembles the "empidine" type adapted for nectar feeding, as in *Rhamphomyia* subgenus *Lundstroemiella.*

The Brachystomatinae differ substantially from these three subfamilies in having the wing-venation very plesiomorphous, with a long anal cell (Figs. 44, 50) of the same type as in the Atelestidae and Hybotinae, but otherwise not found within the Empididae. This indicates that the Brachystomatinae may well have separated from the common empidid line very early. On the other hand, the male genitalia are very apomorphous, with the apparent loss of the epandrium. Taking together the characteristic wing-venation, the structure of the male genitalia and the remarkably long basal article of the antennal style, the genera *Brachystoma* and *Homalocnemis* appear to be correctly classified. The structure of the mouthparts in *Brachystoma* is very remarkable, with their basal maxillar ring and stout maxillae, but unfortunately the mouthparts of *Homalocnemis* could not be studied. The small "*Hybos*-type" prosternum and rather well-developed axillary lobe on the wing in *Homalocnemis* suggest that *Brachystoma* is a more specialised form within the Brachystomatinae.

To summarise the presumed phylogeny of the Empididae, as worked out from the relationships of the recent forms and the probable evolutionary lines, the Oreogetoninae and Brachystomatinae appear to be the most ancient groups, although they have developed in quite different directions. The Empidinae and Hemerodromiinae are closer related, as are the Clinocerinae and Ceratomerinae, and perhaps form a monophyletic subgroup; the partial resemblance of the Hemerodromiinae and Clinocerinae is the result of parallel evolution under similar conditions.

Classification

The suggested new classification of the former family "Empididae", which is now divided into four distinct families, will probably be criticised by some students of dipterology, although the breakdown of the former "Empididae" is not a new idea at all. During the last 20 years it has been pointed out by several authors (e.g. Colless, 1963; Tuomikoski, 1966; Hennig 1970, 1971; Ulrich, 1971; Griffiths, 1972; Chvála, 1981a) that the "Empididae" is a paraphyletic and unnatural unit, particularly with the Dolichopodidae excluded from the family. When discussing my new classification of the former "Empididae" with the leading dipterists working in taxonomy, morphology or phylogeny, I often have been assured that the proposed breakdown of the "Empididae" is rational and long overdue. I therefore hope that this proposed reclassification of the Empidoidea will be generally accepted.

The superfamily Empidoidea, which originally consisted of the families "Empididae" and Dolichopodidae, is here divided into five families, the Empididae (s.str.), Hybotidae, Atelestidae, Microphoridae and Dolichopodidae. Each family is briefly characterised and discussed below.

Family Empididae

The family Empididae, as defined here, is restricted to a group of subfamilies of the former "Empididae" ranked by Hennig (1970) as the subfamily-group Empidoinea. It includes six well-defined subfamilies, the Oreogetoninae, Empidinae, Hemerodromiinae, Clinocerinae, Ceratomerinae and Brachystomatinae, with about 2,100 recent species.

The family may be characterised by apomorphous characters which are part of the ground-plan of the Empidoidea (Orthogenya) and of the whole Eremoneura (Muscomorpha), and which are apomorphous in comparison with the ground-plan of the Brachycera (and Asiloidea). This is partly adapted from Griffiths (1972).

1. Vein M with only 3 branches (M_3 and M_4 fused).
2. Anal cell (cell Cu) shortened, at most slightly longer than 2nd basal cell, apically closed and petiolate (veins Cu_{1b} and 1A with a common apical section, the "anal" vein (A_1)).
3. ♂ genitalia with discrete gonopods expanded dorsally, forming "lateral" lamellae.
4. 9th tergum (epandrium) reduced or forming paired "dorsal" lamellae, never in the form of a large unmodified sclerite.
5. Aedeagus slender, upcurved distally.
6. 9th sternum (hypandrium) forming a hypandrial bridge round the base of aedeagus.
7. Larvae with hypopharyngeal skeleton V-shaped.

The family Empididae includes the primitive groups of the Empidoidea, which possess a full set of characters in their plesiomorphous state, as shown in Fig. 141. Their exclusion from the former "Empididae" is indicated by the following plesiomorphous characters:

1. ♂ genitalia symmetrical and unrotated.

   The same condition is still preserved in the Atelestidae. Very exceptional, true asymmetry of ♂ hypopygium has evolved in *Trichopeza* (Clinocerinae).

2. ♂ genitalia with separate gonopods, originally still bi-articulated, periandrium not developed.
3. Costa running around the wing.

   A character also retained in the Microphoridae but, on the other hand, already lost in some genera of the Empidinae.

4. Sc complete, reaching costa.

   With the same exceptions as in previous character.

5. Radial sector 3-branched, R_{4+5} forked.

   R_{4+5} unforked in *Anthepiscopus* (Oreogetoninae), in the *Rhamphomyia*-group of genera (Empidinae) and in a few genera of the Hemerodromiinae and Clinocerinae.

7. Maxillary lacinia well-developed.

   The lacinia is also present in the Atelestidae and Microphoridae (Microphorini); very rarely reduced, or maxillae completely absent in the Clinocerinae.

The family Empididae, as characterised above, will be included in three further volumes of the 'Fauna ent. scand.' which are now in preparation.

Family Hybotidae

The family Hybotidae, as defined here, is identical with the "Ocydromiinae group of subfamilies" of Tuomikoski (1966). It is a clearly-defined monophyletic group of three subfamilies of the former "Empididae", the Ocydromiinae, Hybotinae and Tachydromiinae, which includes about 1,300 recent species. The presumed phylogeny of the family is given in Fig. 140. It is thought that the Ocydromiinae include the most primitive groups of the family, whilst the Hybotinae and Tachydromiinae, which form a natural subgroup within the family, have more characters in an apomorphous state. A full set of apomorphous characters is present in the Tachydromiinae.

The Tachydromiinae were treated separately in volume 3 of 'Fauna ent. scand.' (Chvála, 1975), and they appear to be the most studied and best-known group of the former "Empididae". The subfamilies Ocydromiinae and Hybotinae are treated in the Systematic part of the present volume.

The family Hybotidae may be characterised by the following ground-plan characters, which are apomorphous in terms of the ground-plan of the Empidoidea.

1. ♂ genitalia with the gonopods connected dorsally, forming a periandrium.
2. ♂ genitalia asymmetrical (symmetrical in the Atelestidae and Cyclorrhapha!).
3. Hypopygium rotated through about 90° towards the right (hypopygium lateroversum), pregenital segments in a normal position.

   A character that is not known in the Eremoneura outside the Hybotidae; in the Microphoridae, the "rotation" of the hypopygium (through 180°) is caused by deflexion and partial rotation of the pregenital segments alone.

4. Aedeagus more or less firmly attached to the inner wall of hypandrium.
5. Maxillary laciniae lost, base of maxillae (stipites) attached to the base of paraphyses (the long sclerotised folds in the anterior wall of labium).
6. Palpi separated from maxillae, situated on a special sclerite ("palpifer").
7. Front tibia with a tubular gland on the inner side near base, its opening visible as a small pit that varies in precise form.
8. Costa ending at tip of R_{4+5} or the first branch of media.
9. Sc abbreviated, not reaching costa.
10. Radial sector 2-branched, R_{4+5} not forked.

Tuomikoski (1966) has listed two further apomorphous characters which I have not personally investigated:

11. Abdominal spiracles with a more or less long tubular opening which is bare (devoid of distinct microtrichia) but distinctly ringed inside. The opening is usually shorter and the internal rings are absent or weakly indicated by rows of microtrichia in the other subfamilies of the former "Empididae" (except some Hemerodromiinae).

12. There is no sclerotised spermatheca in the female, as is probably the rule in most, if not all, other "Empididae".

The distribution of certain characters among the subfamilies Ocydromiinae, Hybotinae and Tachydromiinae of the Hybotidae is shown in Fig. 141.

Family Atelestidae

The family Atelestidae, as defined below, includes two Holarctic genera *Atelestus* and *Meghyperus*, and provisionally also the Chilean genus *Acarteroptera* Coll. (not studied). The family is characterised by the following ground-plan characters which are apomorphous in terms of the ground-plan of the Empidoidea.

1. ♂ genitalia with the gonopods connected dorsally, forming a periandrium.
2. Hypandrial bridge strongly developed and connected to unusually broad periandrial folds.
3. 8th tergum in ♂ in the form of a narrow dorsal band, and 8th sternum (placed ventrally) large.

   This arrangement is also part of the ground-plan of the Cyclorrhapha, but owing to rotation through 180° the narrow 8th tergum is situated ventrally and the large 8th sternum dorsally.
4. Radial sector 2-branched, vein R_{4+5} unforked.
5. Vein M_{1+2} unforked; or in *Meghyperus* the fork of M_1 and M_2 is apical, far beyond the discal cell.

   In other Empidoidea with M_{1+2} forked (Empididae, Microphoridae and most of Ocydromiinae of the Hybotidae), M_1 and M_2 arise separately from the discal cell. This is also the situation in *Bicellaria* (Hybotidae) and some Hemerodromiinae *(Phyllodromia, Hemerodromia)*, but the posterior crossvein closing discal cell is absent. A true apical fork of M_1 and M_2, as in *Meghyperus*, is found in the Platypezidae and, so far as I am aware, in the Empidoidea only in *Chelifera* and *Ceratomerus* (Empididae) and the Chrysosomatinae (Dolichopodidae), but this appears to be the result of convergent development.
6. Costa ending at tip of vein M_1 (or M_{1+2}).
7. Sc incomplete, not reaching costa.
8. Antennal style with a tendency to be 3-articulated.

   The clearly 2-articulated style of *Atelestus* is plesiomorphous within the Eremoneura, but the 3-articulated style of the Nearctic *Meghyperus* or the tendency of the style to be 3-articulated in *Meghyperus sudeticus* is an apomorphous state. Whether or not a 3-articulated style is in the ground-plan of the Cyclorrhapha remains unclear, since the Opetiidae (formerly a subfamily of Platypezidae) still have a 2-articulated style as in the Empidoidea. The exceptional 3-articulated style of some Empididae *(Rhagas, Hesperempis, Hilara, Dryodromia)* has evolved in quite a different way, by division of the short bare terminal part of the style.
9. Abdomen in ♀ "ovipositor-like", the apical five visible abdominal segments very narrow and weakly sclerotised.

10. Proboscis projecting forwards or obliquely forwards.

    The forwardly projecting proboscis obviously indicates predatory habits. Certain Hybotidae, Microphoridae and even the Oreogetoninae (Empididae) have similarly porrect mouthparts. However, this is probably due to parallel development, as an adaptation for hunting other insects, mostly in flight.

11. Hind tibiae (and sometimes also metatarsi) dilated and compressed.

    The similar hind legs of the Platypezidae (together with their modified wing-venation) led to the assignment of *Atelestus* to that family. However, similar adaptations of the hind legs in some Microphoridae (♂ only), Hybotidae and Empididae suggest parallel development: such legs may be an adaptation for holding and/or carrying other insects, or for mating.

12. Acrostichals reduced, biserial (quadriserial in Neartic *Meghyperus*) and well-separated from the dorsocentrals.

    Multiserial acr and dc, evenly distributed on the mesonotum, are a plesiomorphous state within the Empidoidea, present in the Cretaceous *Trichinites* and some recent Hybotidae (including the primitive groups of the Tachydromiinae). Numerous acr (although separated from dc) are also present in the Microphoridae.

Some of the apomorphies given in this list are also present in the Cyclorrhapha and can be regarded as synapomorphies of the Atelestidae and Cyclorrhapha. This agrees well with Griffiths' (1972: 66) statement that "*Atelestus* shows certain strong similarities to some groups of Cyclorrhapha (such as Platypezidae) which cannot be summarily dismissed as due to convergence. The hypothesis that the similarities between *Atelestus* and Platypezidae are retained from a common ancestor of the Ocydromioinea and Cyclorrhapha has much to commend it".

The Atelestidae were formerly included in the Hybotinae (Hybotidae), but they cannot be included in this family because of the following plesiomorphies:

1. ♂ genitalia completely symmetrical and not rotated.

    A symmetrical and unrotated hypopygium is also present in the Empididae but in a more primitive state, as the periandrium is not yet developed. The symmetrical hypopygium of the Atelestidae, with fully developed periandrium, is the ground-plan condition of the Cyclorrhapha (still not rotated to the circumverted position) or the initial stage of the asymmetrical and more or less rotated hypopygium of the Hybotidae (*Trichinomyia, Bicellaria, Syneches*).

2. Aedeagus free.

    In the Hybotidae, the aedeagus is more or less firmly attached to the inner wall of the hypandrium or to the folds linking the inner periandrial wall.

3. Maxillary laciniae present, and palpi attached directly to the maxillae.
4. Legs without ordinary bristles.
5. Front tibiae without a tubular gland.
6. Wings with a well-developed alula and axillary lobe.
7. Antennal style pubescent.
8. ♂ holoptic, ♀ broadly dichoptic.

    This type of dimorphism in the eyes is undoubtedly plesiomorphous within the Empidoidea, clearly retained in the non-specialised forms of the Empidinae but generally lost or less contrasting (in different sexes) in the other families.

9. Anal cell long, at least as long as 2nd basal cell or longer, the vein closing it (Cu_{1b}) convex and joining anal vein at right-angles, or outer angle of anal cell slightly acute.

This character is also present in the Hybotinae and the extinct genus *Trichinites.* I suggest that this type of anal cell is a symplesiomorphy of the Atelestidae and the hybotine branch of the Hybotidae. Hennig (1970: 6) thought that the anal cell of the 'Atelestinae' and Hybotinae might have evolved in two different ways, "bei *Atelestus* und *Meghyperus* durch Verkürzung der hinteren Basalzelle, bei den Hybotinae . . . durch Verlagerung wichtiger Punkte des Flügelgeäders in Richtung nach der Flügelspitze". The same type of anal cell is also present in the Brachystomatinae; however, it is difficult to say at present if this is a symplesiomorphy or a parallel development.

10. Style bi-articulated, with a single small basal article.

The 3-articulated style of Nearctic *Meghyperus* is undoubtedly an apomorphous condition. However, the 2-articulated style of *Atelestus (Acarteroptera* is unknown to me in this respect) is the ground-plan condition of the Empidoidea, whilst the 1-articulated style (or its entire absence) in some Oedaleini, *Ocydromia* (Hybotidae) and Parathalassiini (Microphoridae) is undoubtedly apomorphous by secondary reduction.

11. Ventral intersegmental region of the meso- and metathorax without a sclerotised ridge *(Atelestus).*

The sclerotised ridge named 'Leiste l.L' and interpreted by Ulrich (1971) as a synapomorphy of the Empidoidea families was not found by Ulrich in *Atelestus.* This character is not known for *Meghyperus* (and *Acarteroptera*), and the thoracic skeleton of these two genera needs further study.

Because of the presence of a discal cell, the genera *Meghyperus* and *Acarteroptera* have usually been assigned to the old family "Empididae", either to the Hybotinae or, like other genera of uncertain systematic position, close to Hybotinae. On the other hand, the systematic position of *Atelestus* has been unclear for a long time, particularly because of its somewhat intermediate position between the Platypezidae and "Empididae". Both Walker (1837) and Zetterstedt (1838), when describing this genus (as *Atelestus* and *Platycnema* respectively), included it in the Platypezidae, and with some exceptions this classification was followed until the sixties, when Collin (1961) wrote in his 'British Flies' monograph that "there can be little doubt that *Atelestus* is correctly located in the Empididae and its nearest ally is probably *Meghyperus*". On the contrary, Lundbeck in 'Diptera Danica' placed *Microsania* in the "Empididae" (in 1910) and *Atelestus* (as *Platycnema*) in the Platypezidae (in 1927), although he suggested that the inclusion of *Platycnema* in the "Empididae" would be more natural.

More recently Kessel (1960) has again assigned *Atelestus* to the Platypezidae, and he has been followed by Chandler (1973, 1974). Bährmann (1960), in his study of the male terminalia of the old "Empididae", placed *Atelestus* in the Hybotinae, but Krystoph (1961) removed it from the "Empididae" on the basis of the structure of the mouthparts and suggested that it belonged to the Platypezidae.

Tuomikoski (1966) was the first to separate *Atelestus, Meghyperus* and *Acarteroptera* as "aberrant Empidinae" into a separate taxonomic unit, stating that they "do not belong to the Empididae or to any of the other families hitherto recognized". However, Tuomikoski was wrong to include *Homalocnemis* (Brachystomatinae) and *Ironomyia* (a cyclorrhaphous family) in this group, but he was undoubtedly right in suggesting that "it is not certain whether this group of genera represents a single family or perhaps more, but their removal would undoubtedly leave the Empididae more naturally delimited" and, that in this group "connections to the cyclorrhaphous flies are to be sought".

Kessel & Maggioncalda (1968) also pointed out the close relationship of *Meghyperus* (and the hybotine branch ?) with the Platypezidae, but Griffiths (1972) was the first to give explicit expression to the close phylogenetic relationship between *Atelestus* and Cyclorrhapha after evaluating homologies in the hypopygial structures. Hennig (1976) came to more or less the same conclusion when examining the hypandrial structures of *Atelestus*, suggesting a common ground-plan for the hypandrium of *Atelestus* and Cyclorrhapha (although his hypothesis as to the origin of these structures differed substantially from that of Griffiths). Six years earlier, the same author (Hennig, 1970) had found it impossible to accept that the Cyclorrhapha arose from the "Atelestinae" in view of the 2-articulated style of *Atelestus;* however, this objection can be met by discussion above on the antennal style. Griffiths' (1972: 66) opinion that "the 3-articulated arista condition (in Cyclorrhapha) was probably derived from the 2-articulated condition, not through an independent series of modification" is clearly correct.

Hennig (1970) erected a subfamily Atelestinae for the three genera discussed above and "possibly" also for *Anomalempis*, which has recently been assigned (but not studied by me) to the Brachystomatinae. The subfamily Atelestinae was placed by Hennig in his Ocydromioinea-group of subfamilies in the former "Empididae", and this classification was provisionally accepted by Griffiths (1972), Chvála (1976, 1981a) and Chandler (in litt.).

The "Atelestinae" are now raised to family rank, and are regarded as a monophyletic group of flies very probably sharing a common ancestor with the Cyclorrhapha.

Family Microphoridae

The family Microphoridae, as defined below, includes four widely distributed recent genera, *Microphorus* Macq., *Schistostoma* Beck. (tribe Microphorini), *Parathalassius* Mik, *Microphorella* Beck. (tribe Parathalassiini), and two extinct Cretaceous genera. The family may be fully characterised by the following ground-plan characters which are apomorphous in terms of the ground-plan of the Empidoidea.

1. ♂ genitalia asymmetrical, capsule-like, with hypandrial part considerably enlarged, periandrium developed.
2. 7th and 8th abdominal segments in male deflexed and slightly rotated, consequently hypopygium deflexed and inverted, lying towards the right alongside the abdomen.

   Apparently the hypopygium itself is not rotated, and its position is the result of the deflexion and partial rotation of the two pregenital segments. This inversion and deflexion is also part of the ground-plan of the Dolichopodidae.

According to Griffiths (1972), the postabdomen of *Microphorus* (and its allies) demonstrates one stage in the evolutionary grade found in the Cyclorrhapha, since the inverted position and right-hand direction of 8th segment and hypopygium in *Microphorus* resemble what is found in newly-emerged platypezid males (i.e. the first stage of the hypopygium circumversum of the Cyclorrhapha). However, I would prefer to suggest that the characters in the Microphoridae and Cyclorrhapha have arisen through parallel development because in the Microphoridae the 7th abdominal segment is also slightly rotated (unaffected in the Cyclorrhapha) and, moreover, the rotation of the pregenital segments is obviously not in the ground-plan of the Dolichopodidae. If this were to be regarded as a synapomorphy, it would mean that the Atelestidae must have separated much earlier from the common microphorid-cyclorrhaphous phylogenetic lineage, because their 8th segment is not rotated, and this seems very unlikely. There are many more apomorphies shared by the Atelestidae and Cyclorrhapha, and I suggest that the microphorid lineage separated long ago from the common ancestral stock of the Atelestidae and Cyclorrhapha which still had a symmetrical hypopygium and an unrotated 8th abdominal segment. It is still not known whether the rotation of the pregenital segments in the Microphoridae takes place in the pupal stage or the pharate imago.

3. Base of radial sector (rs) arising opposite humeral crossvein.
4. Radial sector 2-branched, vein R_{4+5} unforked.
5. Basal cells and anal cell remarkably shortened.
6. The vein closing anal cell (vein Cu_{1b}) convex.

The apically rounded anal cell is not developed in *Parathalassius*, where vein Cu_{1b} is almost straight and the anal cell blunt apically as in the Hybotinae. However, the shape of the anal cell in *Parathalassius* may well be a secondary development, as vein Cu_{1b} is already distinctly rounded in the Cretaceous genus *Microphorites*. An anal cell of similar shape is present in some genera of the Clinocerinae and Hemerodromiinae, but this is undoubtedly a parallel development and appears to be an adaptation to basally narrowed wings.

7. Anal vein very short.

Vein A_1 (Cu_{1b} + 1A) is generally very short, not extending beyond the anal cell except in *Parathalassius*, the Nearctic *Microphorus sycophantor* and the Australian *Microphorella iota*, so far as I am aware. The Mesozoic (Cretaceous) *Microphorites extinctus* already has the anal vein very short as in most recent Microphoridae, but the Coenozoic (Baltic amber) *Microphorus rusticus* has a rather long anal vein extending far beyond the anal cell. I believe that the short anal vein may well be a synapomorphy with the Dolichopodidae.

8. Alula not developed.

There are two further characters which may well be synapomorphies of the monophyletic subgroup Microphoridae + Dolichopodidae in terms of the ground-plan of the Empidoidea.

9. Axillary lobe on wing not developed.

A character present in the Lower Cretaceous *Microphorites*, in all the known Upper Cretaceous Microphoridae and Dolichopodidae, and in recent Microphoridae (except Microphorini) and Dolichopodidae. The species of the tribe Microphorini (*Microphorus*, including the Coenozoic *M. rusticus*, and *Schistostoma*) have the axillary lobe well-developed, but this may well be a secondary acquisition by forms with an aerial swarming habit – a secondarily apomorphous condition. It should also be noted that *Microphorites* cannot be regarded as an ancestor of the Parathalassiini alone (based on the undeveloped axillary lobe) because practically all the other characters of *Microphorites* are shared with the tribe Microphorini (see the section on extinct forms).

10. Head broadly dichoptic in both sexes.

Males are holoptic only in the Microphorini and in *Diaphorus* (Dolichopodidae), so far as I am aware. The broadly dichoptic eyes of males are very probably apomorphous in this group of the Empidoidea, and perhaps the comments under no. 9 may also apply here. Unfortunately, the male sex of *Microphorites* is unknown.

The special modification of the ♀ ovipositor, with its cleft and apically spinose 9th tergum (hemitergites, acanthophorites), was considered by Hennig (1971) to be a synapomorphy of the Dolichopodidae and Microphoridae (Parathalassiini). However, this character is widely distributed in the Empidoidea (see the morphology of female abdomen) and appears to be no more than a functional adaptation of significance for oviposition, perhaps in species inhabiting sandy biotopes.

It will be apparent that the list of apomorphies given above, which clearly show the monophyly of the Microphoridae, also apply quite well to the Dolichopodidae (perhaps with some minor exceptions) and demonstrate the common origin of the Microphoridae and Dolichopodidae. The close relationship and the presumed monophyly of these two families as given here is not a new idea. Colless (1963) drew attention to the presumed common origin of the *Microphorus*-group of genera and the Dolichopodidae, as did Tuomikoski (1966), Hennig (1971), who erected the subfamily Microphorinae, and Griffiths (1972). Hennig (1971) contributed a great deal to the elucidation of pointing out the close relationship and similarity between the genera *Microphorella* and *Parathalassius* and the Dolichopodidae. Chvála (1981a) erected the tribe Parathalassiini for these two genera, in connection with his study of the extinct "Microphorinae" and Dolichopodidae, and indicated more precisely the age of the presumed origin of the Dolichopodidae. I am now suggesting that the Parathalassiini share a full set of apomorphies in common with the Dolichopodidae.

The tribe Parathalassiini may be characterised by the following ground-plan conditions which are apomorphous in terms of the ground-plan of the Microphoridae and Microphorini.

1. Antennal style 1-articulated.

The small basal article is also absent in all the fossil Upper Cretaceous Microphoridae *(Cretomicrophorus)* and Dolichopodidae *(Archichrysotus, Retinitus)* – see the correction to Fig. 2 in Chvála (1981a: 231), a footnote on p. 235. However, in the Lower Cretaceous *Microphorites*, all the Microphorini and the Dolichopodidae (!) the style is 2-articulated. This is the only character in which the recent Dolichopodidae differ substantially from both the recent Parathalassiini and the fossil "parathalassiine" Microphoridae and Dolichopodidae.

2. Antennae inserted above middle of head in profile.
3. Clypeus not separated by a membraneous area from face, usually distinctly produced.
4. Maxillary lacinia absent.
5. Labial paraphyses (the long sclerites in anterior wall of labium) firmly connected with the maxillae (Hennig, 1971).
6. The vein closing 2nd basal cell (crossvein tb) incomplete.

   The same character is present in the fossil dolichopodid genus *Retinitus* (but not in *Archichrysotus*), and the gradual loss of this crossvein led to the complete fusion of the 2nd basal cell with the discal cell in the Dolichopodidae.
7. Head in males broadly dichoptic.
8. Eyes microscopically pubescent.
9. 9th tergum in ♀ cleft and apically spinose.

   Although this same female ovipositor is also found in *Schistostoma* and apparently in some Nearctic *Microphorus* (Microphorini), it is part of the ground-plan of the Parathalassiini and the Dolichopodidae.

The distribution of apomorphous characters in the Parathalassiini within the Microphoridae and Dolichopodidae, including the extinct genera, is shown in Fig. 142. A few additional characters have also been included in order to show the apomorphies of the Dolichopodidae in terms of the Microphoridae.

Family Dolichopodidae

The family Dolichopodidae is not discussed here in any depth as it is the largest family of the Empidoidea (over 4,000 recent species) and, as a non-specialist in the group, I am not familiar with all the necessary details. I would just like to point to the presumed origin and separation of the Dolichopodidae from the other families of the Empidoidea (Fig. 140), and to the distribution of the synapomorphies of the family within the monophyletic subgroup of the Microphoridae and Dolichopodidae (Fig. 142).

The family may be roughly characterised by the following ground-plan characters, which are apomorphous in terms of the ground-plan of the Microphoridae and in particular of the Parathalassiini.

1. Costa running only to M_1, absent on posterior wing-margin.
2. Sc very short, attached to and ending in vein R_1.
3. M with 2 branches, vein M_2 absent.

   Except in the Sciapodinae, where the apical fork of M_1 and M_2 is completely developed.
4. The vein closing 2nd basal cell (crossvein tb) completely absent, and consequently discal cell fused with 2nd basal cell.
5. Body metallic or yellow in colour.

These apomorphies are also present in the extinct Upper Cretaceous dolichopodid genera *Archichrysotus* Negr. (except for the character 4) and *Retinitus* Negr.

Life history

Immature stages

The immatures are still very poorly known. Smith (1969) has recently given a review of the published records of the immature stages of the former "Empididae", which included a tentative key to genera of the larvae and some pupae, and Brindle (1973) has given a key to the larvae of four British genera of the Hemerodromiinae. Dyte (1967) has published an important account of the distinctions in the immature stages between the former "Empididae" and the Dolichopodidae, and a comprehensive paper on the morphology and ecology of some edaphic larvae (mainly *Hilara*) was prepared by Trehen (1971).

So far as is known, the majority of species overwinter as larvae, and the pupal stage is rather short, usually a week or two. Larvae live mostly in soil, but also in leaf litter, detritus, dung, under moss, in rotten wood or under bark, and some are aquatic. The larval habitats of the families Hybotidae, Atelestidae and Microphoridae are discussed further in the Systematic part of this volume. Within the Empididae, the larvae of the Hemerodromiinae and Clinocerinae are mostly aquatic, whilst the development of the Brachystomatinae, Ceratomerinae and most Oreogetoninae remains unknown. Larvae of the Empidinae *(Empis, Rhamphomyia, Hilara)* live mostly in soil, but rarely in rotten wood, for instance *Rhamphomyia marginata* (personal observation); Chandler (1972) reared *Hilara lurida* from rotten beech wood. The aquatic larvae of the Hemerodromiinae and Clinocerinae are much better known, and Vaillant (1967) even proposed an ecological zonation of the larvae of *Wiedemannia* (Clinocerinae) in streams and rivers in France.

The larvae are thought to be almost exclusively predaceous, preying on other insects but mainly on the larvae of other Diptera. In captivity, the larvae readily feed on sliced insect larvae but are often cannibalistic. Smith (1968) reported on large numbers of *Rhamphomyia sulcata* larvae occurring in pasture soil under newly-sown crops of barley, but he doubted if the larvae were connected with the crop. It is thought that larvae living in rotten wood or dung (many Ocydromiinae) feed on other insects as well. Aquatic larvae have been repeatedly found to prey on Simuliid larvae: Vaillant (1952, 1953) reported on the larvae of *Wiedemannia* and *Hemerodromia* respectively, and Sommerman (1962) on Nearctic *Oreogeton*. Larvae of the Hemerodromiinae also feed on caddisflies, as Knutson and Flint (1971, 1979) found larvae and pupae of *Neoplasta* and other unidentified genera in pupal cocoons of the trichopterous families Rhyacophilidae and Glossosomatidae.

Only a few modern detailed morphological descriptions of the immature stages of the Empididae and Hybotidae have been published, mainly by B. M. Hobby, K. G. V. Smith, P. Trehen and F. Vaillant. However, there are still so few of them that they cannot yet be used for taxonomic and phylogenetic purposes.

Adults

The bionomics of the adults are very much better known than of immature stages, particularly because of the wide range of different habits and ecology of the adults. The adult biology has been a subject of general interest for many entomologists for the past century, especially the curious courtship of some Empididae connected with the parallel prey presentation (»wedding presents«) of males to females during aerial mating. Individual groups of the Empidoidea differ substantially not only in their mating and/or swarming behaviour but also in their feeding habits, and this has led many students to suggest unnatural groupings of unrelated groups of species with similar habits: the assignment of the Tachydromiinae, Hemerodromiinae and Clinocerinae into a single family is a good example of this.

It is now quite clear that similar behaviour patterns and feeding habits evolved along parallel lines in many unrelated groups, including similar morphological adaptations, and conversely that species of a single monophyletic group can exhibit a wide range of different feeding habits including different morphological adaptations of the eyes, legs or wings (e.g. Ocydromiinae). The mating and/or swarming habits of species of the Empidinae exhibit a wide range of modifications (sometimes very slight), including different feeding habits, activity, flight-period, which all lead to the isolation of related species and, of course, to parallel speciation. A relatively recent evolution of these habits and specialisations to ecological or climatic conditions are phenomena typical of species of *Empis, Hilara, Rhamphomyia* and others. The exclusively predaceous species of the Empididae and Hybotidae do not exhibit such diversified specific behaviour.

In the predaceous species of the Tachydromiinae (Hybotidae) and Hemerodromiinae (Empididae), there are special adaptations for raptorial activity on a solid substrate which have led to the gradual loss of the wings, through stenoptery and brachyptery to aptery. This has also led to a parallel development in their walking activity, as the species move by running very rapidly on foliage, logs or sand, and do not fly. In some brachypterous forms, the adults still move by means of special jumps *(Chersodromia)*, rather like certain Empididae *(Rhamphomyia)* adapted for life in high arctic regions.

Phenology

Flight-periods vary substantially, not only in different zoogeographical regions or climatic zones but also in northern and southern areas of Europe, and there are striking differences even within Scandinavia. In general, the flight-period of species in northern Fennoscandia is shorter and the adults appear later than in southern Scandinavia or Denmark. In southern parts of Europe, in the Mediterranean, the species are on the wing a month or two earlier than in temperate Europe; in the Canary Islands, for instance, the peak of adult seasonal activity is in the winter months.

The flight-period within Scandinavia of the species dealt with in this volume is given in Table 1; any discrepancies with known flight-periods in other parts of northern

Europe, caused by climatic conditions (Atlantic and continental climate) or simply because material from the Scandinavian countries is so sparse, are mentioned in the Systematic part. It should be borne in mind that in Scandinavian species with a long flight-period the two extreme dates are usually found in the southern populations whereas the summer in the north is extremely short. On the other hand, species with a short flight-period, particularly spring or early summer species, occur first in the south but almost a month later, at the end of the flight-period, they may be found in Lapland.

In boreal and temperate Europe the adults of the Empidoidea are, so far as I am aware, exclusively univoltine, but with unusually favourable climatic conditions individuals of a second generation may accidentally appear in late summer or early autumn. Harper (1980), when studying the phenology and distribution of aquatic Hemerodromiinae in North America, reported on 3 generations per year in *Hemerodromia empiformis,* but other Nearctic species studied were univoltine. He considered that males were relatively short-lived whereas females often persisted throughout the summer.

The maximum occurrence of Empidoidea in boreal and temperate Europe is in late spring and early summer, emphasized by the abundant occurrence of many species of the Empidinae. The first species to appear in early spring, with a short flight-period confined mainly to April and May, is *Empis (Platyptera) borealis,* and this is followed by several early spring species of the *Rhamphomyia* (s.str.) *sulcata*-group. There are several other spring or early summer species with a remarkably short flight-period, such as almost all the Oreogetoninae, *Empis* subgenera *Leptempis* and *Polyblepharis,* and *Rhamphomyia* subgenera *Megacyttarus* and *Dasyrhamphomyia.* On the other hand, *Rhamphomyia* subgenus *Amydroneura, R. (Eorhamphomyia) spinipes,* some *Clinocera* and *Wiedemannia* (Empididae) and *Trichinomyia flavipes* (Hybotidae) are typical autumn species, with rather a short flight-period in October and November. There are, however, many species with a remarkably long flight-period, mainly in the Hybotidae, which are on the wing for seven or even eight months in the year and apparently produce only one generation; these include many *Platypalpus, Tachypeza nubila* (v-xi) and *Tachydromia arrogans* (iii-x) in the Tachydromiinae, and *Ocydromia glabricula* (iv-x) and some *Bicellaria* in the Ocydromiinae.

Hibernation in the adult stage has been found in a few species: in the Tachydromiinae in *Platypalpus maculipes* and several *Crossopalpus* and *Stilpon* (Chvála, 1975), and in the Empididae in *Clinocera (Hydrodromia) wesmaelii* (personal observation). However, it is likely that further Clinocerinae will be found to overwinter as adults.

Feeding habits

The adults are either predators (preying on living or dead insects) or flower visitors; the term 'flower visiting species' is deliberately used here instead of the common term 'nectar feeders', as it is by no means clear whether the species visiting flowers imbibe nectar or feed on pollen grains.

Tuomikoski (1952) gave a review of the feeding habits in the Finnish "Empididae". Chvála (1976) suggested that predation (feeding on liquified animal tissue) was the original feeding habit of the Empidoidea, as there were no angiosperms with nectaria secreting sweet substances during the early Mesozoic and the first Empidoidea must therefore have been insectivorous or at least zoophagous. Downes & Smith (1969) suggested that feeding on pollen, which provides a meal as rich in protein as insect prey, could have arisen fairly readily from the insectivorous or scavenger habit. Predation is undoubtedly the basic feeding habit of the Asiloidea and Empidoidea, and the nectar feeding habit (sucking of sweet liquid substances from the nectaria) only evolved from the original predatory activity in certain groups of the Empidoidea, mainly in the Empidinae.

This means that predation is still the main feeding habit in the Empidoidea. Originally, the adults hunted their prey in flight and presumably it is only during the course of evolution that the predatory activity has been transferred from the air to the ground. This evolution was obviously closely connected with aerial swarming and mating, as discussed in the next section. There does not appear to be any marked prey specialisation by the predator: the prey usually consists of the insects that occur most abundantly in the habitat together with the predator, and the prey is often as large as or even larger than the predator itself.

The family Empididae exhibits a wide range of feeding habits with many modifications, although the predatory habit seems to be the basic one, both in primitive forms (Oreogetoninae) and in the very specialised raptorial groups (Hemerodromiinae, Clinocerinae). When erecting the subfamily Oreogetoninae (Chvála, 1976), which undoubtedly includes the most primitive members of the family, I thought that the species were secondary nectar feeders. However, there is strong evidence now that some of them are predators. My original belief was based mainly on the arrangement of the eyes in the two sexes (holoptic in males, dichoptic in females), which is apparently an adaptation for aerial swarming (and mating ?). The rather strong, forwardly-pointing proboscis is probably the most decisive adaptation, for predatory activity. The species are only rarely found in nature, the only record available (Chandler, 1972) is of *Hormopeza obliterata* preying on smoke flies *(Microsania,* Platypezidae). *Rhagas* (adults on tree trunks), *Gloma, Oreogeton* and others are probably predators too, but both *Iteaphila* and *Anthepiscopus* are frequent flower visitors and are armed with conspicuously long mouthparts including palpi. However, it is difficult to say whether they are really nectar feeders or pollen feeders.

The feeding habits of the Empidinae *(Empis, Rhamphomyia, Hilara)* are undoubtedly the best studied and many published records are available, particularly because of the close connexion of feeding with aerial swarming and mating. Generally, both sexes visit flowers during the non-mating period, but males capture insect prey before mating and this prey is given as a "wedding present" to the female during the first mating contact. Females eat the prey during copulation, but there is no evidence that the male itself feeds on the prey it captures. The prey provides the protein-rich meal which is the necessary prerequisite for female ovarian development. It should be noted that the predatory habit has been retained in both sexes in most species of *Empis* subgenera

Xanthempis and *Euempis*, and in some species of *Pararhamphomyia*, whereas *Rhamphomyia* subgenera *Lundstroemiella* and *Holoclera* are exclusively flower visitors and the predatory habit has been completely lost.

The Hemerodromiinae are entirely predaceous, having a series of adaptations for raptorial behaviour and, like the Tachydromiinae, their predaceous activity takes place on the ground, close to water where the larvae develop. The Clinocerinae are also predaceous, except for the monotypic *Dryodromia testacea* which appears to be the only frequent flower visitor. Unlike the Hemerodromiinae, the Clinocerinae also capture their prey in flight or collect both living and dead insects from the surface of the water. No records are available on the feeding habits of the Brachystomatinae, but they appear to be predaceous like the Hemerodromiinae. The Ceratomerinae are thought to be flower visiting species in view of their long *Empis*-like proboscis.

Predation is the basic feeding habit of the Hybotidae. The subfamily Hybotinae includes very specialised forms that hunt their prey in flight, as does *Bicellaria* (Ocydromiinae), and the Tachydromiinae have transferred their predatory activity to the ground, except for a few primitive species (*Platypalpus*) which still catch the prey in flight. Within the Ocydromiinae, the species of *Anthalia*, *Allanthalia* and *Euthyneura* are entirely flower visitors, and clearly at least some of them feed on pollen as Downes & Smith (1969) have shown for the Nearctic *Anthalia bulbosa*.

The Atelestidae are probably also predators, although previously (Chvála, 1976) I suggested that they are nectar feeders. However, no precise field observations are available. Species of *Atelestus* are found locally in large numbers when sweeping low herbage, but they have never been observed on flowers and the forwardly projecting proboscis indicates a predatory habit.

The Microphoridae are undoubtedly predators for the most part, even preying on other insects in flight (? hunting swarms), and *Microphorus* species have been repeatedly observed to feed on dead insects trapped in spiders' webs (Laurence, 1948; Downes & Smith, 1969). However, adults are often found also on flowers, and it is suggested that they feed not only on nectar but also on pollen grains. Laurence (1953) also described the dolichopodid genus *Hercostomus* (Dolichopodinae) as a pollen feeder. Otherwise, the Dolichopodidae are entirely predaceous, some groups catching their prey on the water like the Clinocerinae.

Swarming and mating

Swarming is a well-known phenomenon in the Empidoidea, a very characteristic and undoubtedly sexual behaviour which is even reflected in the popular names of the Empididae - dance flies, "Tanzfliegen". The unmodified swarm is typically stationary and consists of a few to a very large number of individuals, either slowly hovering in the air or flying in wild forward and backward movements. The type of flight, number of swarming individuals, and the size and shape of the swarm are specifically distinctive, and species can often be easily recognised by their characteristic swarm. The swarm is orientated towards some "swarm marker", such as a conspicuous or contrasting object in the landscape, an overhanging branch, a clearing in the tree canopy, a solitary bush or a light-coloured flower, or even a bank, a path or a pool. Orientation towards such swarm markers is visual, as is the orientation towards other swarming individuals or of males to females before mating contact; however, close-range contact between the two sexes in the swarm is thought to be achieved by a contact mating pheromone, as has been shown for instance by Linley & Carlson (1978) in the Ceratopogonidae.

The visual orientation of swarming individuals, as developed in the nematocerous families Simuliidae and Bibionidae, nearly all the swarming Brachycera, and also the Ephemeroptera, is achieved by the highly modified holoptic male eyes, with more or less enlarged upper ommatidia. Generally, this kind of sexual dimorphism, with holoptic males and dichoptic females, indicates a swarming habit.

The organised aerial aggregation of swarming individuals, the so-called synorchesium of Gruhl (1955), is the basic type of swarm in the Empidoidea and the Empidinae in particular. It has been suggested that this type of swarm is the first evolutionary step from the original large unorganised aerial aggregations resulting from mass emergence, so-called synhesmic swarming. Gruhl (1955) differentiated several further types of modified swarming activity: of these polyorchesium, a rhythmic dancing - perching - dancing by several individuals is known in some Empidinae *(Empis, Rhamphomyia),* and monorchesium, a hovering and dancing of isolated individuals, is a common habit in many Ocydromiinae (Hybotidae).

Aerial aggregations are generally regarded as a part of sexual behaviour, as the swarm serves as a meeting-place for the two sexes from different populations and enables the emergent females, particularly from dispersed or sparse populations, to mate immediately. In the course of evolution the whole process of meeting and mating has been transferred secondarily from the air to the ground, especially in the highly predaceous groups of the Empidoidea, and aerial swarming has gradually lost its original function. Within the Empidoidea, the swarming activity has only been preserved in the Oreogetoninae and Empidinae (Empididae), the Ocydromiinae (Hybotidae), and the Atelestidae and some Microphoridae. In other words, in the Hemerodromiinae, Clinocerinae (apparently also in the Brachystomatinae and (?) Ceratomerinae), Hybotinae, Tachydromiinae and the Dolichopodidae, the mating activity has been transferred to a solid substrate and aerial swarming has been completely abandoned. However, of the "swarming" groups just mentioned, so far known only the Empidinae and Oreogetoninae (*Oreogeton*) have preserved the original function of the aerial aggregations, with the two sexes meeting and subsequently mating in the air, or at least performing the initial stages of mating. The other non-empidide forms have already lost their useful ability for long-range recognition in aerial aggregations, since sexual encounters on the ground are based merely on specific short-range recognition.

There is an extensive literature on swarming habits, particularly of the Empidinae as this group has achieved the highest degree of development of aerial swarming within the Empidoidea or even the Diptera, with a parallel predatory activity leading to a very curious courtship - a ritualised prey transfer which has a high specific significance. A paper on the swarming and mating habits of Finnish "Empididae" was published by Tuomikoski (1939), and there are more recent papers with complete bibliographies on this subject by Downes (1969, 1970) and Chvála (1976, 1980). Papers summarizing the evolutionary trends within the former "Empididae", although with different conclusions, were published by Poulton (1913), Tuomikoski (1939), Kessel (1955) and Chvála (1976).

The Empidinae exhibit a wide range of different types of swarming and mating behaviour, but there are even some groups, such as *Empis* subgenus *Xanthempis* or *Rhamphomyia* subgenus *Lundstroemiella,* which have already transferred their mating activity to a solid substrate and do not form swarms. Generally, the males of the Empidinae *(Empis, Rhamphomyia, Hilara)* capture an animal prey which is given to the female at the moment of direct contact in the aerial mating swarm. The secondarily nectarivorous females feed on it during copulation, obviously receiving a supplementary source of the protein-rich food necessary for ovarian development. A large fresh insect prey undoubtedly serves as a food for female ovarian maturation but this is hardly likely to be the case when the prey is very small or is repeatedly used (becoming no more than an empty husk), or if copulation is very short (even several dozen seconds in some *Empis* species). The repeated use of an empty prey without edible contents, or of its equivalent (an inanimate object, such as a flower-seed), no longer has any food value, and the prey presentation has lost its original meaning and can only be acting as a stimulus for mating or as an object facilitating better long-range specific recognition of the hovering male. The prey presentation has been ritualised and the object that is being transferred, either the remains of an insect enclosed in a silken web (many *Hilara* species) or simply an empty silken web (the so-called balloon-flies), still has the purpose of attracting the female. The value of the "prey" as a protein-rich food has gradually been lost, but a sufficient amount of protein can have been stored for the adult from the larval stage or, in the "flower visiting" species, can be replaced by the protein-rich pollen grains.

Apart from the Empidinae, the Oreogetoninae are the only members of the Empididae that are known to swarm in aerial aggregations. However, it is surprising to note that no mating in aerial swarms has been observed in other swarming Empidoidea, the Hybotidae (Ocydromiinae), Atelestidae and Microphoridae. A list of the non-empidine species known to produce aerial swarms has been given by Chvála (1980: 13). The loose aerial aggregations of rapidly-flying Clinocerinae over water are undoubtedly only hunting swarms (aggregations of individuals searching for prey above water), which is a common habit also found in some Empidinae *(Hilara, Rhamphomyia* subgenus *Megacyttarus)* and many Dolichopodidae.

The purely male or female synorchesic swarms of the non-empidine forms, like the almost motionless hovering of isolated individuals (monorchesia), without any mating, where swarming and mating (transferred to the ground) are two independent habits, appear at first sight to be a ritual behaviour without any plausible significance. One suggested function, to activate the male gonads before mating, was refuted by Chvála (1980) in the case of the very rare but unusually large swarming aggregations of the common *Bicellaria nigrita.* Another unlikely explanation is that individuals in swarms find protection against predators, because it is precisely these swarming individuals that are the most frequent victims of other hunting predators, often of hunting Empidoidea.

Pajunen (1980) suggested that the swarming of males could be interpreted as a kind of territorial behaviour. He suggested that the males defend topographically fixed areas (territories) against other males, by aggressively motivated displays and even by fighting, thus advertising the occupancy of the site. This may perhaps explain the fixed monorchesic flight of a single individual at one spot, with conspicuous reactions to approaching flying insects, but not the large synorchesia of almost motionless swarming individuals (as in *Bicellaria nigrita*) which are not attracted by nearby flying insects or by each other. Chvála (1980) suggested another explanation for these ostensibly meaningless, large, male, aerial aggregations. They are thought to be a relict form of behaviour on the part of species that originally mated in aerial aggregations but have subsequently transferred their mating activity to the ground. I have coined the term "relict swarm" for swarms where this type of behaviour has been retained although no longer has any obvious significance.

Zoogeography

This section only discusses the species dealt with in the present volume, and so the Empididae are not considered any further.

Of the 18 genera treated, the 3 genera of the Hybotinae are best represented outside the Holarctic region, mostly in the Oriental, but also in the Afrotropical and Neotropical regions; this also applies to the ocydromiine genus *Oropezella*, and partly to *Ocydromia*. The rest of the genera of the Ocydromiinae, and of the Atelestidae and Microphoridae, are almost entirely Holarctic in distribution, *Leptodromiella* and *Atelestus* being exclusively Palaearctic. Eight species (*Trichinomyia flavipes*, *Trichina clavipes*, *Bicellaria spuria*, *Anthalia schoenherri*, *Allanthalia pallida*, *Ocydromia glabricula*, *Leptopeza flavipes* and *L. borealis*) were considered by Melander (1965) to be Holarctic in distribution, but I have not had the opportunity of studying any Nearctic material to confirm this or to establish whether there are any further Scandinavian species that are actually Holarctic.

As in the earlier 'tachydromiine' volume (Chvála, 1975), there are six distribution-patterns (A–G) for Scandinavian species and these are shown in Figs. 143 and 144. The distribution-pattern of the 48 recorded Scandinavian species (and of a further four species whose occurrence there is very likely) is summarised in Table 1. The type of distribution also shows the supposed origin of species.

A – Atlantic elements, extending from England to Denmark and the extreme south of Sweden. About 15% of the Scandinavian fauna belong here.

B – Species of southern distribution within a larger area covering Denmark, the southern parts of Norway, Sweden, and the Baltic coast of Finland including NW of European USSR, reaching approximately 62° N. About 19% of species recorded.

C - Species with the same distribution-pattern as B, but extending further northwards into the central and northern parts of Fennoscandia, to approximately 65–66° N; absent in the extreme north and in the Kola Peninsula. A large group of species, about 21% of the fauna.

D - Widely distributed species throughout Scandinavia up to and including the extreme north (Lapland) and the Kola Peninsula. The largest group of species, about 23% of the fauna, consisting entirely of the Ocydromiinae and *Bicellaria* in particular.

E - Northern species, covering a large area including the extreme northern and central parts of Fennoscandia, absent in the south of Norway, Sweden and in Denmark, but often reaching the Baltic coast of Finland. A rather small group of species, about 10% of the fauna.

F - Extreme northern species, distributed within a smaller area covering Norwegian, Swedish and Finnish Lapland and including the Kola Peninsula. Only a few species, 6% of the fauna.

G - Continental eastern and central European species, extending through the NW parts of the European USSR to the eastern parts of Fennoscandia, sometimes probably along the Gulf of Bothnia as far as southern Sweden. Absent in western Fennoscandia, Denmark and Great Britain. Like F, a very small group with 6% of the fauna.

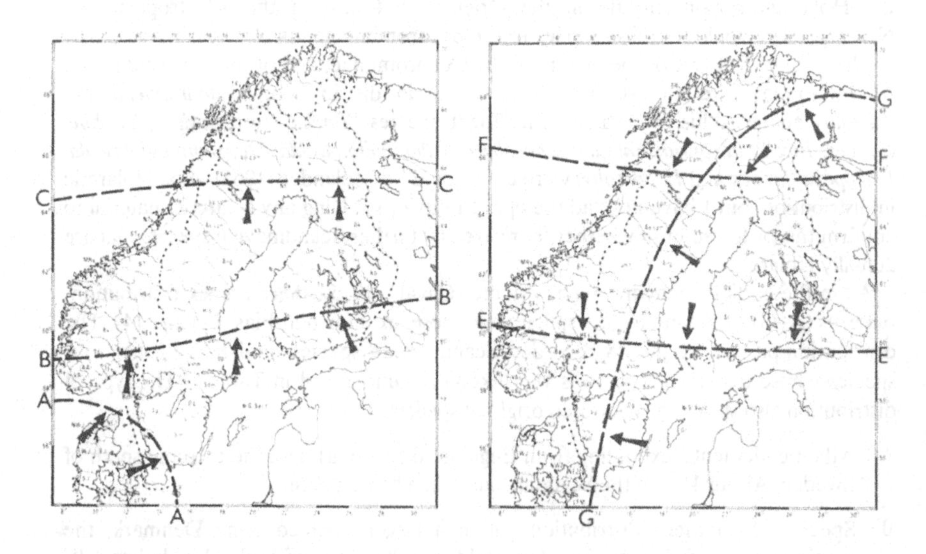

Fig. 143. Distribution-pattern of Fennoscandian species. - Groups A–C.

Fig. 144. Distribution-pattern of Fennoscandian species. - Groups E–G.

A list of species with their distribution-patterns (symbols A–G) is given in Table 1. The expected pattern of distribution of species not yet found in Scandinavia, or where another type is excepted on the basis of the distribution in adjacent countries, is given in brackets. About 70% of the Scandinavian species also occur in Great Britain, about 80% in Central Europe, but only about 40% in southern Europe. In the adjacent USSR, over 60% of the Scandinavian fauna occurs in the north-western part. The distribution of species in the central area of the European USSR has been completed for me by Dr. V. G. Kovalev, but this type of distribution, as given in Table 1, does not correspond exactly with the "central" area of the 'Opredelitel nasekomych' (Bei-Bienko, 1969) as it extends much further southwards, up to and including the Caucasus, i.e. the W, C and SW areas of the 'Opredelitel'.

Of the 39 Scandinavian species also found in Central Europe, at least three (*Trichinomyia fuscipes, Allanthalia pallida* and *Leptopeza borealis*, but probably also *Ocydromia melanopleura)*, are very probably boreoalpine in distribution, and a further three (*Bicellaria pilosa, B. austriaca* and *Euthyneura gyllenhali)* are thought to be boreomontane. Although about 20% of the Scandinavian species do not occur in temperate Central Europe (mostly distribution types E and F), the fauna here is at least as rich as in Scandinavia, mainly due to the presence of further, more southern species of *Bicellaria, Oedalea, Microphorus* and others. On the other hand, the British fauna is poorer than that of Scandinavia, and only seven British species (*Syneches muscarius, Bicellaria nigrita, B. halterata, Oedalea apicalis, Euthyneura myricae, E. halidayi* and *Atelestus dissonans)* have not yet been found in Scandinavia.

Table 1	Type of distribution (cf. Figs. 143, 144)	Great Britain		European USSR			Europe			Asia	Flight period in Denmark and Fennoscandia
		England	Scotland	north-west	north	central	central	south	south-west		
Hybos											
1. grossipes	C	+	+	+	+	+	+				V-IX
2. culiciformis	B (C)	+	+	+		+	+	+	+	+	VI-IX
3. femoratus	B (C)	+	+	+	+	+	+	+	+		VI-IX
Syndyas											
4. nigripes	B-C	+		+		+	+				VII-VIII
Syneches muscarius	(A)	+				+	+	+	+		(VI-IX)
Trichinomyia											
5. flavipes	B	+		+		+	+				VIII-X
6. fuscipes	E			+	+		+				VI-VIII
Trichina											
7. clavipes	B-C	+	+	+		+	+	+	+		VI-IX
8. bilobata	C	+	+	+		+	+	+			VI-VIII
9. elongata	D	+	+	+	+	+	+	+	+		V-IX
10. opaca	B	+	+		+	+	+	+			IX
11. pallipes	C (D)	+		+		+	+				VI-IX

Table 1	Type of distribution (cf. Figs. 143, 144)	Great Britain		European USSR			Europe			Asia	Flight period in Denmark and Fennoscandia
		England	Scotland	north-west	north	central	central	south	south-west		
Bicellaria											
12. simplicipes	D	+	+	+	+		+				V-IX
13. pilosa	D	+	+	+	+		+				V-VIII
14. austriaca	A-B					+	+	+			VI-VIII
15. subpilosa	D	+	+	+	+	+	+	+			V-VII
16. bisetosa	F-E	+	+	+						+	VII
17. spuria	D	+	+	+	+	+	+	+	+		V-X
18. mera	A	+					+				VI-VIII
19. sulcata	D		+	+	+	+	+	+	+	+	V-X
20. vana	A-B	+				+	+				VI-IX
21. intermedia	C	+	+	+		+	+	+			VI-IX
22. nigra	D	+	+	+	+	+	+	+	+		V-IX
nigrita	(A)	+				+	+	+			(V-VII)
Oedalea											
23. stigmatella	C	+	+	+		+	+	+			V-VIII
24. freyi	F			+							?
25. zetterstedti	D		+		+	+	+	+			V-VII
26. holmgreni	C	+	+	+			+		+		VI-VII
27. ringdahli	F (C)	?									VI
28. tibialis	B	+					+		+		VII
29. flavipes	A	+	+			+	+				V-VIII
kowarzi	(G)			+			+				(VI)
30. oriunda	A	+									VI
31. hybotina	C (D)			+	+	+	+				VI-VII
Euthyneura											
32. myrtilli	D	+	+	+	+	+	+				V-VIII
33. albipennis	E	?			+						VI-VII
34. gyllenhali	C (D)	+	+	+	+	+	+	+			V-VIII
Anthalia											
35. schoenherri	E			+	+						VI-VII
Allanthalia											
36. pallida	E			+	+						VII
Ocydromia											
37. glabricula	D	+	+	+	+	+	+	+	+	+	IV-X
38. melanopleura	C-D	+	+	+		+	+				V-IX
Leptopeza											
39. flavipes	D	+	+	+	+	+	+	+	+	?	V-IX
40. borealis	E		+	+	+	+	+				VI-VII
Leptodromiella											
41. crassiseta	G					+					VI
Oropezella											
42. sphenoptera	A	+				+	+	+			VI-IX
Atelestus											
43. pulicarius	B	+	+	+		+	+	+	+		VII
dissonans	(A)	+					+				(VI-VIII)
Meghyperus											
44. sudeticus	G (B)			+			+	+		+	VI-VII

Microphorus										
45. holosericeus	C-D	+	+	+		+	+	+	+	V-VII
46. anomalus	A	+					+	+	+	VI-VIII
47. crassipes	B	+	+	+			+	+	+	VII
Microphorella										
48. praecox	G			+			+			VI

SYSTEMATIC PART

Key to the families of the Empidoidea

1 Vein R_{4+5} forked; if not, then either prosternum large (fused with episterna) and metapleura usually bristled, or costa running around the wing. First antennal segment bristled, at least with a few bristly hairs beneath. Male hypopygium symmetrical and unrotated, with discrete gonopods (side lamellae) .... **Empididae**

– Vein R_{4+5} not forked; prosternum in the form of a small sclerite separated by membrane from the episterna (except Dolichopodidae), costa ending at wing-tip (except Microphoridae) and metapleura always bare. First antennal segment very small, without bristles. Male hypopygium with gonopods connected dorsally, forming periandrium 2

2 (1) Wings with alula. Male hypopygium symmetrical and unrotated, female abdomen remarkably narrowed apically, ovipositor-like. Hind tibiae (or also metatarsi) laterally compressed and dilated in both sexes .... **Atelestidae** (p. 228)

– Alula on wings not developed. Male hypopygium asymmetrical, rotated towards right or deflexed 3

3 (2) Basal cells moderately large, anal cell differently shaped or even absent. Radial sector originating well beyond humeral crossvein. Front tibiae with a sense organ. Male hypopygium along the longitudinal axis or upturned .... **Hybotidae** (p. 86)

– Basal and anal cells conspicuously small, anal cell usually rounded apically. Radial sector originating opposite humeral crossvein. Front tibiae without a sense organ and male hypopygium deflexed 4

4 (3) Discal cell present, emitting 3 veins to wing-margin, veins M_1 and M_2 arising independently from discal cell. Costa running around the wing. Body black or greyish .... **Microphoridae** (p. 243)

– Discal cell fused with 2nd basal cell, M with 2 branches, vein M_2 absent (M rarely forking apically into M_1 and M_2). Costa ending at M_1. Body generally metallic or yellow .... **Dolichopodidae**

Family Hybotidae

The family Hybotidae is a well defined monophyletic group consisting of the subfamilies Ocydromiinae, Hybotinae and Tachydromiinae, the 'Ocydromiinae group of subfamilies' of Tuomikoski (1966), and includes the more specialised forms of the former "Empididae". The family is fully discussed in the General part and is characterised by several apomorphies: male hypopygium more or less asymmetrical and rotated towards right (hypopygium lateroversum), periandrium developed, aedeagus rather short and stout (never long and slender), more or less firmly attached to hypandrium; front tibiae with a tubular gland; radial sector with 2 branches, R_{4+5} not forked; maxillary laciniae absent and palpi connected to the maxillary stipites by a special sclerite, the 'palpifer'.

The separation of the Hybotidae as a family was first proposed by Fallén (1816, as the family 'Hybotinae'), and this was followed by Meigen (1820), Macquart (1834 - Hybotides) and Zetterstedt (1842 - Hybotidae), until Schiner's (1860) reclassification. Originally, the Hybotidae included (with some minor differences) only the recent subfamilies Hybotinae and Ocydromiinae, and Tachydromiinae were recognised as a distinct family covering also the Hemerodromiinae and Clinocerinae - the Tachydromiae of Fallén (1816).

Key to subfamilies of Hybotidae

1 No discal cell, vein M unbranched. Anal cell (cell Cu) usually absent; if present, then only very small and anal vein scarcely visible. Small predaceous species, 1-5 mm in length ...... **Tachydromiinae** (see vol. 3 of *Fauna ent. scand.*, 1975).

– Discal cell present; if absent *(Bicellaria)*, then vein M branched into M_1 and M_2. Anal cell present 2

2 (1) Anal cell large, as long as or longer than 2nd basal cell. Only 2 veins from discal cell, vein M_2 absent. Hind femora more or less enlarged and/or long bristled beneath .... **Hybotinae** (p. 86)

– Anal cell much shorter than 2nd basal cell. Three veins from discal cell, in the tribe Ocydromiini M_1 abbreviated or absent. Hind femora slender; if stout *(Oedalea)*, then with only small spines beneath towards tip and discal cell emitting 3 veins .... **Ocydromiinae** (p. 110)

SUBFAMILY HYBOTINAE

Diagnosis. Head holoptic in both sexes (except *Lamachella* ♀), proboscis horizontal in position, forwardly directed and heavily sclerotised including small apical labellae (except *Afrohybos*). Thorax distinctly arched and prosternum in the form of a small sclerite separated by membrane from episterna (Fig. 41). Wings always with a more or less developed axillary lobe, no alula, discal cell present and emitting 2 veins (M_2 absent) to wing-margin. Anal cell large, at least about as long as, or longer than basal cells (shorter in *Stenoproctus* and *Lamachella*). Radial sector rather short, extremely short in *Syndyas* and *Euhybus*, longer in *Syneches, Stenoproctus, Afrohybos* and *Parahybos*. Antennal segment 3 (Figs. 145–147) small, ovate, with a long filiform 2-articulated terminal arista (except *Acarterus*). Hind femora more or less thickened and remarkably bristled (except *Acarterus* and *Parahybos*). Male hypopygium still almost symmetrical in *Syneches, Parahybos, Stenoproctus* and *Lamachella*, which are regarded as the most primitive groups of the subfamily.

Classification. The Hybotinae in the sense here proposed includes the following genera: *Hybos* Meig., *Syndyas* Loew, *Euhybus* Coq., *Afrohybos* Smith, *Ceratohybos* Bezzi, *Lactistomyia* Mel., *Syneches* Walk., *Parahybos* Kert., *Lamachella* Mel., *Stenoproctus* Loew and *Acarterus* Loew. The last-mentioned genus, with a single South African (Cape) species, has the characteristic hybotine wing venation, but the simple hind legs without bristling and the conical 3rd antennal segment with distinctly shorter arista distinguish it from the other genera of the subfamily. The genera *Meghyperus* Loew and *Acarteroptera* Coll., usually assigned to the Hybotinae, have been transferred to the family Atelestidae.

Originally, the subfamily Hybotinae was a very heterogeneous group of flies, including 'archaic' forms of the "Empididae" with the plesiomorphous large anal cell. Melander (1928) was the first, although he was not always followed by subsequent authors, to restrict the Hybotinae almost to the present concept and to exclude the subfamily Brachystomatinae (including *Homalocnemis*), but he still included the Tasmanian *Ironomyia* and *Sciadocera* (cyclorrhaphous Hypoceran families) within the Hybotinae. On the other hand, for some obscure reason he placed *Stenoproctus* and *Lamachella* in the Ocydromiinae. Collin's (1961) unnatural concept of the Hybotinae, which included all the non-tachydromiine forms of the former "Empididae" (including Oreogetoninae and Microphoridae) with a small *Hybos*-like prosternum, is, of course, unacceptable.

Distribution. The subfamily is mainly distributed, and most rich in species, in the tropics and subtropics of the Oriental and Neotropical regions, where at least two-thirds of the known species occur. The genera *Hybos, Syndyas* and *Syneches* are worldwide in distribution but *Syndyas* is absent in South America. *Parahybos* is exclusively Oriental and *Euhybus* is known only from the American continent. Two genera, *Lactistomyia* and *Ceratohybos*, are Neotropical but one species of the former genus has also been described from the Oriental region. Four genera, *Stenoproctus, Lamachella, Acarterus* and *Afrohybos*, with only a few species, are entirely Afrotropical as the single *Stenoproctus* described from Nepal may well be a species of *Syneches*.

Only six species in three genera of the Hybotinae are known from Europe, and four of these have been found in Scandinavia.

Key to genera of Hybotinae

1 Anal cell much longer than basal cells (in European species); radial sector rather short, at most as long as the distance from humeral crossvein to its origin (Figs. 148, 149). Wings clear or uniformly clouded, not maculated 2
– Anal cell shorter, as long as basal cells; radial sector long, much longer than the distance from humeral crossvein to its origin (Fig. 150). Wings maculated (in European species) .... *Syneches* Walker (p. 104)
2 (1) Basal section of vein M (separating basal cells) distinct, radial sector rather longer (Fig. 148). Antennae inserted at or below middle of head in profile. Hind tibiae and metatarsi simple (Fig. 151) .... *Hybos* Meigen (p. 88)
– Basal section of vein M (separating basal cells) scarcely visible, radial sector very short (Fig. 149). Antennae inserted above middle of head in profile. Hind tibiae at tip and hind metatarsi thickened (Fig. 174) .... *Syndyas* Loew (p. 99)

Genus *Hybos* Meigen, 1803

Neoza Meigen, 1800: 27. Suppressed by I. C. Z. N., 1963: 339.
Type-species: *Musca grossipes* Linné, 1767 (des. Coquillett, 1910).
Hybos Meigen, 1803: 269.
Type-species: *Hybos funebris* Meigen, 1804 (des. Curtis, 1837) = *Hybos grossipes* (Linné, 1767).

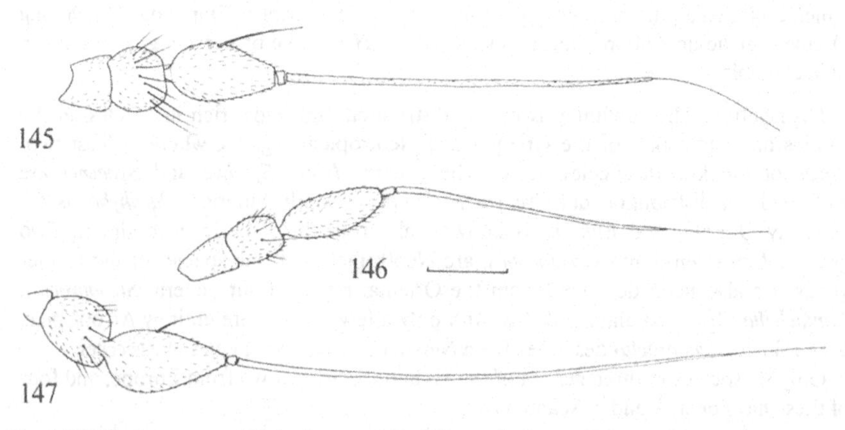

Fig. 145–147. Antennae of Hybotinae. – 145: *Hybos culiciformis* (Fabr.); 146: *Syndyas nigripes* (Zett.); 147: *Syneches muscarius* (Fabr.). Scale: 0.1 mm.

Head rather small, almost circular in frontal view, holoptic in both sexes. Upper facets more or less enlarged, sometimes very conspicuously so, but in some non-Palaearctic species all facets equally small, a character not depending on the sex. Antennae inserted at or below middle of head in profile, face short and not very broad. One or two pairs of ocellar bristles and a row of distinct postocular bristles. Antennae small, 1st

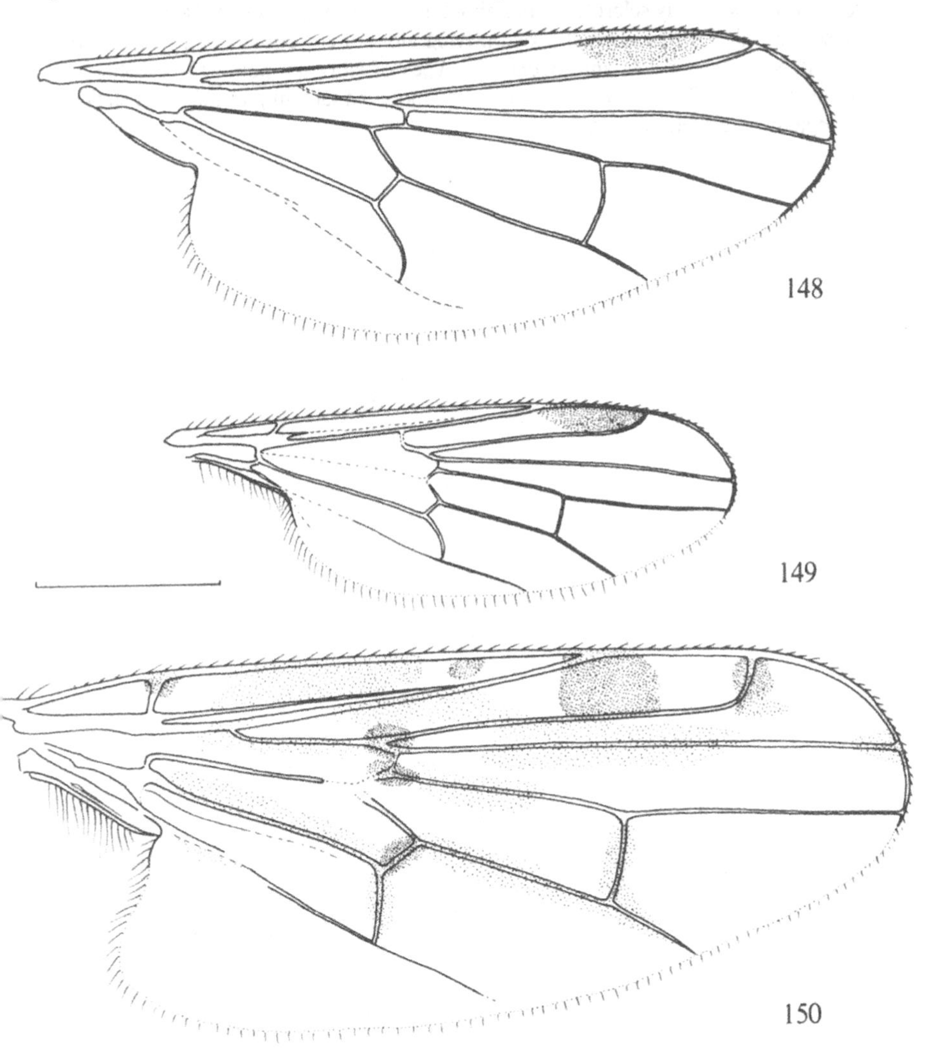

Figs. 148–150. Wings of Hybotinae. – 148: *Hybos grossipes* (L.); 149: *Syndyas nigripes* (Zett.); 150: *Syneches muscarius* (Fabr.). Scale: 1 mm.

segment indistinctly separated from the 2nd, the latter with a circlet of short preapical bristles. 3rd segment short ovate, often with 1 or 2 bristles above or below; terminal arista very long and slender, at least twice as long as antenna, finely pubescent on the somewhat stouter basal part, leaving apical quarter to extreme tip bare and hair-like, basal article very small. Mouthparts (Figs. 152–155) projecting forwards or slightly upwards, proboscis heavily sclerotised and about as long as head is deep. Palpi as long as proboscis, slender, in the normal position lying along the proboscis and with several long, widely spaced bristles anteroventrally. Apex of labrum 3-pointed, hypopharynx with numerous small teeth before tip at sides, and even labium and the small slender labellae heavily sclerotised (Fig. 155).

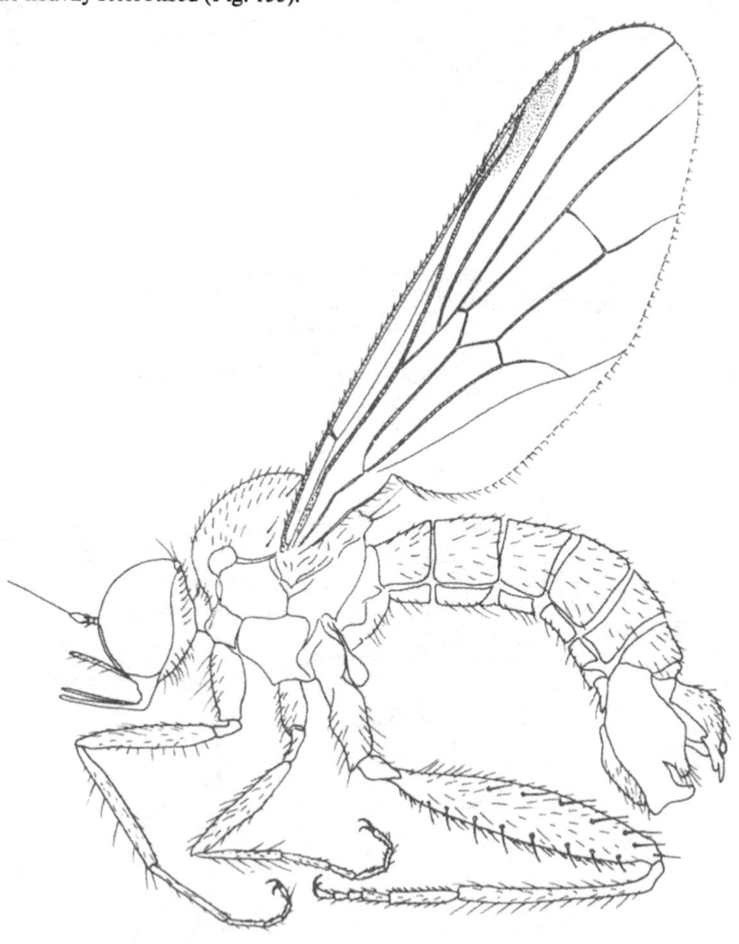

Fig. 151. Male of *Hybos culiciformis* (Fabr.). Total length: 3.3 – 5.6 mm.

Thorax distinctly arched above, prescutellar depression somewhat concave and bare. Prosternum in the form of a small isolated saddle-shaped sclerite (Fig. 41), humeri small but distinct. Acr 2- to 6-serial, dc uniserial, similar hairs also present at sides; large bristles reduced to several notopleurals, a postalar, last prescutellar pair of dc, and a single apical pair of scutellars, other scutellar bristles reduced to small marginal hairs.

Wings (Fig. 148) clear or uniformly clouded, large, with a distinct axillary lobe, alula not developed, wing entirely covered with microtrichia. Costa running to tip of M_1, costal ciliation rather long but no distinct costal bristle. Sc close to R_1, joining it far beyond radial bifurcation. Radial sector rather short, usually shorter than the distance between humeral crossvein and its origin. Costal stigma long-elliptical, when compared with *Syndyas*, longer spaced basad of R_1 and not reaching tip of R_{2+3}. Veins R_{4+5} and M_1 slightly diverging. Discal cell large, emitting 2 veins to wing-margin, M_2 absent. Anal cell longer than the equally long basal cells, the vein closing it distinctly convex and reaching vein A (which is very fine on its basal part) nearly at right angles, or the outer angle obtuse (Fig. 51). Basal section of vein M, separating basal cells, always distinct.

Legs rather long and slender, with specifically distinct hairs and bristles that also differ between the sexes. Hind femora more or less incrassate and lengthened, usually armed with conspicuous spine-like bristles. Hind tibiae and tarsi slender.

Abdomen long, semicylindrical in cross-section and slightly curved downwards, consisting of eight visible segments, no marginal bristles. 8th tergum mostly hidden beneath

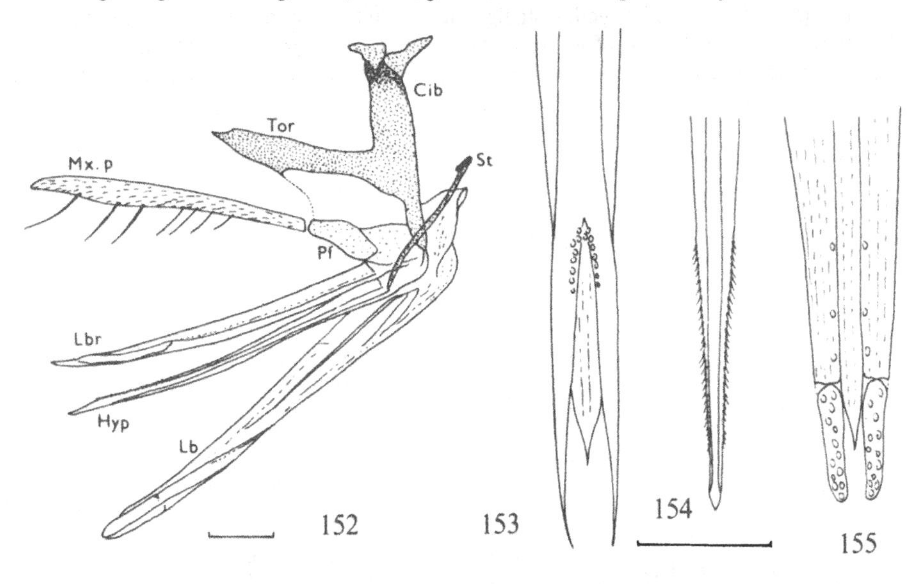

Figs. 152–155. Mouthparts of *Hybos culiciformis* (Fabr.) ♂. – 152: lateral view; 153: tip of labrum, ventral view; 154: tip of hypopharynx, ventral view; 155: tip of labium, dorsal view. Scale: 0.1 mm. For abbreviations see p. 13.

the 7th in male, distinct and slightly produced at sides in female. 1st and 8th tergum narrowed in male, shorter than corresponding sterna. Hypopygium (Fig. 151) rather large, sometimes very conspicuously so, asymmetrical and rotated through almost 90° to the right, so that hypandrium lies on the left side. Periandrial lamellae with conspicuous, specifically distinct terminal appendages. Hypandrium large with lateral margins folded inwards, forming a firm ring round a short stout aedeagus. Aedeagal apodeme about as long as aedeagus, cerci very small. Female abdomen (Figs. 156, 157) distinctly blunt-ended, sterna small and broadly separated by membraneous parts, 8th sternum enlarged, partly covering small slender cerci from beneath.

Distribution. The genus is mainly distributed in the Oriental region, with 37 known species; 8 species are known from South America, 2 from Africa, and 4 have been described from Micronesia and New Guinea. The genus is only poorly represented in the northern hemisphere, 1 species being Nearctic, 3 European, and 7 are from east Asia (Japan, China). Three Indoaustralian species were placed by Frey (1953) in the subgenus *Pseudosyneches* Frey. Of the three Scandinavian *Hybos* species, *grossipes* and *culiciformis* are often misidentified.

Biology. Adults are predaceous in both sexes; flower visiting and swarming have not been observed, and mating activity has been transferred to a solid substrate. Both males and females hunt insects in flight, more often other small Diptera and other available insects, and with their strong chitinised proboscis they are also able to attack Hymenoptera or even small beetles (staphylinids). Adults prefer rather moist and shady biotopes, sitting on the tips of leaves of bushes and low vegetation, and waiting for nearby flying insects. European species occur rather in late summer; immature stages unknown.

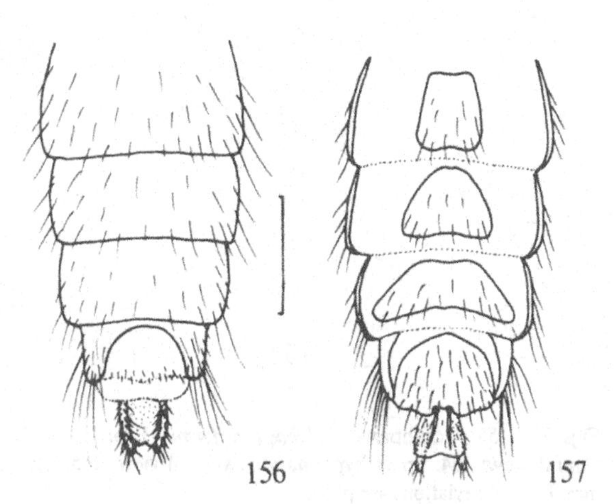

Figs. 156, 157. Female abdomen of *Hybos culiciformis* (Fabr.), – 156: dorsal view; 157: ventral view. Scale: 0.5 mm.

Key to species of *Hybos*

1 Mesonotum polished black, leaving only narrow sides and a median stripe greyish dusted. Anterior four legs yellow. ♂: fore tibiae and tarsi and hind legs with long pale hairs; genitalia rather small .... 3. *femoratus* (Müller)

– Mesonotum uniformly thinly grey or brownish dusted, at most subshining. Anterior four legs blackish 2

2 (1) Large thoracic bristles yellowish. Mid femora with a row of ad bristles, mid tibiae with several strong av bristles. Meso-

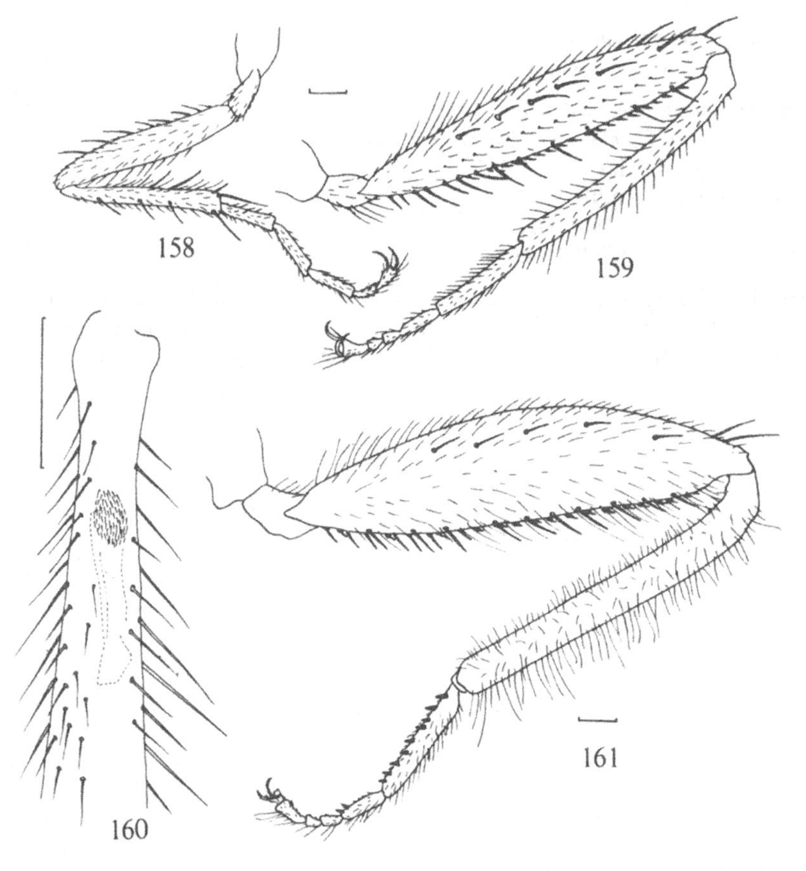

Figs. 158–161. Legs of *Hybos*. – 158: *culiciformis* (Fabr.) ♂, fore leg in posterior view; 159: same species, hind leg in anterior view; 160: same species, fore tibia with tubular gland in ventral view; 161: *grossipes* (L.) ♂, hind leg in anterior view. Scale: 0.2 mm.

notum finely brownish dusted. Hind metatarsus with only short bristles beneath, and at least anterior metatarsi translucent brownish. ♂: fore tibiae and tarsi rather short pubescent; genitalia very large .... 2. *culiciformis* (Fabricius)

– Large thoracic bristles black. No ad bristles on mid femur and no strong av bristles on mid tibia. Mesonotum very thinly grey dusted, rather subshining. Hind metatarsus with short but strong black spines beneath and legs quite black. ♂: fore tibiae and tarsi with very long black hairs; genitalia scarcely broader than tip of abdomen .... 1. *grossipes* (Linné)

1. ***Hybos grossipes*** (Linné, 1767)
   Figs. 41, 51, 148, 161, 162–165.

Musca grossipes Linné, 1767: 988.
Asilus culiciformis Gmelin, 1790: 2900 (nec Fabricius, 1775).
Empis clavipes Fabricius, 1794: 403.
Hybos funebris Meigen, 1804: 240.
Hybos pilipes Meigen, 1820: 349.
Hybos claripennis Strobl, 1893: 43.

Legs wholly black, hind femora conspicuously thickened and hind metatarsi with short black spines beneath. Large thoracic bristles black. Anterior four tibiae and metatarsi in male with conspicuously long, black bristly hairs at sides; hypopygium rather small.

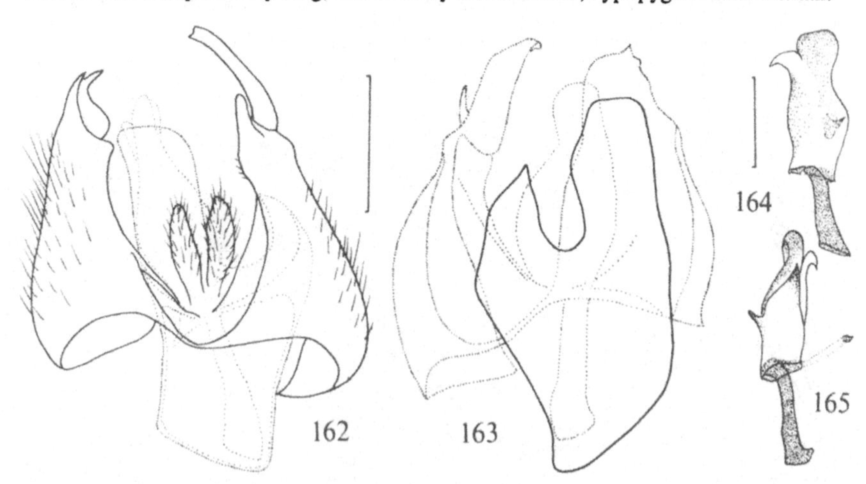

Figs. 162–165. Male genitalia of *Hybos grossipes* (L.). – 162: periandrium with cerci, hypandrium and aedeagus dotted; 163: hypandrium, periandrium and aedeagus dotted; 164, 165: aedeagus in two different views. Scale: 0.2 mm.

♂. Antennae black, arista more than twice as long as antenna, its thicker basal part finely pubescent, apical quarter bare and hair-like. Mesonotum finely dark brownish grey dusted, margins and pleura greyish. All thoracic hairs and bristles black, acr irregularly 4-serial, dc uniserial and as long as acr, ending in one pair of long prescutellars about as long as 2 notopleurals, 1 postalar and 1 pair of scutellars; 6 additional scutellars very small and hair-like.

Wings (Fig. 148) quite clear to very dark brown clouded (var. *funebris*), squamae with several long dark bristly hairs. Stigma long ovate, very distinct especially in the clear-winged specimens. Halteres pale. Legs quite black, hind femora very elongated and thickened, considerably convex above, and quite polished. Fore femora with long pv bristly hairs, about as long as femur is deep, mid femora with similar pv and av bristly hairs and several shorter ad hairs on basal third. Fore tibiae with 2 long dorsal bristles, a circlet of preapical bristles and a row of very long posterior bristle-like hairs becoming longer towards tip; similar bristle-like hairs on both sides on basal two tarsal segments, the longest hairs on metatarsus quite four-fifths of its length. Mid tibia with 3 strong dorsal bristles and long lateral hairs, 2nd tarsal segment with only the usual short hairs. Hind femora (Fig. 161) spinose beneath, tibiae everywhere clothed with long dark hairs, metatarsus ventrally with 6–9 black spines mixed with short rufuous pubescence, and also with two similar spines at tip of 2nd tarsal segment.

Abdomen uniformly thinly brownish dusted, subshining black from some angles, dorsum with short hairs, sides and venter clothed with longer and paler hairs. Hypopygium (Fig. 162) rather small, when viewed from above (left lamella dorsal due to rotation) not deeper than abdomen.

Length: body 3.5–5.5 mm, wing 4.0–5.5 mm.

♀. Resembling male very closely but anterior tibiae and tarsi without long lateral bristly hairs, although the short pubescence is still longer and denser than in *culiciformis* and the not so thickened hind femora are still deeper and more shining. The short ventral spines on hind metatarsus as in male.

Length: body 3.2–4.6 mm, wing 3.5–5.2 mm.

Distribution and biology. Rather uncommon but widespread in Denmark and S. Sweden, more common towards the north beyond 66°N, up to the polar circle. The hybotine species with the most northern distribution, in Finland to ObN and Ks, in Norway to NT. - Europe, the most southern genuine record is from the Romanian mountains; probably absent in the south. In temperate Europe more common in colder areas and at higher altitudes. - End May to late September, but mainly in July and August.

2. ***Hybos culiciformis*** (Fabricius, 1775)
Figs. 145, 151, 152–158, 166–170.

Asilus culiciformis Fabricius, 1775: 796.
Hybos vitripennis Meigen, 1820: 348.
Hybos rufitarsis von Roser, 1840: 53.

Hybos infuscatus Zetterstedt, 1842: 234.
Hybos rufitarsis Zetterstedt, 1849: 2994 (nec von Roser, 1840).

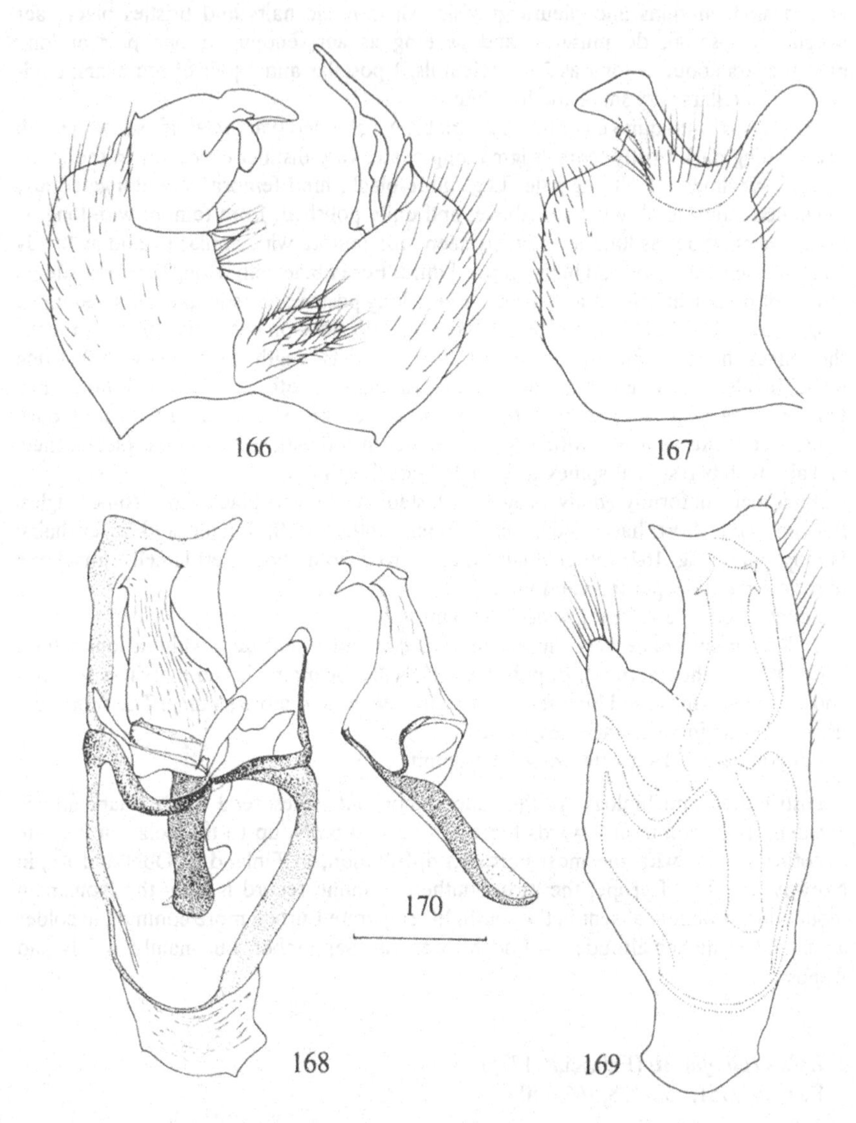

Figs. 166–170. Male genitalia of *Hybos culiciformis* (Fabr.). – 166: periandrium with cerci; 167: right periandrial lamella; 168: hypandrium in ventral view, with hypandrial bridge and aedeagus; 169: hypandrium dorsally, aedeagus dotted; 170: aedeagus. Scale: 0.3 mm.

Legs with tips of anterior tibiae and at least base of mid metatarsi dark brownish. Hind femora elongate but less stout than in *grossipes,* hind metatarsi not spinose beneath. Large thoracic bristles pale. Anterior legs with rather short hairs in male, and hypopygium very large and globose.

♂. Head and antennae as in *grossipes,* 3rd ant.s. often with a similar small dorsal bristly hair. Mesonotum more brownish dusted but also slightly subshining in dorsal view. Small thoracic bristles blackish, acr irregularly 6-serial throughout, dc uniserial, but the large bristles (the same number as in *grossipes*) more or less yellowish. Apical scutellars always pale, other marginal bristles more numerous (4–5 pairs) and slightly longer, usually one-third length of the strong apical pair.

Wings as in *grossipes,* also varying in colour from quite clear to very darkened (var. *infuscatus*), stigma always distinct and of the same shape as in *grossipes.* Halteres pale but the long bristly hairs on squamae pale. Legs not quite so black, all 'knees' and tips of tibiae brownish, mid tibia near tip and at least basal part of mid metatarsus translucent brownish, in very pale specimens (*rufitarsis*) dorsum of mid tibia and anterior tarsi more or less dark reddish. The bristling on fore legs almost as in *grossipes* but mid femur with a complete row of long black ad bristles (about 6 bristles as long as femur is deep), mid tibia with more numerous dorsal bristles and with about three av bristles much longer than tibia is deep. Fore tibiae and metatarsi with the lateral hairs much shorter and less densely placed compared with *grossipes,* the hairs on metatarsus not much longer than metatarsus is deep. Mid tibiae and tarsi with short hairs, no long pubescence as in *grossipes.* Hind femora duller black and less thickened than in *grossipes,* with similar bristles but obviously more elongate. Hind metatarsus with only short bristly hairs beneath, no distinct spines.

Abdomen dull black when viewed from above, in lateral view terga with distinct grey dusting anteriorly, posterior parts contrasting polished black. Pubescence mostly pale and apparently longer at sides and on venter than in *grossipes.* Hypopygium (Fig. 166) very large and globose, much broader than abdomen when viewed from above, and remarkably greyish dusted.

Length: body 3.3–5.6 mm, wing 3.6–5.6 mm.

♀. Very much like male but hind femora more slender and the pv spines near base absent, replaced by long yellowish hair-like bristles along the whole length of femur. Females of *culiciformis* and *grossipes* differ mainly in the bristling and pubescence on legs, as described under the male sex.

Length: body 3.4–4.8 mm, wing 4.0–5.5 mm.

Distribution and biology. Very common in Denmark, S. Sweden and along the Finnish Baltic coast, extending to 61° N, very rarely up to 63° to Ångermanland (1 ♀). A species of rather southern distribution in Scandinavia; not yet found in Norway, but it should occur there in the south. - Europe, including extreme south (Spain, Corsica, Yugoslavia, Greece) and Turkey. - End June to September, mainly from late July to early September, in C. Europe to middle of October.

3. ***Hybos femoratus*** (Müller, 1776)
Figs. 171–173.

Asilus femoratus Müller, 1776: 2135.
Hybos flavipes Meigen, 1804: 241.
Hybos fumipennis Meigen, 1820: 349.

Anterior four legs and hind tarsi yellow. Mesonotum extensively polished black with a dull grey median stripe, all hairs and bristles on body and legs pale.

♂. Antennae black, 3rd segment with a distinct dorsal bristle or two, arista about 2.5 times as long as antenna, apical fifth bare. Mesonotum polished black, narrow margins, notopleural and prescutellar depressions and a narrow median stripe silvery-grey dusted, thoracic pleura greyish pollinose. Acr 4-serial, situated on the median greyish stripe, dc uniserial, with last prescutellar pair strong. All hairs and bristles on thorax pale, one pair of strong scutellars with 5–6 pairs of minute hairs at sides.

Wings clear or intensely darkened *(fumipennis)*, stigma very long-elliptical and always indistinct, faintly brownish to yellowish, halteres pale. Extreme base of wing between basal stems of veins R and Cu anterior to basal cells devoid of microtrichia. Anterior four legs (not coxae) and hind tarsi yellow to yellowish brown. Fore and mid femora with conspicuously long av and pv bristly hairs beneath, at least as long as femur is deep. Anterior four tibiae with 2–3 strong dark bristles above and with very long pale hairs laterally and ventrally, the hairs scarcely shorter than those of *grossipes*; similar

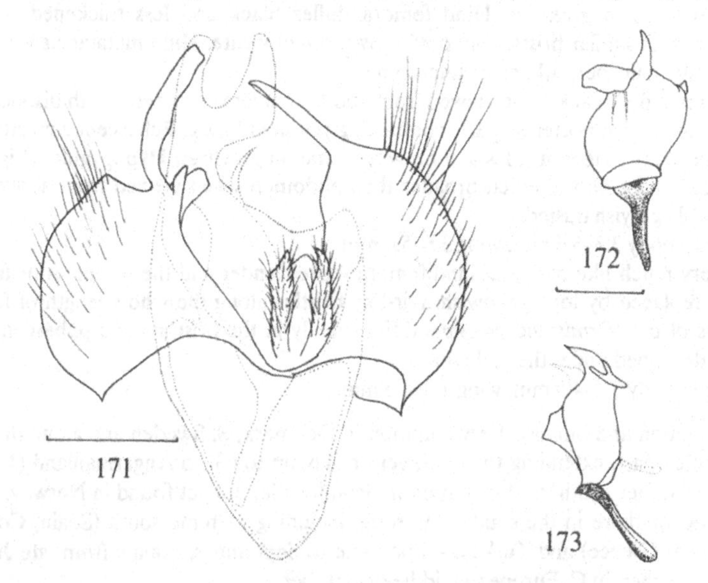

Figs. 171–173. Male genitalia of *Hybos femoratus* (Müll.). – 171: periandrium with cerci, hypandrium and aedeagus dotted; 172, 173: aedeagus in two different views. Scale: 0.2 mm.

long hairs also on fore metatarsus, mid metatarsus with less numerous but longer hairs. Hind femora very thickened (almost as in *grossipes*); in addition to the usual short black spines beneath with longer av and pv spines, and 2–3 still longer ad spines before tip. Femur everywhere clothed with long outstanding pale hairs almost as long as femur is deep. Hind tibiae covered with similar long pale hairs, those at sides almost 3 times as long as tibia is deep. Hind metatarsus beneath with short pale hairs and 4–5 short but strong black spines, and similar spines also on 2nd segment at tip beneath.

Abdomen slender, uniformly very thinly grey dusted, almost subshining black, sides and venter with sparse, very long pale hairs. Hypopygium (Fig. 171) broader than abdomen, periandrial lamellae thinly dusted and very convex, hypandrium polished with a remarkably overlapping bifurcated appendage.

Length: body 3.3–4.7 mm, wing 3.6–5.0 mm.

♀. Resembling male but legs with shorter hairs everywhere, especially on anterior four tibiae and metatarsi, and hind femora obviously less thickened.

Length: body 3.0–4.6 mm, wing 3.4–5.3 mm.

Distribution and biology. Very common in Denmark and S. Sweden, north to approximately 63° N, in Finland to Ta, St and Kb, in Sweden to Hls., in Norway to HO. – Throughout Europe including the south, N. Africa (Egypt). – June to September, mainly in July and first half of August, in temperate Europe by end of May.

Genus *Syndyas* Loew, 1857

Syndyas Loew, 1857: 369.

Type-species: *Syndyas opaca* Loew, 1857 (des. Coquillett, 1903).

Sabinios Jones; 1940: 273.

Type-species: *Sabinios jovis* Jones, 1940 (orig. des.).

Head very much as in *Hybos*, holoptic in both sexes with upper facets more or less enlarged, but antennae inserted above middle of head in profile and face consequently longer and much narrower, almost linear. Ocellar tubercle prominent, anterior pair of ocellar bristles long. 3rd antennal segment ovate, without a bristle above. Arista much longer than antenna and its thicker part always bare, basal article distinct. Mouthparts (Figs. 175–178) as in *Hybos*, directed forwards and heavily sclerotised including labium, labellae somewhat shorter and broader. Palpi similarly slender and scarcely shorter than proboscis, apically with 1 or 2 bristles, hypopharynx with less numerous (8–9) and more distinct teeth before tip, and labrum distinctly serrate anteroventrally towards tip.

Thorax more or less arched above, prosternum *Hybos*-like, large thoracic bristles confined to 2 notopleurals and a pair of apical scutellars. Other thoracic bristles hair-like and numerous, usually as long as the large bristles, acr 2- to 6-serial, dc generally uniserial with additional hairs at sides, and 3 to 5 pairs of small hairs on scutellum lateral to the strong apical pair.

Wings (Fig. 149) clear or uniformly clouded, costal stigma (if visible) short ovate to almost circular, occupying the costal section between tips of veins R_1 and R_{2+3}, and the whole apex of cell R_1. Vein R_1 shorter than in *Hybos* and R_{2+3} more abruptly bent towards costa, sometimes very conspicuously so. Radial sector very short, much shorter than in *Hybos*. Wings only seldom completely covered with microtrichia (as for instance in *nigripes*), more often at least costal cell and partly also basal cells devoid of microtrichia and consequently paler. No distinct costal bristle. Discal cell emitting 2 veins to wing-margin, basal section of vein M, separating basal cells, and basal part of

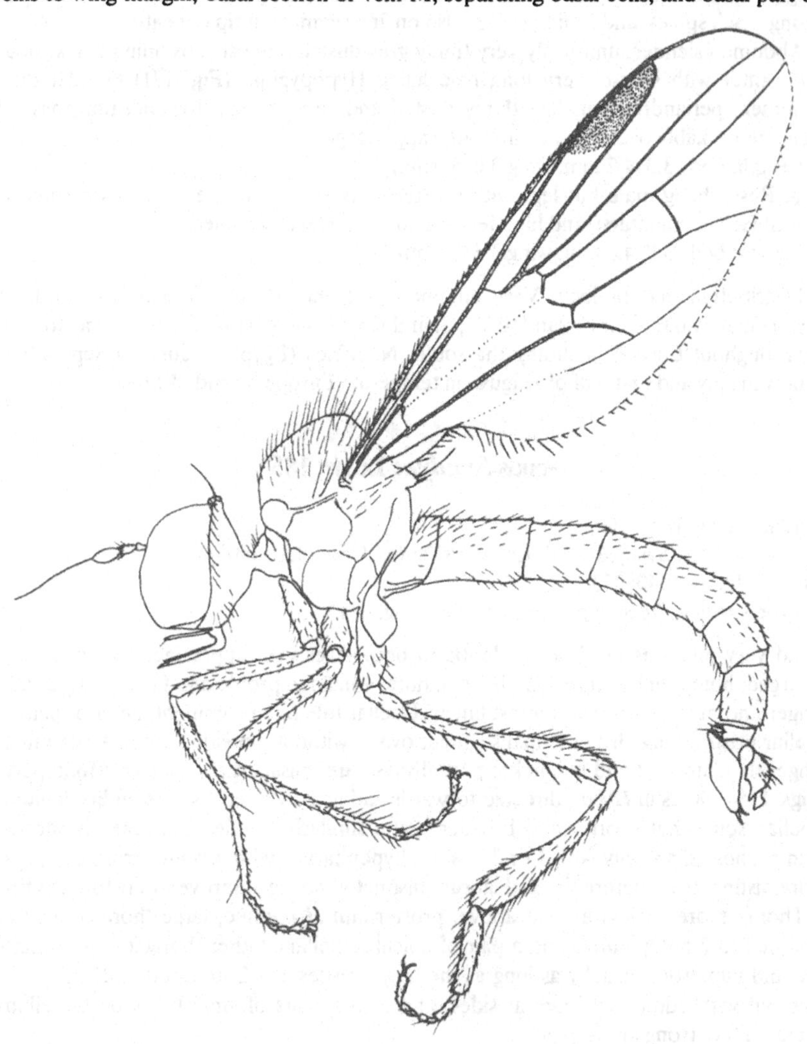

Fig. 174. Male of *Syndyas nigripes* (Zett.). Total length: 3.0–3.3 mm.

anal vein very indistinct, in the form of a whitish fold. Anal cell about as long as or longer than basal cells, axillary lobe developed.

Legs with specifically distinct hairs and bristles, anterior legs slender and simple, fore tibiae with a tubular gland sometimes ending in a conspicuous tubercle (Fig. 179). Hind femora more or less thickened (at least towards tip) and armed with spine-like bristles beneath, hind tibiae and hind metatarsus considerably swollen.

Abdomen long and rather slender, clothed with hairs only, 5th and following segments slightly curved downwards. Basal and 8th sternum in male (Fig. 184) longer than corresponding terga, 8th segment partly concealed within the much larger 7th segment. Hypopygium (Fig. 187) asymmetrical and rotated to the right through about 90°. General structure as in *Hybos* with similar specifically distinctive appendages on both periandrial lamellae, but cerci longer, hypandrium often distinctly convex and aedeagus with a flat, leaf-shaped apodeme. First sternum in female (Fig. 185) longer than corresponding tergum as in male but abdomen telescopic and pointed apically, much narrower than in *Hybos*, both 8th tergum and sternum equally elongated, cerci long and slender.

Distribution. The genus is best represented in Africa with 14 known species; 8 species have been described from the Oriental region, 1 or 2 from New Guinea (a revision is needed), 6 from North America and only 2 from Europe. In the northern hemisphere the genus is evidently more abundant in North America (see Teskey & Chillcott, 1977) as only one species is known from boreal Europe and Scandinavia.

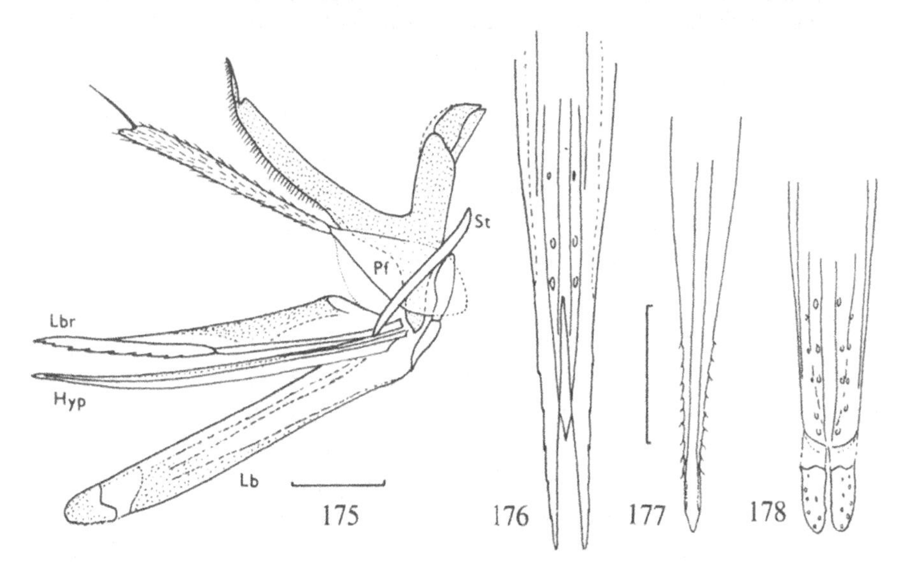

Figs. 175–178. Mouthparts of *Syndyas nigripes* (Zett.) ♂. - 175: lateral view; 176: tip of labrum, ventral view; 177: tip of hypopharynx, ventral view; 178: tip of labium, dorsal view. Scale: 0.1 mm. For abbreviations see p. 13.

Biology. Adults are predators in both sexes with habits similar to those of *Hybos* species; no swarming activity has been observed and mating has been transferred to a solid substrate. The immature stages are unknown, but the adults occur in the temperate and cooler zones of the northern hemisphere and appear to be restricted to cold sphagnum bogs.

4. ***Syndyas nigripes*** (Zetterstedt, 1842)

Figs. 146, 149, 174–190.

Ocydromia nigripes Zetterstedt, 1842: 240.

A small black *Hybos*-like species with hind femora very slender on basal half, hind tibiae towards tip and hind metatarsi distinctly swollen.

♂. Antennae (Fig. 146) black, 3rd segment clothed with minute hairs only; arista nearly twice as long as antenna, its thicker basal part bare, apical quarter hair-like. Palpi black, long and slender, with 1 or 2 black preapical bristles. Thorax polished black on mesonotum, pleura thinly grey dusted. Large thoracic bristles (2 notopleurals, apical pair of scutellars) black, other bristles almost as long but fine, hair-like and pale: acr 2-serial, dc uniserial, with numerous similar hairs at sides, and 4–5 pairs of hair-like marginal scutellars half length of the strong apical pair.

Wings more or less faintly brownish clouded, costal stigma dark brown and elliptical, occupying the whole apex of cell R_1. The vein separating basal cells very indistinct and

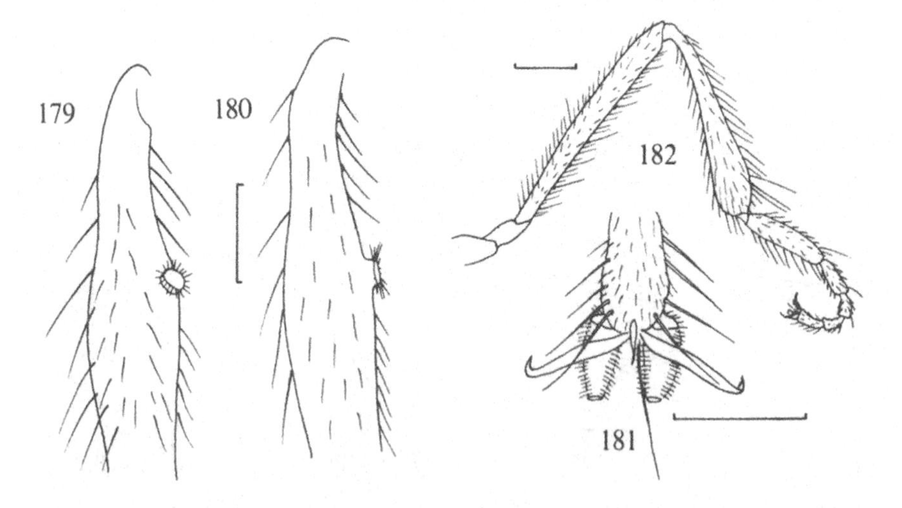

Figs. 179–182. Legs of *Syndyas nigripes* (Zett.) ♂. – 179: base of fore tibia with tubular gland in anteroventral view; 180: the same in anterior view; 181: praetarsus of mid leg; 182: hind leg in anterior view. Scale: 0.1 mm.

whitish. Halteres black, squamae dark grey. Legs subshining black, anterior two pairs slender, hind legs elongated with femur slightly dilated towards tip, but tibia on apical half and whole metatarsus very thickened. Fore tibiae swollen in basal third with a distinct sense organ (Fig. 179) and metatarsus also covered with long pv hairs. Mid tibia with about 7 dorsal bristles, 2–3 preapical bristles, the ventral one very long (not much shorter than metatarsus), mid metatarsus with 2 long pv bristles and one still longer dorsal bristle. Hind femora with long hairs dorsally on basal third, pv along the whole length, and with a row of about 8 long spine-like ventral bristles arising from small warts, 4 similar black ad bristles before tip. Hind tibiae with long black ad and pd bristles arranged in irregular pairs, metatarsus with several rather long av spine-like bristles and 2 preapical ones.

Abdomen long and slender, shining black and clothed with long pale hairs at sides and below. Hypopygium (Fig. 187) not broader than end of abdomen, dulled by brownish dust and short pubescence, the small bifurcated appendage of right lamella overlapping the whole hypopygium.

Length: body 3.0–3.3 mm, wing 2.7–3.1 mm.

♀. Resembling male very closely but abdomen gradually tapering, more downcurved and very pointed.

Length: body 2.6–3.0 mm, wing 2.7–3.0 mm.

Holotype identification. Described by Zetterstedt (1842) from a single ♂ from "Ölandia, Jul. 1819". In the Dipt. Scand. Coll. at Lund there is a pair under *Ocydromia*

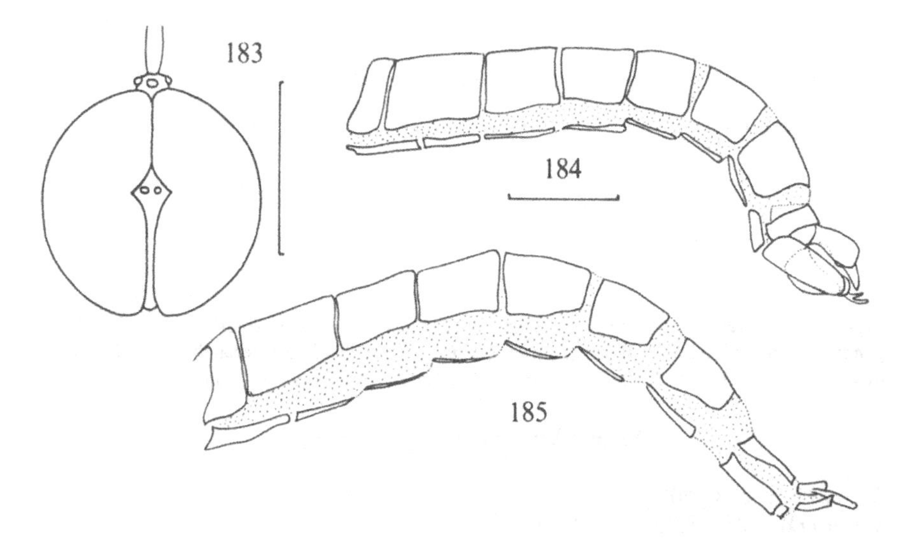

Figs. 183–185. *Syndyas nigripes* (Zett.). 183: ♂ head, anterior view; 184: ♂ abdomen; 185: ♀ abdomen. Scale: 0.5 mm.

nigripes, the ♂ labelled "Ölandia", the ♀ "Smol. Skänfiö". The ♂ is undoubtedly the holotype of *Syndyas nigripes* (Zett.) and it was labelled accordingly in 1977.

Distribution and biology. Rare in S. Sweden (only 5 specimens from Öl., Sm. and Nrk.); rather common in Finland along the Baltic coast from Al to Russian Karelia, scarce northwards to Oa, Tb, Sb, the northernmost record in ObN (Pissavara naturpark, 2 ♀). Not yet recorded from Denmark and Norway. – From England along the North Sea coast through the Netherlands, Schleswig-Holstein and Poland to NW and central European USSR, and in C. Europe (Czechoslovakia, Hungary, Austria, Switzerland). – July to middle of August, locally common in marshy biotopes and peat-bogs in particular.

Note. The second SW European species, *S. subsabinios* Chv., is larger (body 3.7–4.0 mm), wings with distinctly spherical stigma, costal cell and most of basal cells devoid of microtrichia, and legs yellow on anterior four tibiae and all tarsi.

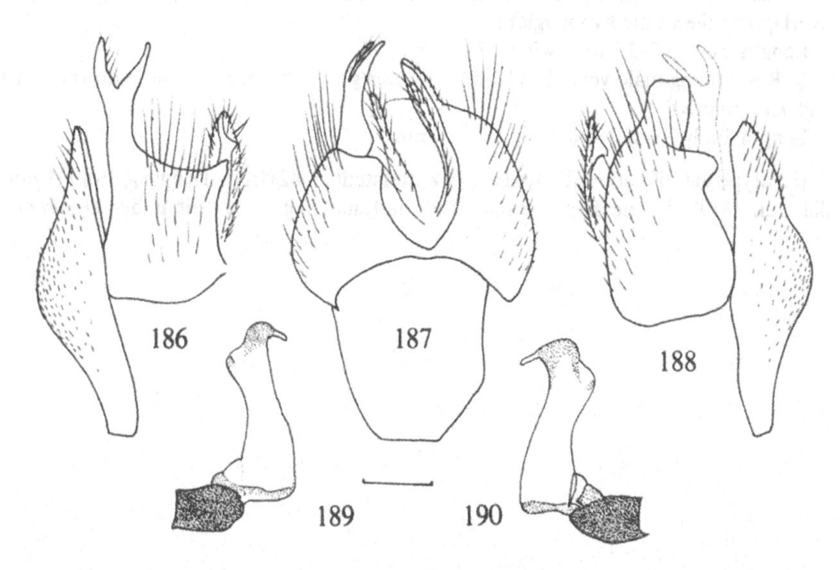

Figs. 186–190: Male genitalia of *Syndyas nigripes* (Zett.). – 186: right periandrial lamella with hypandrium and cercus; 187: hypopygium (aedeagus omitted); 188: left periandrial lamella with hypandrium and cercus; 189, 190: aedeagus in two different views. Scale: 0.1 mm.

Genus *Syneches* Walker, 1852

Damalis Fabricius, 1805: 147.*)
Acromyia Bonelli MS; Latreille, 1809: 305.
Type-species: *Acromyia asiliformis* Bonelli MS (mon.) = *Syneches muscarius* (Fabricius, 1794).

Syneches Walker, 1852: 165.
Type-species: *Syneches simplex* Walker, 1852 (mon.) (see the 'Note' on p. 108).
Pterospilus Rondani, 1856: 152.
Type-species: *Asilus muscaria* Fabricius, 1794 (mon.).
Harpamerus Bigot, 1859: 306. (? subgenus)
Type-species: *Harpamerus signatus* Bigot, 1859 (mon.).
Epiceia Walker, 1860: 149. (? subgenus).
Type-species: *Epiceia ferruginea* Walker, 1860 (mon.).

Eyes meeting on frons in both sexes for a long distance, head distinctly flattened anteriorly above, upper part of occiput slightly concave; in frontal view head almost circular. Antennae inserted below middle of head in profile and eye-facets very enlarged on more than upper half of eyes. Ocellar tubercle very prominent, anterior pair of ocellar bristles (placed more behind) distinct. Face short and rather broader, antennal excision of variable shape, nearly acute or only shallow. 2nd antennal segment with a circlet of preapical bristles, 3rd segment short ovate, with one or several conspicuous bristles above; arista bare and rather slender, at least three times as long as antenna, extreme tip hair-like. Mouthparts very much as in *Hybos*, proboscis (Fig. 193) slender and heavily sclerotised included small labellae, pointing forwards or slightly upwards, as long as head is deep. Palpi (Fig. 194) with similar bristling to *Hybos* but much shorter, at most half length of proboscis; palpifer bristled.

Thorax more or less arched above, often very humped, but flattened behind. Strong bristles as in *Hybos* but scutellars more numerous, usually 5 or more pairs of strong marginal bristles mixed with small hairs. Small thoracic bristly hairs always much smaller than in *Hybos*, sometimes almost invisible, acr 4-serial (or more numerous), dc uniserial.

Wings (Fig. 150) large, membrane entirely covered with microtrichia, axillary lobe strongly developed, axillary excision usually rectangular, no alula. Wings either uniformly clouded with a more or less distinct large elongate *Hybos*-like costal stigma or, less often, maculated as in the western Palaearctic *S. muscarius*. Costa reaching vein M_1, finely short-bristled, no distinct costal bristle. Vein Sc very close to vein R_1 and joining it far beyond radial bifurcation. Vein R_{2+3} sometimes abruptly bent towards costa, radial sector very long, not much shorter than the basal section of vein M separating basal cells. Veins R_{4+5} and M_1 parallel or convergent, sometimes very

*) Fabricius (1805) erected this genus for 4 species, 2 Neotropical "Empididae" *(curvipes, quadricincta)* and 2 Oriental Asilidae *(planiceps, myops)*. Westwood (1835) designated the first of these, *D. curvipes* (now *Syneches curvipes*), as type-species of *Damalis*, although Wiedemann (1828: 415) had already used the name *Damalis* for the two asilid species and had transferred the "empidids" to the genus *Hybos*. All subsequent authors have followed Widemann, not Westwood, and *Damalis* has generally been accepted as a genus of the Asilidae and has never been used in the "Empididae". For this reason, and for nomenclatural stability, Hull (1962: 53) proposed the asilid *Damalis planiceps* Fabricius as a new type-species for *Damalis*; however, no application was sent to the I. C. Z. N. The generally accepted nomenclature following Hull (1962) is accepted here, and at the same time the I. C. Z. N. has been requested (Smith & Chvála, 1982) to designate *Damalis planiceps* Fabricius, 1805, as type-species of *Damalis* Fabricius.

strongly so *(Harpamerus, Epiceia)*. Anal cell not longer than basal cells, the vein closing it reaching anal vein at about right-angles, latter distinct throughout, seldom faint or abbreviated at tip. Discal cell large, often very elongated, emitting two veins to wing-margin.

Legs simple on anterior two pairs, specifically bristled especially on tibiae, fore tibia

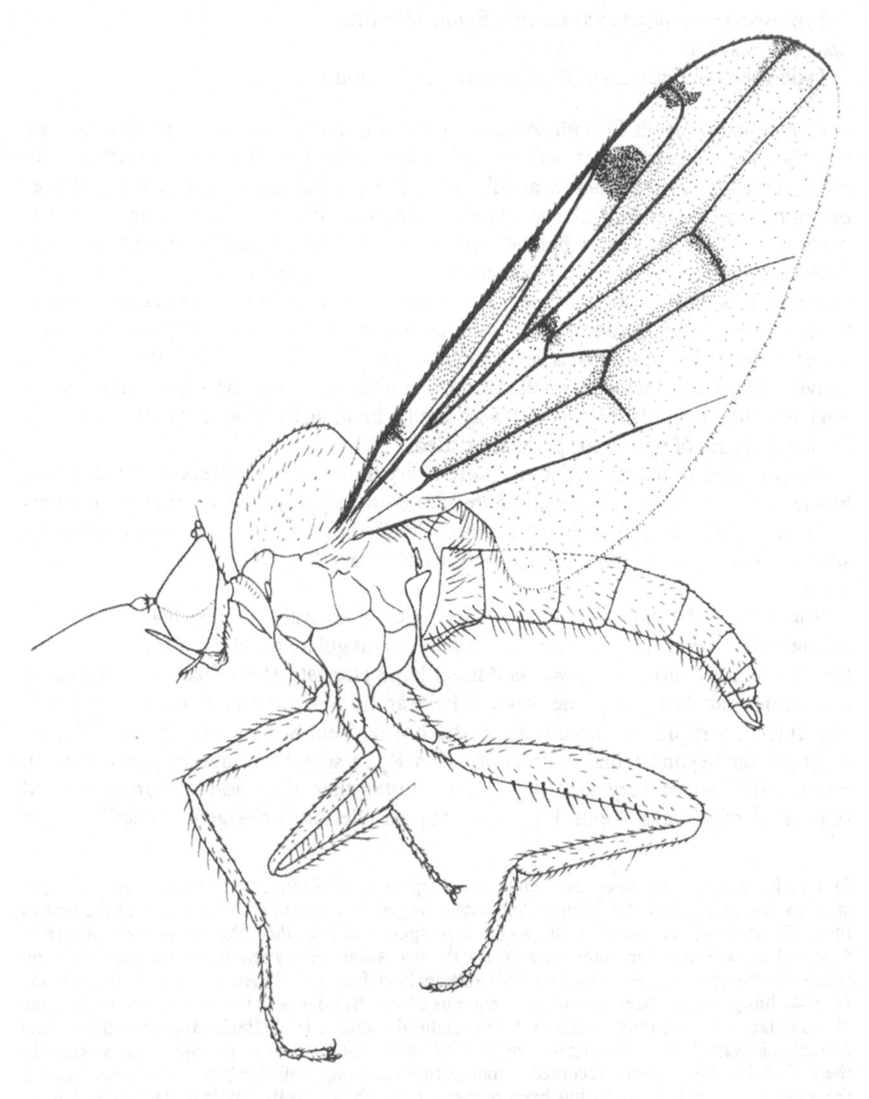

Fig. 191. Male of *Syneches muscarius* (Fabr.). Total length: 4.6–5.3 mm.

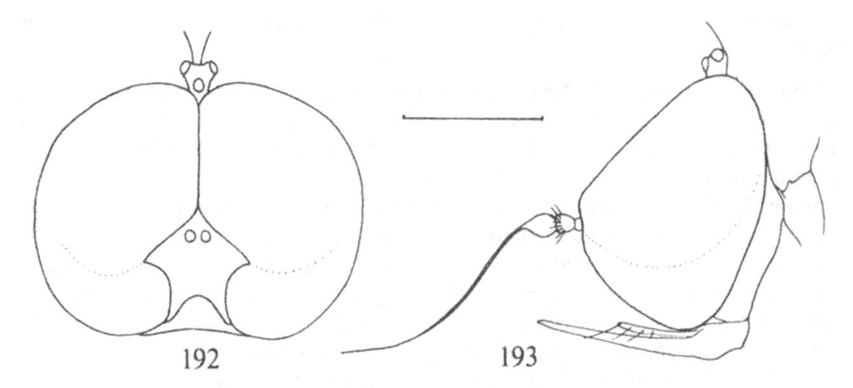

Figs. 192, 193. Head of *Syneches muscarius* (Fabr.) ♂. – 192: anterior view; 193: lateral view. Scale: 0.5 mm.

with a distinct "sense organ" near base and often slightly dilated there. Hind legs with elongate and more or less thickened femora which are variously armed beneath with bristles, simple spines, or spines arising from special warts or tubercles. Hind tibiae towards tip and metatarsi rarely swollen.

Abdomen long and rather slender, apically tapering and slightly curved downwards in both sexes, clothed with hairs only. 8th segment in male unmodified, almost as long as 7th and partly visible. Hypopygium (Fig. 196) almost symmetrical, in some species only slightly rotated to the right, in others practically through 90°. Periandrial lamellae narrowly connected anteriorly, rather simple, without conspicuous dorsal appendages, cerci slender. Inner wall of periandrium forming a distinctly bristled periandrial fold which is narrowly but firmly connected with hypandrial bridge, hypandrium simple and usually flat. Aedeagus short and stout, laterally with large flattened sclerites (postgonites) originating at base of aedeagus. The structure of the male postabdomen (unmodified 8th segment, almost symmetrical hypopygium) is clearly the primitive type in the Hybotinae. Female abdomen telescopic, often dorsoventrally flattened posteriorly and ending in two long slender cerci.

Distribution. The species of *Syneches* are the most primitive of the subfamily Hybotinae and this is also reflected in their distribution-pattern in the southern hemisphere. The genus is distributed mainly in the Oriental (38 species) and Neotropical (29 species) regions; 8 species are Afrotropical and 7 Indoaustralian. Twelve species are known from North America and 10 from the Palaearctic, 9 species being eastern Palaearctic (Japan, Ussuri region) and only one European.

Biology. The adults, generally large flies, often much over 5 mm in length, are amongst the largest species in the subfamily or even the family, and are predators in both sexes. The predaceous behaviour was fully described for the Nearctic *S. thoracicus* (Say) by Wilder (1974). No swarming and/or mating activity in the air have been observed, and reproductive activity will undoubtedly be very much as in *Hybos* species. Immature stages unknown.

Note. For many years the Nearctic *Syneches phthia* (Walker, 1849) was considered to be the type-species of the genus, based on the presumed synonymy *simplex* = *phthia*. However, according to Wilder (1974), *S. simplex* Walker, 1852, is a valid species and *S. phthia* is a synonym of *S. thoracicus* (Say, 1823).

Some Oriental and Australian *Syneches* species have been separated into the subgenera *Harpamerus* and *Epiceia* on the basis of the strongly converging veins R_{4+5} and M_1, and the differently bristled, tuberculose and/or spined hind femora. However, these subgeneric characters do not wholly fit species from other regions, as has also been pointed out by Smith (1969) for the Afrotropical species, and some of the differential features are very slight and vary from species to species. For this reason, it is probably best for the moment not to accept any subgeneric divisions.

Syneches muscarius (Fabricius, 1794)
Figs. 147, 150, 191–203.

? *Empis avidus* Harris, 1780: 151.
Asilus muscaria Fabricius, 1794: 390.
Stomoxys asiliformis Fabricius, 1794: 395.
Asilus hybos Lamarck, 1816: 404.

Large blackish species with yellow tarsi and hind tibiae, wings conspicuously brown maculated with anal cell as long as basal cells.

♂. Antennae (Fig. 147) black on basal segments, 3rd segment yellowish including a very long bare terminal arista. Proboscis yellowish, as long as head is deep and directed upwards. Palpi yellow and slender, half length of proboscis and armed with several long

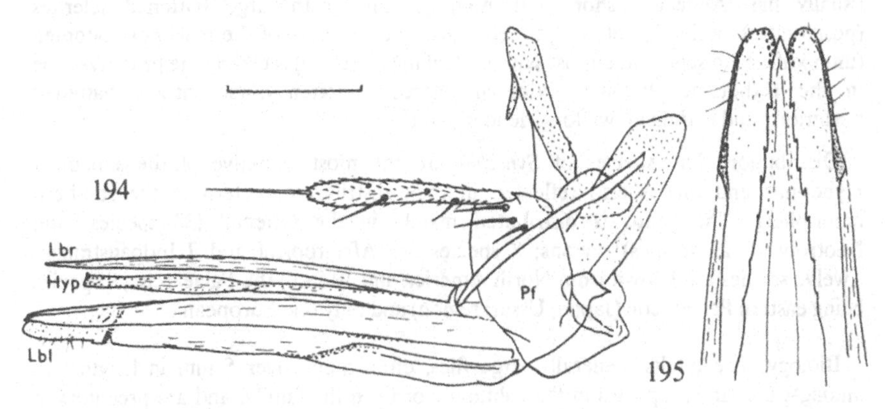

Figs. 194, 195. Mouthparts of *Syneches muscarius* (Fabr.) ♂. – 194: lateral view; 195: tip of labium, dorsal view. Scale: 0.1 mm. For abbreviations see p. 13.

black bristles. Thorax black on dorsum, dulled by brownish dust, pleura translucent brownish, humeri, postalar calli and a narrow base of scutellum yellow. Thoracic hairs and bristles black, acr small and broadly 4-serial, separated from the still smaller uniserial dc. Scutellum with 8–10 pairs of distinct marginal bristles, inner 2 pairs stouter and longer.

Wings (Fig. 150) large, uniformly faintly brownish clouded and conspicuously dark brown maculated: a large ovate patch at tip of R_1, small patches at tips of Sc, R_{2+3} and R_{4+5}, and all crossveins, bifurcations and most of longitudinal veins distinctly clouded. Legs blackish, anterior four femora and tibiae often somewhat brownish, hind tibiae with a broad black ring at base, otherwise legs yellow including tarsi. Mid femora with long av and pv black bristly hairs, anterior tibiae black bristled above, mid tibia with a

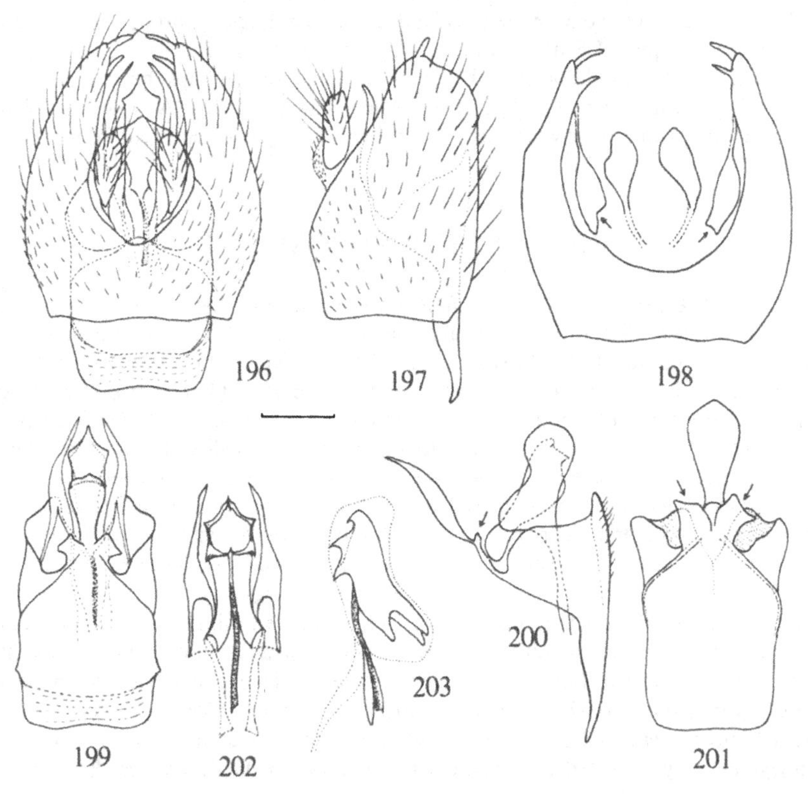

Figs. 196–203. Male genitalia of *Syneches muscarius* (Fabr.). - 196: dorsal view; 197: left periandrial lamella with cercus and hypandrium; 198: periandrium with periandrial folds (indicated by arrows) and cerci; 199: hypandrium with aedeagus in dorsal view; 200: the same in lateral view; 201: hypandrium in dorsal view (arrows indicating attachments of periandrial folds with hypandrial bridge); 202: aedeagus in ventral view; 203: the same in lateral view, postgonite dotted. Scale: 0.1 mm.

very long black dorsal bristle near base and a ventral one at tip. Hind femora slightly thickened, armed with long ventral and still longer av black spines arising from small warts, and long thin pv bristly hairs.

Abdomen subshining blackish brown, uniformly thinly brownish grey dusted and clothed with long pale hairs on venter and at sides. Hypopygium (Fig. 196) small and compact, not broader than terminal abdominal segments, rotated through 90° towards right but almost symmetrical.

Length: body 4.6–5.3 mm, wing 4.7–5.1 mm.

♀. Resembling male in all respects, but wings with a larger and deeper blackish patch at tip of vein Sc and abdomen narrowly tapering; cerci very long and slender.

Length: body 4.3–5.0 mm, wing 4.8–5.0 mm.

Distribution and biology. Widespread in Europe from England through Belgium east to western European USSR, C. Europe, south to Spain and Italy. Not yet found in Scandinavia but it should occur in Denmark (Jutland) and perhaps the extreme south of Sweden. – End June to middle of September, mainly in July. A typical grassland species; rather common in wet meadows at suitable biotopes, often swept from grasses.

SUBFAMILY OCYDROMIINAE

Diagnosis. The members of this subfamily are much more heterogeneous than the more specialised and rather homogeneous Hybotinae. Head both holoptic and dichoptic, sometimes almost the same in both sexes, or females very broadly dichoptic. Antennae originally with a rather short 2-articulated style (Trichinini), which may rarely be very shortened, reduced to one article or quite absent (Oedaleini), or even very long and filiform (Ocydromiini). Mouthparts usually small and short, rarely long and/or projecting (Oedaleini), labium always with well-developed soft labellae with distinct pseudotracheae. Palpi, compared with the Hybotinae, small ovate and flattened, at most short cylindrical (Trichinini). Thorax arched above, with a small *Hybos*-like prosternum. Wings with or without axillary lobe, no alula, discal cell present (except for *Bicellaria* and *Hoplocyrtoma*) and emitting 3 veins to wing-margin, rarely 2 or with a short stump of M_1 (Ocydromiini); vein M_2, compared with the Hybotinae and Tachydromiinae, always present. Anal cell well-developed but shorter than basal cells. Legs generally simple, hind legs rarely raptorial. Male hypopygium asymmetrical (nearly symmetrical in *Trichinomyia, Stuckenbergomyia* and *Bicellaria*), more or less rotated towards right and often directed upwards. Female abdomen telescopic, or with 8th segment very narrowed and lengthened, ovipositor-like (Oedaleini, Ocydromiini).

Classification. The Ocydromiinae, as characterised above, include 17 recent genera which are classified in the present paper into three fairly well defined tribes, based mostly on the structure of antennae, wing venation and male genitalia. The characteristics of the tribes are given in the systematic part below. The first serious attempt to classify the Ocydromiinae was made by Tuomikoski (1966) who separated three groups of genera ("perhaps tribes", as stated on p. 285), viz., the *Oedalea-*, *Leptopeza-* and *Trichina*-groups. My own recently proposed subdivision is in general very similar to that of Tuomikoski, except for the position of the genus *Ocydromia* which was placed by Tuomikoski in his *Trichina*-group on the basis of the "dorsiventral" 3rd antennal segment and the short cylindrical palpi with a sensory hollow at the truncate tip. However, even these two characters are not confined to this group of genera, and all other characters of *Ocydromia* (the structure of male genitalia and aedeagus in particular) are, on the contrary, characteristic of the *Leptopeza*-group (tribe Ocydromiini in my classification).

As to the phylogeny of the subfamily Ocydromiinae, the tribe Trichinini is clearly the most primitive, as it has a full set of plesiomorphous characters. The other forms of the subfamily appear to have evolved along two parallel lines, one with a shortened antennal style in combination with forwardly-projecting mouthparts (tribe Oedaleini), and the other with an antennal style lengthened into a filiform arista in combination with a reduction or entire loss of vein M_1 (tribe Ocydromiini).

The tribe Trichinini, as defined here, includes 5 genera: *Trichinomyia* Tuomik., *Trichina* Meig., *Bicellaria* Macq., *Hoplocyrtoma* Mel., and probably also the Afrotropical genus *Edenophorus* Smith, originally described into the '*Microphorus*-group' of genera. The tribe Oedaleini includes the genera *Oedalea* Meig., *Euthyneura* Macq., *Anthalia* Zett., *Allanthalia* Mel., and very probably also the genus *Stuckenbergomyia* Smith (syn. *Stuckenbergia* Smith, nec Tshernyshev). The tribe Ocydromiini includes 7 genera: *Ocydromia* Meig., *Leptopeza* Macq., *Leptodromiella* Tuomik., *Oropezella* Coll., *Hoplopeza* Bezzi, *Scelolabes* Phil. and *Pseudoscelolabes* Coll.

Distribution. The Ocydromiinae are world-wide in distribution, although the majority of genera are mainly Holarctic. This applies particularly to the tribes Trichinini and Oedaleini: of the former tribe, only about a dozen species known as "*Trichina*" are Neotropical, the occurrence of *Bicellaria* in the Orient (Formosa) needs verification, and the genus *Edenophorus* is exclusively Afrotropical; of the tribe Oedaleini, all the genera are Holarctic except for the provisionally included Afrotropical genus *Stuckenbergomyia* and a single Oriental species of *Oedalea*. On the other hand, the tribe Ocydromiini is only poorly represented in the Holarctic region (9 species in 4 genera); most species of *Leptopeza, Ocydromia* and *Oropezella* are known from the southern hemisphere (at least 20 species), and the genera *Hoplopeza* and *Scelolabes* from South America and *Pseudoscelolabes* from New Zealand only. However, these southern hemisphere species of the tribe Ocydromiini need urgent generic revision and the occurrence of further taxa may be expected, even at the generic level.

About fifty five species of the Ocydromiinae in eleven genera are known from Europe, and of these thirty eight have been found in Scandinavia.

Key to genera and tribes of Ocydromiinae

1 Discal cell emitting 2 veins to wing-margin, sometimes a short stump of a third upper vein M_1 (Fig. 495). Antennal

style long, arista-like, much longer than 3rd segment (Fig. 499). Proboscis very short, pointing downwards (**tribe Ocydromiini**) 2

– Discal cell emitting 3 veins to wing-margin, vein M_1 complete, rarely vein M_2 abbreviated or discal cell absent. Antennal style at most as long as 3rd segment 5

2 (1) 3rd antennal segment ovate with arista supra-terminal (Figs. 499, 500). Legs without distinct bristles. Mesonotum almost bare, 2-serial acr inconspicuous. Males holoptic, females narrowly dichoptic .... *Ocydromia* Meigen (p. 206)

– 3rd antennal segment pointed with arista terminal (Figs. 501–504). Legs with at least several distinct bristles on mid tibiae dorsally 3

3 (2) Mesonotum with multiserial acr, head holoptic in both sexes. Wings with well-developed axillary lobe, discal cell rather broad and short, about twice as long as deep (Fig. 496). Females with conspicuously narrowed ovipositor-like 8th abdominal segment (Fig. 540) .... *Leptopeza* Macquart (p. 212)

– Mesonotum almost bare with inconspicuous biserial acr, eyes dichoptic in both sexes, even if very narrowly. Axillary lobe slightly developed or absent, discal cell longer. 8th abdominal segment in female short, not ovipositor-like 4

4 (3) Antennae inserted at about middle of head in profile (Fig. 542), or very slightly above middle, face linear. First antennal segment short and arista pubescent (Fig. 503). Axillary lobe slightly developed, discal cell about 3.5 times as long as deep (Fig. 497), a distinct costal bristle .... *Leptodromiella* Tuomikoski (p. 219)

– Antennae inserted much above middle of head in profile and eyes touching below antennae. First antennal segment longer than deep and arista bare (Fig. 504). Axillary lobe absent, discal cell long and slender, about 5 times as long as deep (Fig. 498), costal bristle absent .... *Oropezella* Collin (p. 223)

5 (1) Antennal style as long as, or not conspicuosly shorter than, conical 3rd segment (cf. Fig. 278). Small mesonotal hairs few in number, acr 2-serial, dc 1-serial. Proboscis short and pointing downwards (**tribe Trichinini**) 6

– Antennal style very short (rarely absent), much shorter than the conspicuously lengthened or leaf-shaped 3rd segment (cf. Figs. 380, 456). Mesonotal hairs numerous, acr and dc multiserial or acr at least 4-serial. Proboscis often lengthened and projecting forwards (**tribe Oedaleini**) 8

6 (5) Discal cell absent, medial fork of veins M_1 and M_2 with its upper branch (M_1) abbreviated at base (Fig. 206). Eyes holoptic in both sexes. Legs blackish, last tarsal segment not com-

pressed .... *Bicellaria* Macquart (p. 134)
- Discal cell present. Eyes holoptic in males, narrowly dichoptic in females. Legs often yellowish, last tarsal segment laterally compressed 7

7 (6) Hind tibiae slender. Wing stigma shortened, not extending to tip of vein R_{2+3} (Fig. 204). At least 3 pairs (6–10 bristles) of distinct scutellar bristles .... *Trichinomyia* Tuomikoski (p. 115)
- Hind tibiae dilated towards tip. Wing stigma large, extending to tip of R_{2+3} (Fig. 205). Usually only 2 pairs of scutellar bristles, rarely 6–8 scutellars .... *Trichina* Meigen (p. 121)

8 (5) Hind femora thickened and armed with numerous short black spines beneath towards tip. 3rd antennal segment very long, strap-shaped (Figs. 380–391). Proboscis short, pointing almost downwards (Figs. 376, 377). Larger species, about 3–4 mm in length .... *Oedalea* Meigen (p. 163)
- Hind femora rather slender and not spinose beneath to-

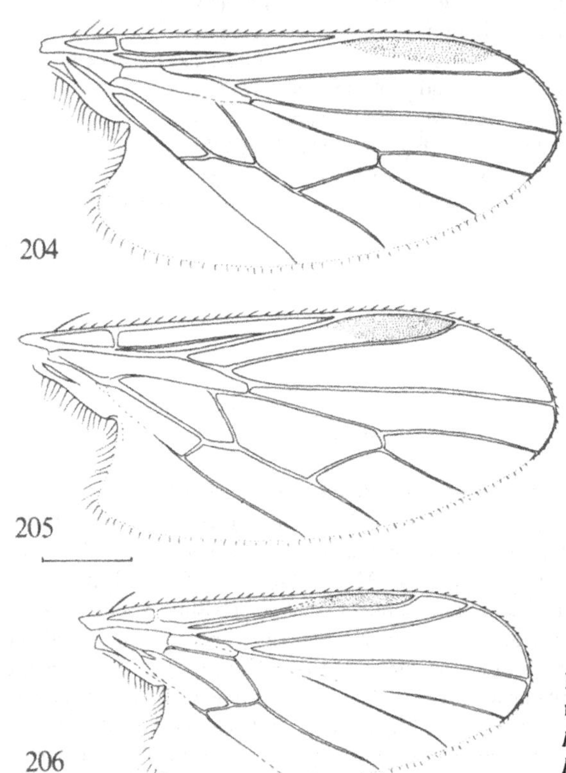

Figs. 204–206. Wings of Trichinini. – 204: *Trichinomyia flavipes* (Meig.); 205: *Trichina clavipes* Meig.; 206: *Bicellaria spuria* (Fall.). Scale: 0.5 mm.

wards tip. 3rd antennal segment short and broad, ovate to leaf-shaped. Smaller species 9

9 (8) Proboscis long, usually longer than head is high, projecting diagonally or slightly obliquely forwards. 2nd basal cell very broad towards tip and the vein closing it almost perpendicular (Fig. 372). Posthumeral bristle absent. Rather smaller polished black species, 2–2.5 mm in length. Mesonotal hairs short and hair-like (*Anthepiscopus* of the Oreogetoninae is larger, with longer bristly dc and costa continuous on posterior wing-margin) .... *Euthyneura* Macquart (p. 187)

– Proboscis very small or quite concealed. 2nd basal cell narrower towards tip and the vein closing it very oblique (Figs. 374, 375). A distinct posthumeral bristle. Very small yellow or black species, about 1.5–2 mm in length 10

10 (9) Antennae with a very small point-like style (Fig. 477) and proboscis very short. Vein M_2 (in European species) abbreviated at tip, not reaching the wing-margin (Fig. 374). Males holoptic, females broadly dichoptic. A small black species with dark legs .... *Anthalia* Zetterstedt (p. 195)

– Antennae without style (Fig. 478) and proboscis entirely concealed in the mouth-cavity. Vein M_2 not abbreviated, very rarely disappearing just before tip (Fig. 375). Head broadly dichoptic in both sexes. A small yellowish to dark brown species with pale legs .... *Allanthalia* Melander (p. 200)

Tribe Trichinini

Males holoptic, females narrowly dichoptic (holoptic in *Bicellaria*), face generally narrow in both sexes. Antennae moderately long, 3rd segment conical, terminal style about half as long or shorter than 3rd segment, finely pubescent with a bare tip. Proboscis very short and pointing downwards. Mesonotal bristles fine and sparse, acr biserial, dc uniserial, large bristles in full number. Discal cell emitting 3 veins to wing-margin, or medial fork of M_1 and M_2 very fine and discal cell absent (*Bicellaria, Hoplocyrtoma*). Hind legs often raptorial, elongated, ordinary bristles absent in the *Trichina*-complex. Male genitalia with a tendency to be symmetrical, aedeagus very small with a bifid coaxial apodeme, without appendages but covered laterally by leaf-shaped postgonites. Female abdomen short telescopic, not ovipositor-like. Generally highly predaceous species.

Genus *Trichinomyia* Tuomikoski, 1959

Trichinomyia Tuomikoski, 1959: 103.
Type-species: *Trichina flavipes* Meigen, 1830 (orig. des.).

Small, about 2–3 mm long, black species with legs rather short and slender, devoid of ordinary bristles. Head placed low on thorax; eyes in ♂ meeting on frons for a long distance, with upper facets very enlarged, more or less separated on frons in ♀ with facets almost equally small. Face narrow in both sexes. Ocellar and vertical bristles small and fine, occipital pubescence not coarser. Antennae inserted at about middle of head in profile. Basal segment very small and indistinctly separated from the globular 2nd seg-

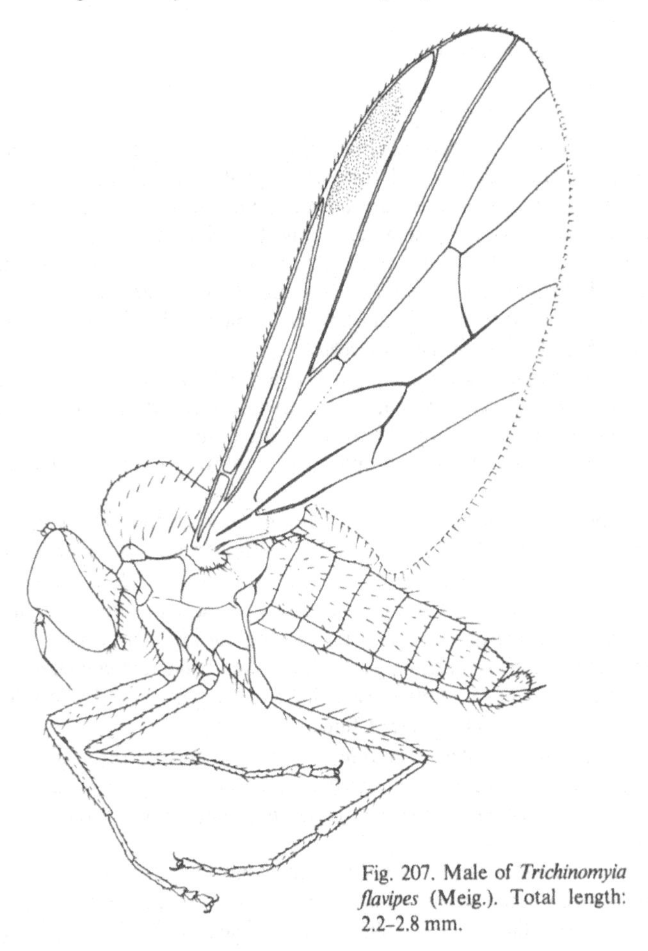

Fig. 207. Male of *Trichinomyia flavipes* (Meig.). Total length: 2.2–2.8 mm.

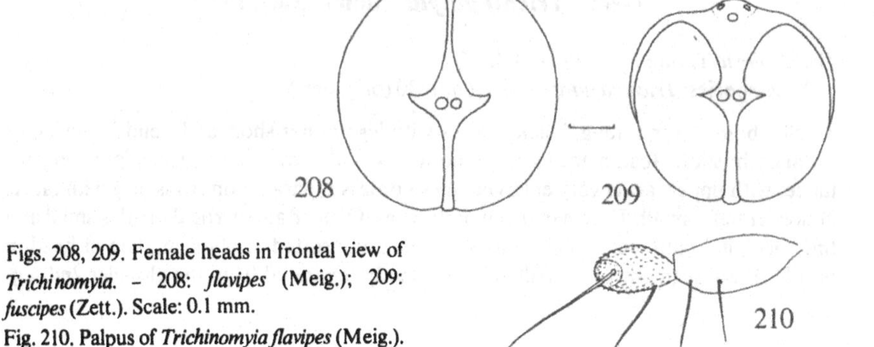

Figs. 208, 209. Female heads in frontal view of *Trichinomyia*. – 208: *flavipes* (Meig.); 209: *fuscipes* (Zett.). Scale: 0.1 mm.
Fig. 210. Palpus of *Trichinomyia flavipes* (Meig.). Scale: 0.1 mm.

ment, which is armed with a preapical circlet of minute hairs. 3rd segment long, almost evenly arched above and below and clothed with equally long hairs on both sides; style often shorter than 3rd segment, its bare terminal part well-separated from the basal pubescent part. Mouthparts very small and almost hidden in the mouth-cavity, very much as in *Trichina*. Palpi (Fig. 210) ovate with one or more bristles at tip, situated on a bristled palpifer. Labrum short and very convex above, hypopharynx as long as labrum and stylet-like, labium with large fleshy labellae armed with fine sensory hairs and pseudotracheae. Maxillary stipites as in *Trichina*.

Thorax rather longer than deep and distinctly arched above, all thoracic hairs and bristles fine. Acr closely biserial, small, dc uniserial, longer posteriorly or at least last prescutellar pair bristle-like. Large bristles in full number: at least 1 humeral, 1 posthumeral, 1 intrahumeral, 2–4 notopleurals, 1 supra-alar, 1 postalar and 6–10 scutellars.

Wings broad but rather long, iridescent, darker in ♂ and clearer in ♀, wholly covered with microtrichia. Stigma distinct and rather long, although not extending to tip of vein R_{2+3}. Vein Sc closely approaching R_1 but fading away half way along, costa ending at tip of vein M_1, vein R_{4+5} almost straight and ending right in the wing-tip. Three veins from discal cell, which is large and has its upper hind corner (from which M_1 and M_2 arise) very produced. 2nd basal cell very broad, 1st basal cell and anal cell equally slender. Vein A rather fine but complete, reaching the wing-margin. Axillary lobe very large, axillary angle almost rectangular. A fine costal bristle.

Legs rather short and slender, sometimes all femora and tibiae moderately stout, hind tibiae not dilated. Legs entirely covered with fine hairs only, without distinct bristles. Mid femora at most finely pubescent beneath and last tarsal segment of all pairs laterally compressed.

Abdomen rather slender, scarcely broader near base but distinctly tapering in ♀, more or less laterally compressed. Abdominal pubescence fine, hind marginal hairs more distinct. ♂ genitalia small and almost symmetrical, rotated towards right, and whole hypopygium rather vertical in position with remarkably projecting aedeagus. Periandrial lamellae broadly connected anteriorly and almost symmetrical, without conspicuous processes but distinctly serrate above. Cerci slender and moderately large, symmetrical hypandrium not overlapping lamellae and more or less quadrate, upper margin straight. Aedeagus conspicuously long, straight, with a simple (not bifurcated) coaxial apodeme which is loosely attached to hypandrium. Tip of aedeagus with flat wing-appendages, hypandrial bridge developed.

Distribution. An exclusively Holarctic genus. Two species are western Palaearctic and at least two others are Nearctic. Although the genus has not yet been recorded from North America, *T. flavipes* is presumably Holarctic, and at least *Trichina basalis* Mel. and *T. pullata* Mel. are almost certainly species of *Trichinomyia*.

Biology. The adults occur in damp shaded biotopes on low herbage, often at forest margins, and the flight-period is short. The adults are often swept on low vegetation and from grasses; males may be found hovering in swarms (Lundbeck, 1910) but no precise records of swarming and the mating habits are available.

Note. *Trichinomyia* species are very probably the most primitive members of the subfamily or even of the whole family in view of the many plesiomorphous characters (antennae, wing-venation, simple legs, almost symmetrical ♂ genitalia). Only a few recent species are known, and it is suggested that speciation took place a very long time ago.

Key to species of *Trichinomyia*

1 Larger, body 2.2–3.1 mm, mesonotum subshining black owing to a weak covering of brownish dust, legs yellowish. Style as long as 3rd ant.s. (Fig. 211). ♀: frons (Fig. 208) very narrow, halteres light brown .... 5. *flavipes* (Meigen)

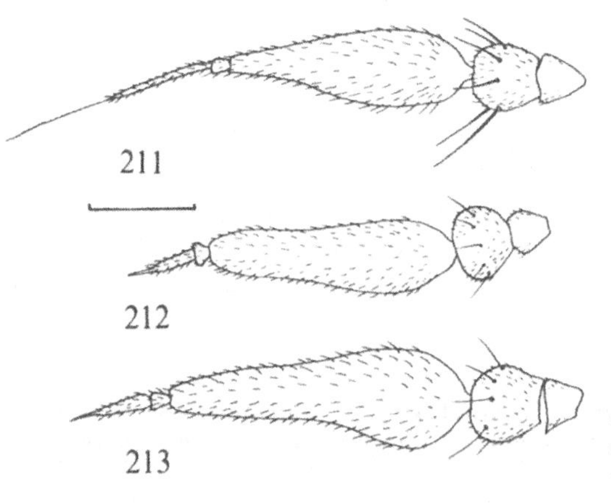

Figs. 211-213. Antennae of *Trichinomyia*. – 211: *flavipes* (Meig.) ♂; 212: *fuscipes* (Zett.) ♂. 213: same ♀. Scale: 0.1 mm.

- Smaller, body 1.6–2.2 mm, mesonotum polished black, legs blackish. Style much shorter than 3rd ant.s. (Fig. 212). ♀: frons (Fig. 209) polished black and very broad, halteres blackish .... 6. *fuscipes* (Zetterstedt)

5. ***Trichinomyia flavipes*** (Meigen, 1830)

Figs. 204, 207, 208, 210, 211, 214–218.

Trichina flavipes Meigen, 1830: 336.

Larger (2.2–3.1 mm) with yellowish simple legs and mesonotum slightly dulled by brownish microscopic pile. Antennal style as long as 3rd segment, bare terminal part as long as the pubescent basal part.

♂. Head black with occiput and a narrow face (as deep as 2nd ant. s.) thinly brownish dusted, eyes holoptic with upper facets very large. Antennae (Fig. 211) black, style as long as 3rd segment, its bare terminal part long, occupying apical half of style. Palpi and proboscis very small, scarcely visible, palpi with 2–3 black terminal hairs. Thorax subshining black, pleura and whole of mesonotum covered with a thin brownish dusting. Thoracic hairs and bristles black including 6–10 scutellars.

Wings faintly brownish grey clouded, halteres and squamae blackish, latter with long black fringes. Stigma faintly brownish, ending before tip of vein R_{2+3}. Legs rather long and slender, unicolorous yellowish, leaving only apical 2 or 3 tarsal segments brownish. Pubescence pale, hind femora with rather long hairs above and below.

Abdomen subshining blackish brown, covered with long black hairs, hind marginal hairs more bristle-like. Genitalia (Fig. 214) small, black, periandrial lamellae evenly

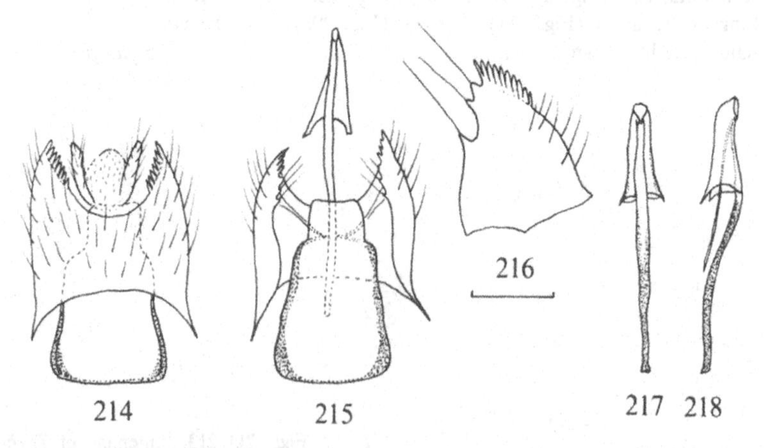

Figs. 214–218. Male genitalia of *Trichinomyia flavipes* (Meig.). – 214: dorsal view, aedeagus omitted; 215: ventral view; 216: left periandrial lamella; 217, 218: aedeagus in two different views. Scale: 0.1 mm.

serrate above and quite simple on dorsal edges. Aedeagus with a small winged-process close to the ovate aperture, hypandrium narrowed apically.

Length: body 2.2–2.8 mm, wing 2.6–3.0 mm.

♀. Eyes very narrowly separated on frons, the latter not broader than anterior ocellus, upper facets slightly enlarged towards vertex. Thoracic bristles rather brownish, including 6–10 scutellars. Wings almost clear, halteres greyish to dirty yellow, paler than in ♂. Abdomen with finer and shorter brownish pubescence, telescopic and ending in two slender ovate cerci.

Length: body 2.2–3.1 mm, wing 2.6–3.3 mm.

Distribution and biology. Southern parts of Scandinavia; Denmark, in Sweden north to Upl., in Finland to Ta, but rather common only in Denmark and S. Sweden. – England, eastwards to NW of European USSR, a common species in C. Europe but probably absent in the south. – Mid August to October, a typical late summer and autumn species, most often in September.

6. ***Trichinomyia fuscipes*** (Zetterstedt, 1838)
Figs. 209, 212, 213, 219–222.

Trichina fuscipes Zetterstedt, 1838: 540; 1842: 256 *(Microphora)*.
Trichina nigripes Strobl, 1898: 206.
Trichina elongata Haliday; Frey, 1913: 59.

Smaller (1.6–2.2 mm) with legs blackish and mesonotum entirely polished black. Antennae longer, style at most one-third as long as 3rd segment, bare terminal part (at least in ♂) short.

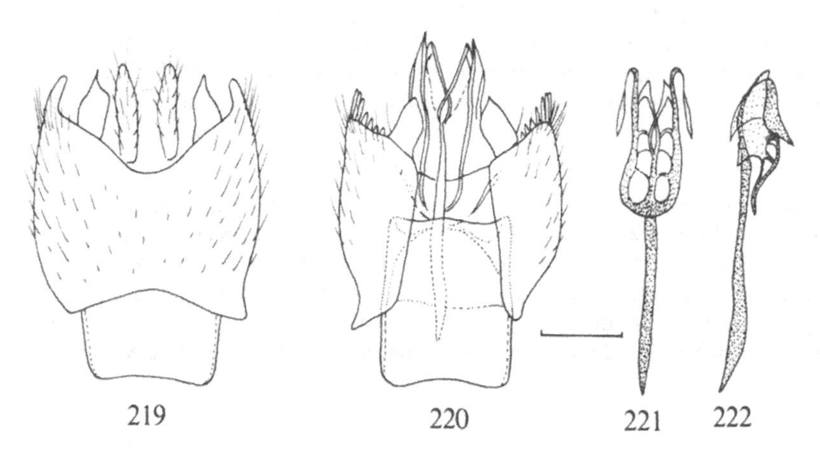

Figs. 219–222. Male genitalia of *Trichinomyia fuscipes* (Zett.). – 219: dorsal view, aedeagus omitted; 220: ventral view; 221, 222: aedeagus in two different views. Scale: 0.1 mm.

♂. Eyes meeting on frons with upper facets very enlarged, occiput and face rather densely brownish grey dusted. Antennae (Fig. 212) black, 3rd segment not so much narrowed at tip as in *flavipes* and apparently longer, covered only with microscopic pubescence, longer hairs absent. Style much shorter than 3rd segment, its bare terminal part very short. Palpi and proboscis black, hardly visible, palpi with a single black bristle-like terminal hair. Thorax deep black, mesonotum wholly polished, thin greyish dusting only in notopleural depression and on a triangular patch in front of scutellum. All thoracic hairs and bristles black, 6–10 scutellars.

Wings rather short and deeper brown clouded, veins black, stigma as in *flavipes*. Halteres and squamae black. Legs uniformly dark brown to blackish brown, somewhat shorter and stouter than in *flavipes*, all tibiae and femora with brownish pubescence, hind femora with a row of distinct pale av hairs.

Abdomen black in ground-colour but rather densely brownish grey dusted from some angles, pubescence black, short and very fine. Genitalia (Fig. 219) resembling those of *flavipes* but both periandrial lamellae dorsally with small lobe-like projections anteriorly and posteriorly, hypandrium shorter and almost regularly rectangular, aedeagus with a more complicated armature at tip.

Length: body 1.6–2.2 mm, wing 2.0–2.3 mm.

♀. Resembling male but wings clearer and eyes broadly separated on frons, with all facets equally small. Frons (Fig. 209) polished black and triangularly widening above.

Length: body 1.6–2.1 mm, wing 2.0–2.3 mm.

Lectotype designation. In Zetterstedt's Ins. Lapp. Coll. at Lund there are 2 ♀ of *Trichinomyia fuscipes* under the name "*Trichina fuscipes* ♀"; one labelled "Stensele" is hereby designated as lectotype of *Trichinomyia fuscipes* (Zett.) and was labelled as such in 1977; the second ♀ is labelled "Gransele ♀ 56 Holmgr.". A third female of *T. fuscipes* is in the Dipt. Scand. Coll. under "*Microphora fuscipes* ♀", labelled "Gaernes, Vaerdal. Norveg. 30.6.-12.7.40", which, like the Gransele ♀, does not belong to the type-series.

Trichina nigripes Strobl, although having 3rd ant.s. more evenly cylindrical, is undoubtedly conspecific with the northern *Trichinomyia fuscipes* (Zett.). I have studied the holotype ♀, described by Strobl from "Scheibleggerhochalpe bei Admont" (19.5.) and labelled by Morge in 1961 as "Typen-Exemplar", and another ♀ taken by Strobl on 18.6. in "Styria".

Distribution and biology. Not uncommon in northern Scandinavia, absent in the south and in Denmark. Mainly in Lapland, in Norway south to NTi, in Sweden to Jmt. and Hrj. approximately 62°N, frequent in Finnish Lapland and the Kola Peninsula (Lr), south to ObN and Ks, the most southern locality being in Alandia (Finnström 1 ♂, leg. Frey; *T. elongata* Hal. det. Becker). – Rare in the Alps; apparently a boreoalpine species in distribution. – June to first half of August, in the Alps as early as May.

Genus *Trichina* Meigen, 1830

Trichina Meigen, 1830: 335.
Type-species: *Trichina clavipes* Meigen, 1830 (des. Rondani, 1856).
Pipistrellus v. Gistl, 1848: xi (unjustified replacement name for *Trichina*).
Type-species: isogenotypic with *Trichina*.
Oedaleopsis Tuomikoski, 1959: 109 (as subgenus) – **syn. n.**
Type-species: *Oedalea pallipes* Zetterstedt, 1838 (orig. des.).

Generally small, body length about 2.5–3 mm, polished black species (Fig. 223) with yellowish legs; thoracic bristles, halteres and legs usually paler in females. Eyes holoptic in males with upper facets enlarged, narrowly dichoptic in females with all facets equally small. Face very linear in both sexes. A distinct ocellar tubercle; anterior pair of ocellars and upper postocular occipital bristles coarser than other pubescence. Head higher than deep, antennae inserted at about middle of head in profile. Antennae black, basal segment very small, 2nd segment globular with a circlet of small preapical bristles; 3rd segment elongate, broad at base and narrowing towards tip, with lower edge convex and dorsum almost straight, covered with only short pubescence. Style 2-articulated, at most slightly more than half length of 3rd segment, somewhat thickened and pubescent, leaving short hair-like terminal part bare. Mouthparts (Figs. 224, 225) small and almost concealed in the mouth-cavity except for large labellae which bear distinct pseudotracheae and fine sensory bristles. Maxillae reduced to slender stipites, and one-segmented ovate palpi situated on the finely bristled palpifers. Labrum produced above but sclerotised on its lower edge only, firmly joined with tips of maxillae laterally near base, hypopharynx stylet-like and heavily sclerotised.

Thorax arched above, mesonotum more or less brightly shining black with scattered hairs, acr narrowly biserial and distinctly diverging, dc uniserial and equally fine except for a strong bristle-like prescutellar pair. Large bristles in full number: 1 humeral, 1 intrahumeral, 3–4 notopleurals, 1 anterior supra-alar, 1 postalar and 4–8 scutellars. Pronotum and prothoracic episterna finely bristled, prosternum small, *Hybos*-like.

Wings (Fig. 205) large with a conspicuous axillary lobe, axillary excision rectangular, and wings entirely covered with microtrichia; a distinct costal bristle. Costa reaching vein M_1, Sc running close to R_1 and fading away before reaching costa, radial sector rather short, shorter than half distance between its origin and humeral crossvein. Costal stigma distinct, extending right up to tip of vein R_{2+3}. Discal cell elongate and emitting 3 veins to wing-margin, basal cells equally long but 2nd basal cell much deeper. Anal cell shorter, the vein closing it straight but slightly recurrent, anal vein reaching wing-margin. Squamae with long fringes.

Legs rather short and slender, hind tibiae dilated towards tip and last tarsal segment on all pairs distinctly compressed laterally. Ordinary bristles on legs practically absent, hind trochanter beneath often with a spine in males. Fore tibiae somewhat spindle-shaped with the conspicuous opening of a tubular gland beneath near base.

Abdomen slender, slightly stouter near base, covered with fine hairs, hind marginal hairs longer and bristle-like. 8th tergum in ♂ very narrowed and only slightly concealed within 7th segment, 8th sternum unmodified. Hypopygium rather small, distinctly asymmetrical and rotated towards right through 90°, often more or less upturned. Periandrial lamellae usually broadly connected anteriorly, armed with conspicuous and specifically distinct dorsal appendages, left lamella in particular. Aedeagus short and stout, almost circular, with a short coaxial apodeme, loosely attached to a rather slender but remarkably asymmetrical hypandrium; hypandrial bridge rarely developed. Hypandrium firmly connected by a special process with only right periandrial lamella near tip. Cerci short and finely pubescent. Female abdomen evenly tapering, 8th segment very narrowed but not lengthened, cerci small and slender.

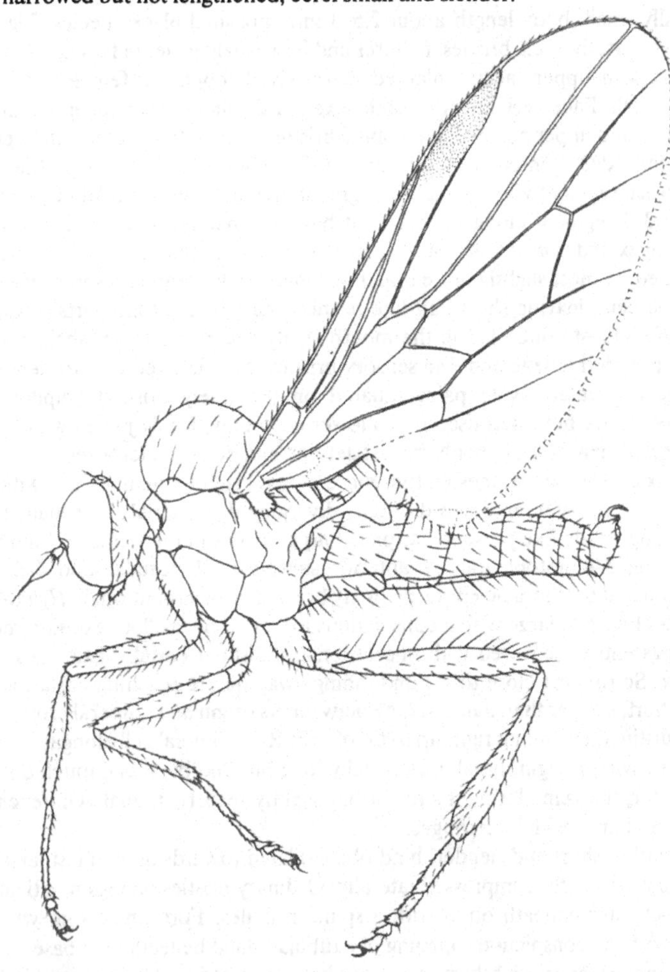

Fig. 223. Male of *Trichina clavipes* Meig. Total length: 2.4–3.0 mm.

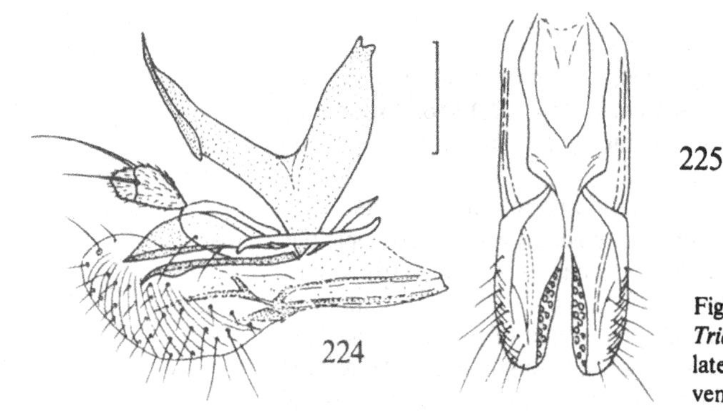

Figs. 224, 225. Mouthparts of *Trichina clavipes* Meig. ♂. – 224: lateral view; 225: tip of labium, ventral view. Scale: 0.1 mm.

Distribution. A Holarctic genus, represented in Europe with certainty by six species if *T. lissonota* Bezzi is only a darker-legged form of *T. clavipes*, as was suggested by Strobl (1910: 72). Five species also occur in northern Europe and one, *T. unilobata* Chv., is south-west European. *T. fumipennis* Frey was described from Japan and I have seen another undescribed species close to *clavipes* from Mongolia. Of the seven *Trichina* species recorded from North America (Melander, 1965), only *T. nura* Mel. definitely belongs to this genus, apart from the supposedly Holarctic *T. clavipes*, and perhaps another two, whilst the others are *Trichinomyia* species. The twelve South American "*Trichina*" species described by Collin (1933) from Chile and Argentina can hardly belong to this genus, as was pointed out by Tuomikoski (1959) and later indirectly by Collin himself (1961) who made no mention of the genus *Trichina* from the southern hemisphere.

Biology. Adults are predaceous in both sexes and occur for a long period during summer, inhabiting rather shaded and moist biotopes; they are often swept from low herbage in deciduous forests. Small swarms or individually hovering specimens of both sexes are often observed at the tips of branches or along forest margins from sunset almost to dusk, but mating activity has probably been transferred to a solid substrate.

Note. The subgeneric classification proposed by Tuomikoski (1959) is not accepted here. For details, see the 'Synonymy' under *T. pallipes*.

Key to species of *Trichina*

1 Antennal style (Figs. 226–232) slender, bristle-like. Legs bicolored or extensively blackish. Usually 4 scutellar bristles, rarely 6–8 *(opaca)*. Proboscis and palpi blackish 2

– Antennal style (Fig. 233) laterally compressed and very broad, as deep as tip of 3rd segment. Legs extensively yellow, 6–8 scutellar bristles. Proboscis and palpi yellow .... 11. *pallipes* (Zetterstedt)

2 (1) Eyes meeting on frons – ♂♂ 3
– Eyes separated on frons, even if narrowly – ♀♀ 6
3 (2) Thorax entirely shining black, front part of mesonotum without microscopic pile. 3rd ant.s. (Fig. 226) short, style more than half length of 3rd segment, its bare terminal part not more than 1/5 of style .... 7. *clavipes* Meigen
– At least front part of mesonotum with microscopic pile 4
4 (3) Only front part of mesonotum covered with microscopic pile, rest of disc polished. Hind trochanters with a black spine beneath. 4 scutellar bristles 5
– Whole of mesonotum dulled by a slight microscopic pile. Hind trochanters without a black spine beneath. 6–8 scutellar bristles. 3rd ant.s. (Fig. 231) rather broadened, style at least half length of 3rd segment. Legs extensively darkened .... 10. *opaca* Loew
5 (4) 3rd ant.s. (Fig. 227) rather short, style quite half length of 3rd segment. Left lamella of genitalia (Fig. 240) with 2 almost equally long and stout processes. Legs yellow with hind coxae and trochanters, apical half of hind femora, apical 2/3 of hind tibiae and all tarsi contrastingly dark .... 8. *bilobata* Collin
– 3rd ant.s. (Fig. 229) longer and narrower, style less than half length of 3rd segment. Left lamella of genitalia (Fig. 246) with one long projecting process, the second posterior one very small. Legs extensively darkened .... 9. *elongata* Haliday
6 (2) Front part of mesonotum polished black, without microscopic pile 7
– Front part of mesonotum dulled by microscopic pile 8
7 (6) Thorax entirely polished black including notopleural depression. 3rd ant.s. short, style more than half length of 3rd segment, its bare terminal part not more than 1/5 of style. Eyes very narrowly separated, frons narrower than anterior ocellus .... 7. *clavipes* Meigen
– Notopleural depression dulled by microscopic pile. 3rd ant.s. (Fig. 228) longer, style at most half length of 3rd segment and its bare terminal part about 1/3 length of style. Eyes more broadly separated, frons at least as deep as anterior ocellus and distinctly deeper than face .... 8. *bilobata* Collin
8 (6) 3rd ant.s. (Fig. 230) long and narrow, style at most 1/3 length of 3rd segment. Legs with conspicuous yellow pattern. 4 scutellar bristles .... 9. *elongata* Haliday
– 3rd ant.s. (Fig. 232) distinctly broadened, style longer, slightly more than 1/3 length of 3rd segment. Legs extensively darkened or at least the pale parts more yellowish brown. Usually 6 scutellar bristles .... 10. *opaca* Loew

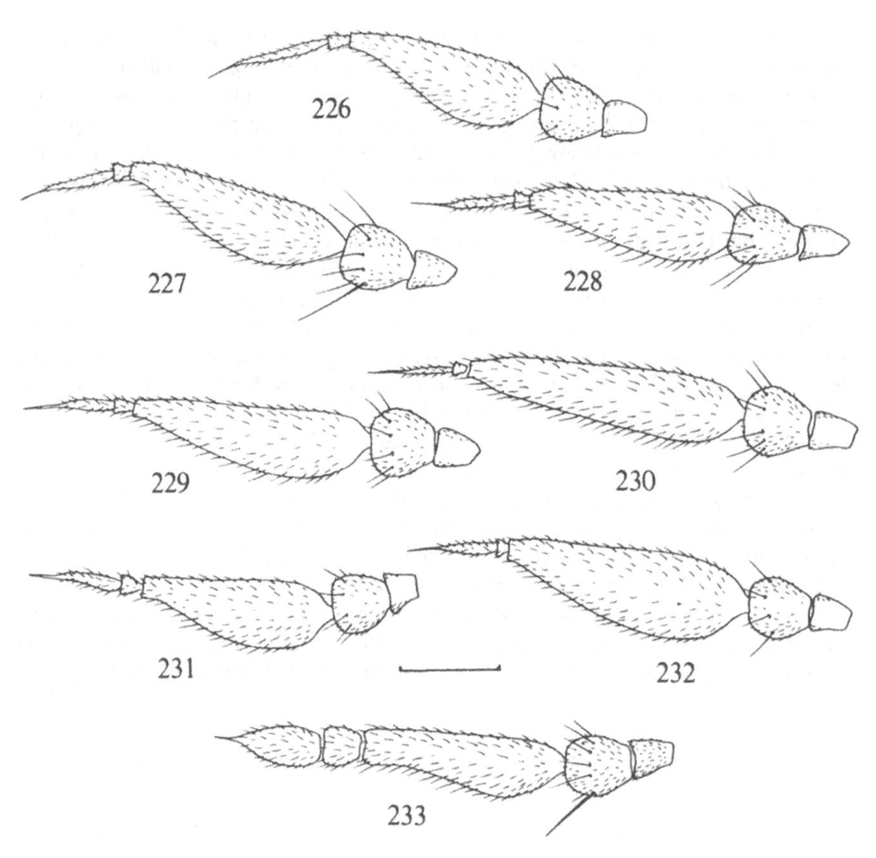

Figs. 226–233. Antennae of *Trichina*. - 226: *clavipes* Meig. ♂; 227: *bilobata* Coll. ♂; 228: *bilobata* Coll. ♀; 229: *elongata* Hal. ♂; 230: *elongata* Hal. ♀; 231: *opaca* Loew ♂; 232: *opaca* Loew ♀; 233: *pallipes* (Zett.) ♂. Scale: 0.1 mm.

7. ***Trichina clavipes*** Meigen, 1830
Figs. 205, 223–226, 234–238.

Empis minuta Fallén, 1816: 32 (nec Fabricius, 1787: 365 = Phoridae).
Trichina clavipes Meigen, 1830: 336.
? *Trichina lissonota* Bezzi, 1899: 144.

Mesonotum brightly shining black in both sexes, even right in front and in notopleural depression. Antennae short, style longer than half length of 3rd segment. Frons in ♀ extremely narrow.

♂. Eyes meeting on frons for a long distance, upper facets slightly enlarged, face greyish black and very linear. Antennae (Fig. 226) black, 3rd segment rather short with long style (more than half length of 3rd segment), its terminal fifth bare and hair-like. Mesonotum very brightly shining black, even right in front and in notopleural depression, pleura completely polished, but prescutellar depression, scutellum and metanotum thinly greyish. Thoracic bristles yellowish brown, 4 scutellar bristles.

Wings very faintly brownish, brown stigma extending to tip of vein R_{2+3}. Halteres and squamae blackish, latter with brownish fringes. Legs mostly yellowish, leaving hind coxae, apical half of hind femora and tibiae, and practically all tarsi blackish brown; posterior metatarsi often yellowish. Hind femora slightly thickened, hind tibiae distinctly swollen on the black apical half. Fore femora with rather long yellow av and pv hairs, mid tibiae with a dark ad bristle in basal third and another preapical one, hind femora with outstanding long pale bristly hairs dorsally near base and a row of pale av bristly hairs not much shorter than femur is deep. Hind tibiae with dark bristly hairs dorsally becoming longer and more bristle-like towards tip. Hind trochanter with a black spine beneath.

Abdomen subshining black at sides, dorsum velvety-black when viewed from above, brownish dusted in anterior view; covered with rather long, densely set yellowish brown hairs, marginal hairs longer. Genitalia as in Figs. 234–238.

Length: body 2.4–3.0 mm, wing 2.6–3.2 mm.

♀. Frons very linear, narrower than anterior ocellus, all eye-facets equally small. Mesonotum entirely polished black as in ♂. Wings clearer and both squamae and halteres pale yellowish. Legs much more yellowish, hind coxae and femora entirely

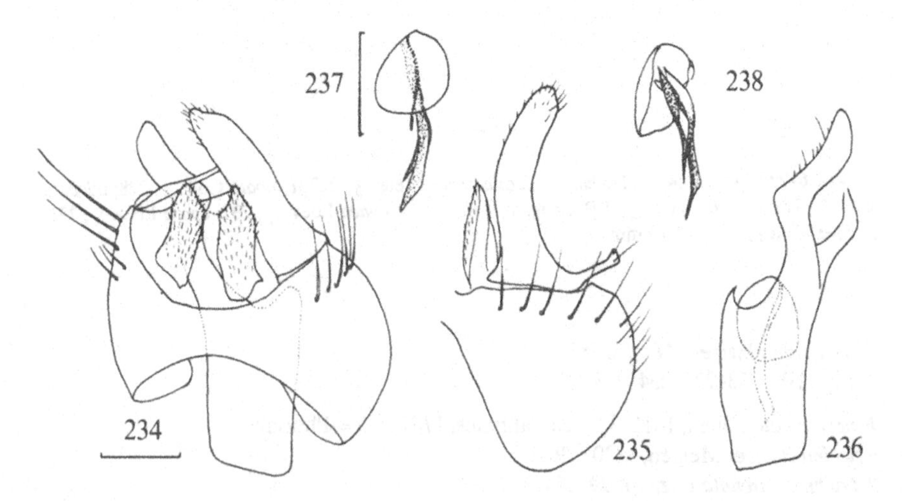

Figs. 234–238. Male genitalia of *Trichina clavipes* Meig. – 234: dorsal view; 235: left periandrial lamella with cercus; 236: hypandrium with aedeagus; 237, 238: aedeagus in two different views. Scale: 0.1 mm.

yellow, hind tibiae blackish on apical third only, and basal segments of all tarsi paler. Hairs on legs generally shorter and paler, no ad bristle on mid tibia and a black ventral spine on hind trochanters. Abdomen evenly narrowing posteriorly and extensively shining black even on dorsum, abdominal pubescence shorter and rather sparse.

Length: body 2.3–3.0 mm, wing 2.6–3.2 mm.

Distribution and biology. A species with a southern distribution in Scandinavia, rather common in Denmark and S. Sweden, north to Ög., very common along the Finnish Baltic coast, rarely extending to 64° N (ObS); not yet recorded from Norway. - Widespread and common in temperate Europe, also recorded from Spain (Strobl, 1899) and North America (Melander, 1965). - Mid June to August, rarely to September, in England as early as May.

8. ***Trichina bilobata*** Collin, 1926

Figs. 227, 228, 239–243.

Trichina bilobata Collin, 1926: 213.

Very much like *clavipes*, having similarly shaped antennae, but thorax in ♂ mostly black bristled and mesonotum thinly dusted in front. Female with broader frons, mesonotum wholly polished even in front but notopleural depression dusted.

♂. Antennae (Fig. 227) very much as in *clavipes*, style about half length of 3rd segment, scarcely longer. Thorax polished black but anterior part of mesonotum and notopleural depression dulled by brownish grey pile. Thoracic bristles darker, almost black, arranged as in *clavipes* including 2 pairs of strong scutellars and 1 pair of prescutellar dc.

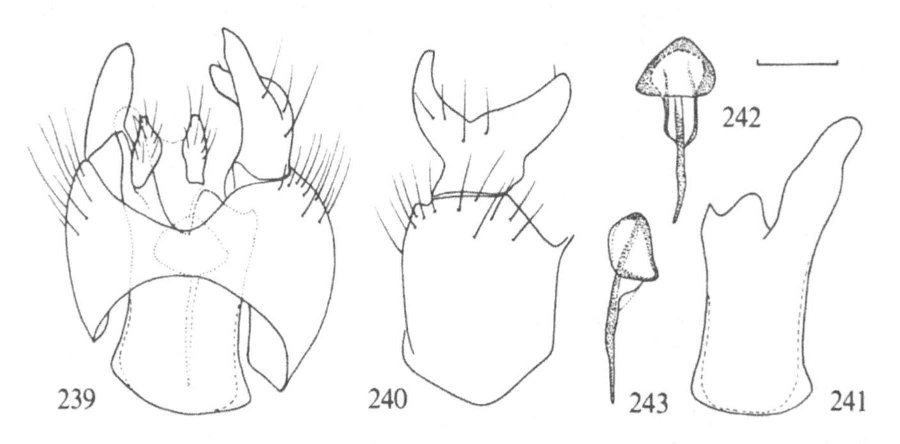

Figs. 239–243. Male genitalia of *Trichina bilobata* Coll. - 239: dorsal view; 240: left periandrial lamella; 241: hypandrium; 242, 243: aedeagus in two different views. Scale 0.1 mm.

Wings rather shorter, more or less tinted brownish, halteres and squamae blackish as in *clavipes* but latter with darker fringes. Legs of similar colour as *clavipes*, perhaps the yellow and dark coloration more contrasting, hind tibiae and femora often darkened on more than apical half. Fore femora with shorter hairs beneath, fore tibiae with a dark dorsal bristly hair at about middle, and the av bristly hairs on hind femora, although equally long, darker and more bristle-like. Hind tibiae more incrassate on apical half but similarly bristled above. Hind trochanters with a black spine.

Abdomen black, venter more translucent brown, subshining, but dorsum densely brownish to brassy yellow dusted even when viewed from the side. Pubescence apparently darker than in *clavipes*. Genitalia (Figs. 239–243) very distinctive because of the dorsal appendage of left lamella, which consists of two almost equally long finger-like processes, and the projecting appendage of right lamella.

Length: body 2.4–3.1 mm, wing 2.5–2.8 mm.

♀. Frons distinctly broader than in *clavipes*, slightly deeper than anterior ocellus, face much narrower. Antennae (Fig. 228) with 3rd segment perhaps slightly longer than in ♂, style as long as or slightly shorter than half length of 3rd segment. Thoracic bristles paler, yellowish brown, and mesonotum brightly shining right up to front edge as in *clavipes* ♀, but notopleural depression greyish pollinose as in *bilobata* ♂. Halteres and squamae (with concolorous fringes) yellowish. Legs much paler than in ♂, mostly yellow, leaving at most only apical half of hind tibiae and apical three segments of all tarsi blackish brown. Hind femora with pale av bristly hairs and hind trochanters without the black spine beneath. Abdomen tapering and extensively polished black even on dorsum, pubescence pale, sparse and much shorter than in male.

Length: body 2.3–2.9 mm, wing 2.3–2.8 mm.

Note. *T. bilobata* could easily be mistaken for the south-west European *T. unilobata* Chv. which has similarly shaped antennae, but the bare terminal part of the style occupies its whole apical half, the legs are extensively blackish in male, hind femora (dilated on apical third) are armed with very long black av bristles, and ♂ hypopygium possesses only a single long process on left lamella.

Distribution and biology. With the same type of distribution as *clavipes* in the south but extending further northwards, nearly up to the polar circle, in Sweden to Nb., in Finland to ObN and Ks. – Widely distributed and common in C. Europe, south to Yugoslavia (Coe, 1962). – June to early August, in England and temperate Europe as early as May.

9. ***Trichina elongata*** Haliday, 1833

Figs. 229, 230, 244–248.

Trichina elongata Haliday, 1833: 158.
Trichina clavipes Meigen; Lundbeck, 1910: 187.
Trichina picipes Tuomikoski ♂ (nec ♀); Frey, 1956: 596.

Mesonotum narrowly brownish grey pollinose in front and in notopleural depression in both sexes, 3rd ant.s. elongate and slender, style less than half length of 3rd segment. Legs in ♂ extensively brownish, hind femora with a row of short black av spine-like bristles.

♂. Occipital hairs and bristles black except for a few pale hairs above mouth-opening. Antennae (Fig. 229) black, 3rd segment considerably long and slender, style apparently less than half length of 3rd segment, its bare terminal part not very much shorter than the pubescent thicker basal part. Thorax polished black, but anterior fifth of mesonotum and whole of notopleural depression brownish grey pollinose. Thoracic bristles black, 4 scutellars, outer pair smaller.

Wings uniformly brownish, squamae and halteres black. Legs almost uniformly brownish, at most light brown on "paler" parts, fore tibiae rather slender and very darkened except for base. Hind femora blackish brown, at most basal third paler, rather like basal half of the apically swollen and darkened hind tibiae. Mid femora with a row of small blackish pv bristly hairs, no distinct dorsal bristle on mid tibiae, and hind femora with a row of rather short black av spine-like bristles. Hind trochanters with only a small, rather indistinct black spine beneath.

Abdomen subshining at sides but dull above, covered with long but sparse dark bristly hairs. Genitalia (Figs. 244–248) with a prominent long finger-like anterior process on left lamella, the posterior process very small, slender and pointed.

Length: body 2.5–3.2 mm, wing 2.6–3.3 mm.

♀. Frons rather narrow, about as deep as anterior ocellus (frons as in *bilobata*, broader than in *clavipes*), 3rd ant.s. apparently longer than in ♂ and style consequently shorter (Fig. 230). Mesonotum pollinose anteriorly as in ♂ but all bristles much paler, brownish yellow. Wings not so infuscated, almost clear, squamae brownish yellow with pale fringes, halteres yellowish with base of stalk blackish. Legs much paler than in ♂,

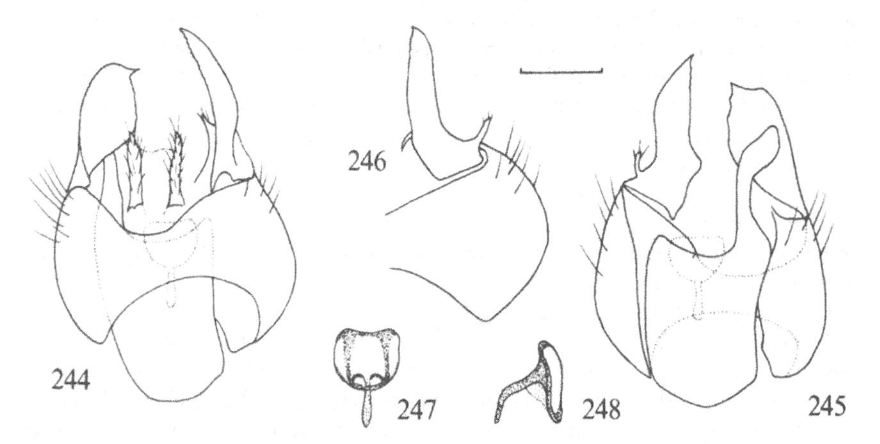

Figs. 244–248. Male genitalia of *Trichina elongata* Hal. - 244: dorsal view, aedeagus dotted; 245: the same in ventral view, cerci omitted; 246: left periandrial lamella; 247, 248: aedeagus in two different views. Scale: 0.1 mm.

coloured as in males of *clavipes* or *bilobata*, the bristles on posterior four femora beneath paler and thinner, hind trochanters without a black spine. Abdomen tapering and almost polished, with sparse shorter pubescence than in ♂.

Length: body 2.3–3.1 mm, wing 2.5–3.0 mm.

Distribution and biology. Widespread in Scandinavia to extreme north, in Norway to Fn, in Sweden to Nb. and in Finland to Li and the Kola Peninsula. - Very common in Europe, south to France, Romania and the Caucasus, in warmer regions preferring the mountains. - May to September, but mainly in June and July, in England to the beginning of October.

10. ***Trichina opaca*** Loew, 1864
Figs. 231, 232, 249–254.

Trichina opaca Loew, 1864: 40.
Trichina picipes Tuomikoski, 1935: 99 – **syn.n.**

Mesonotum entirely microscopically pollinose in ♂, only on front part in ♀. Antennae with 3rd segment rather broadened, style not half as long as 3rd segment. Legs extensively darkened, almost blackish in ♂.

♂. Antennae with 3rd segment rather broad (Fig. 231), apparently deeper than in all previous species, style shorter than in *clavipes* and even *bilobata*, its apical third bare. Mesonotum entirely thinly brownish pollinose and consequently rather dulled, bristles in full number, black; prescutellar pair of dc very strong, 6 scutellar bristles, the outer (third) pair sometimes very small.

Wings conspicuously darkened brown, veins blackish, squamae including fringes and halteres deep black. Legs extensively blackish to blackish brown, "knees" and extreme base of tibiae often paler. Legs covered mostly with hairs, fore tibiae rather slender, mid tibiae without distinct dorsal bristles and no spine on hind trochanters beneath. Hind femora with equally thin dark bristly hairs above and below, those near base above the longest. Hind tibiae slender on basal third but very dilated towards tip, no distinct bristles above.

Abdomen covered with rather long dark bristly hairs, somewhat subshining at sides but very dull black above. Genitalia (Figs. 249–254) very small, in general closely resembling those of *elongata* but the finger-like anterior process on left lamella smaller and the posterior one, on the contrary, larger.

Length: body 2.5–3.0 mm, wing 2.5–2.9 mm.

♀. Frons rather broader, at least as deep as anterior ocellus or broader, greyish dusted, face very linear. Antennae (Fig. 232) even broader than in ♂ and style obviously shorter. Mesonotum remarkably unlike that of ♂, rather thinly but distinctly greyish (not brownish) dusted in front and in notopleural depression, otherwise polished. Thoracic bristles pale, 2 or 3 pairs of scutellars, rarely 4 pairs. Wings almost clear, squamae (including fringes) and halteres pale. Legs considerably variable in colour, paler than in ♂ but still extensively darkened, sometimes anterior four coxae, trochanters and femora, hind coxae, and base of hind femora and tibiae, almost yellowish brown; fore tibiae always almost blackish. All legs with much paler hairs and hind tibiae less dilated apically. Abdomen almost polished, with shorter and paler pubescence.

Length: body 2.5–3.2 mm, wing 2.5–3.0 mm.

Holotype identifications. *T. opaca* was described by Loew (1864) from a single male "in der Nähe des am Fusse des Altvaters gelegenen Dorfes Waldenburg" (i.e. Bělá near Domašov in the Jeseníky Mts., northern Moravia, Czechoslovakia). There are 2 ♂ in Coll. Loew in Berlin; one ♂ bears an original Loew label "*Trichina opaca* m.", "No. 10563" and "St. Mz" - obviously not the holotype as it very probably comes from the Swiss St. Moritz; the second ♂ has only a label "Kowarz". Both males are conspecific and belong to *T. opaca* Loew as described above.

The female sex was described much later by Tuomikoski (1935) under the name of *T. picipes*; the holotype ♀ (now headless) was examined by me in Helsinki in 1980. It is labelled "Helsingfors, Spec. typ. 4897, *Trichina picipes* Tuomik." and is undoubtedly conspecific with *T. opaca* Loew. There is another ♂ of *picipes* (= *opaca*) in the Helsinki Museum, taken by Frey at Sveaborg (Helsinki). However, the male of "*T. picipes*" as described by Frey (1956: 596) in "Lindner" from Utsjoki, Lapland ("Arista kurz, etwa ¼ so lang wie das 3. Fühlerglied, . . . Hypopygium relativ gross" - and keyed on p. 593 as "Mesonotum vorn in der Mitte kahl"), which Collin (1961: 285) suggested might be a distinct species, is actually a male of *elongata* as is proved by Frey's material in Helsinki.

Distribution and biology. Very rare in Scandinavia, known only from the above two specimens from S. Finland (N) from the vicinity of Helsinki. - Widely distributed but overlooked species in Europe, known to me so far from Great Britain, Finland, Ger-

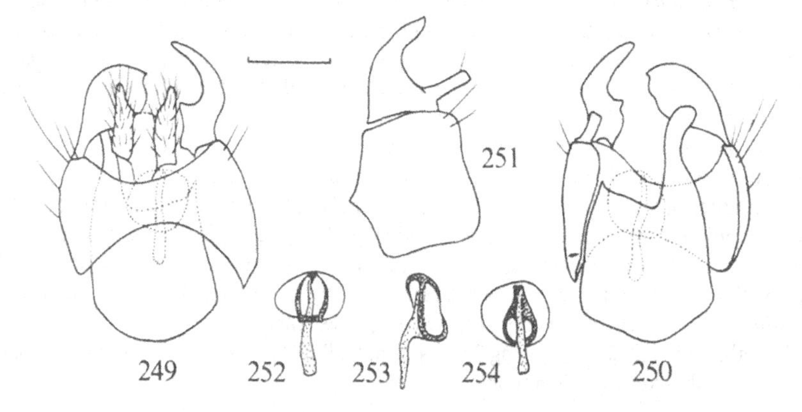

Figs. 249–254. Male genitalia of *Trichina opaca* Loew. - 249: dorsal view, aedeagus dotted; 250: the same in ventral view, cerci omitted; 251: left periandrial lamella; 252–254: aedeagus in three different views. Scale: 0.1 mm.

man Democratic Republic, Czechoslovakia, Switzerland and the Caucasus; Kovalev (in litt.) has also recorded it from the north of European USSR. - September, in England as early as late April, in C. Europe mainly from May to July.

11. ***Trichina pallipes*** (Zetterstedt, 1838)
Figs. 233, 255–260.

Oedalea pallipes Zetterstedt, 1838: 538.
Trichina clavipes var. *sexsetosa* Frey, 1913: 58.

Antennal style very broad, looking like a continuation of 3rd segment. Legs extensively pale yellow in both sexes, mesonotum broadly pollinose anteriorly in ♂, wholly polished in ♀.

♂. Antennae (Fig. 233) black, 3rd segment long and almost evenly slender; style very compressed, as deep as 3rd segment, basal article large and only extreme tip of style bristle-like. Palpi and proboscis pale yellow. Thorax broadly brownish grey pollinose on anterior third of mesonotum and on a median stripe right up to middle, all bristles dark, rather small and fine; 6–8 scutellars the most prominent bristles.

Wings faintly brownish clouded, squamae (including fringes) and halteres blackish. Legs mostly pale yellow, apical half of the apically thickened hind tibiae and apical three segments of all tarsi brown. Hind coxae and a preapical ring on hind femora sometimes brownish. No distinct bristles, hind femora with long yellowish brown hairs above about as long as femur is deep, ventrally with similar but shorter hairs. Fore tibiae slightly spindle-shaped and quite yellow, hind trochanters beneath at most with a pale bristly hair, no spine.

Abdomen almost shining at sides but very dull on dorsum, covered with rather long dark hairs at sides and below. Genitalia (Figs. 255–260) very small, resembling those of other *Trichina* species but left lamella with posterior process longer, anterior process (very long in other species) rather short and broad. Lamellae only narrowly connected anteriorly and hypandrium broader than usual, with well developed hypandrial bridge.

Length: body 2.3–3.0 mm, wing 2.0–2.9 mm.

♀. Frons polished black and relatively deep, distinctly broader than anterior ocellus, in this respect resembling *T. opaca* ♀. Mesonotum wholly polished even in front, only humeri anteriorly and notopleural depression finely greyish pollinose. Thoracic bristles pale including 6–8 scutellars. Wings practically clear with paler veins, squamae very pale with short whitish fringes, halteres yellowish. Legs even paler than in ♂, only apical third of hind tibiae and last tarsal segments brownish. Abdomen extensively polished black with much shorter, sparse pale pubescence.

Length: body 2.2–2.6 mm, wing 2.0–2.4 mm. Collin (1961) erroneously gave the body length in both sexes as 1.75 mm only.

Lectotype designation. There are 2 specimens of *T. pallipes* in Ins. Lapp. Coll. at Lund, both labelled "Lapponia": a male, labelled *Oedalea minuta,* and a female, labelled *Oedalea pallipes;* the female was labelled in 1977 and is herewith designated as lectotype of *Trichina pallipes* (Zett.). In the Dipt. Scand. Coll. there are a further 1 ♂ (as *minuta*) and 2 ♀ (as *pallipes*) from Gotland.

T. clavipes var. *sexsetosa* Frey, 1913 was correctly synonymised with *pallipes* by Tuomikoski (1935). In the Finnish Coll. in Helsinki there are 4 ♂ and 8 ♀ of *pallipes* under the name var. *sexsetosa* from Karislojo (Ab), Runsala (Ab) and Helsinge (N); one male from Karislojo (Spec. typ. No. 4417) and another from Helsinge (Spec. typ. No. 4418) have Frey's original type labels. No lectotype has been selected, but since Frey (1913: 38) described it from Jomala (Al), Helsinge (N), Yläne (St) and Kexholm (Kl), the male from Helsinge (Palmén, No. 4418) could be selected as lectotype if necessary.

Distribution and biology. Rather a rare species in Scandinavia (36 specimens examined), commoner in S. Finland. In Denmark only 1 ♂ from Funen (Lohals, 13.9. 1926, Lundbeck), in Sweden from Gtl., Nrk., Nb. and T. Lpm., in Finland north to Om, also Russian Karelia. - Uncommon species in Europe, recorded from England (Collin, 1961), Schleswig-Holstein (Kröber, 1958), central parts of European USSR (Kovalev, in litt.), and recently found in some numbers in the lowlands of S. Moravia, Czechoslovakia. - June to mid September.

Synonymy. The subgenus *Oedaleopsis* Tuomikoski, 1959, with *T. pallipes* as type-species, is regarded here as a new synonym of *Trichina* Meig. The three subgeneric features given by Tuomikoski for *T. pallipes* are in my opinion nothing but specific characters: (1) the elongate 3rd ant.s., with a broad laterally compressed style compared with the slender style of *clavipes* and others, is a characteristic specific feature in, for instance, *Oedalea* species (*hybotina* versus *zetterstedti*); (2) 6–8 scutellar bristles are also present in *T. opaca;* (3) the absence of the bristle-like hairs on the fore coxae in *pallipes* is simply part of the general reduction in the bristling on the thorax and legs in *pallipes*. Moreover, the general structure of the male genitalia (including the aedeagus) in *pallipes* is exactly the same as in other *Trichina* species; the only substantial difference seems to be the presence of the hypandrial bridge, but the aedeagus is still only loosely

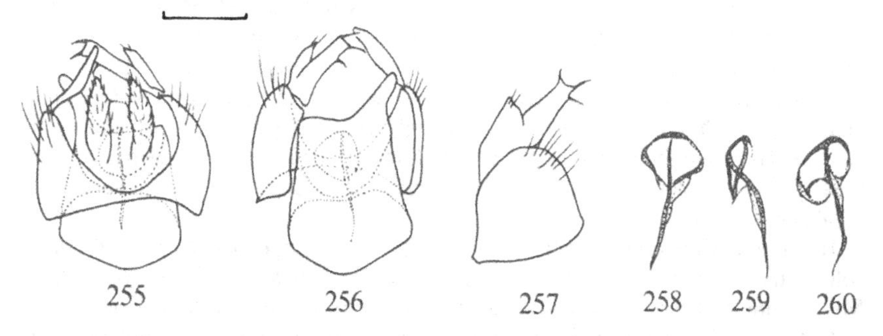

Figs. 255–260. Male genitalia of *Trichina pallipes* (Zett.). - 255: dorsal view, aedeagus dotted; 256: the same in ventral view, cerci omitted; 257: left periandrial lamella; 258–260: aedeagus in anterior, lateral and posterior views. Scale: 0.1 mm.

attached to the hypandrium as in other species and, if for instance the related *Hybos* species are considered, the presence or near absence of a hypandrial bridge are not treated as more than a specific character. Studies of the male terminalia in Nearctic species might also be useful.

Genus *Bicellaria* Macquart, 1823

Bicellaria Macquart, 1823: 155.

Type-species: *Bicellaria nigra* Macquart, 1823 (mon.) = *Bicellaria spuria* (Fallén, 1816).

Cyrtoma Meigen, 1824: 1.

Type-species: *Cyrtoma atra* Meigen, 1824 (des. Westwood, 1840) = *Bicellaria spuria* (Fallén, 1816).

Enicopteryx Stephens, 1829: 264 (catalogue name).

Type-species: not designated (3 *Bicellaria* spp. included).

Calo v. Gistl, 1848: viii (unjustified replacement name for *Cyrtoma*).

Type-species: isogenotypic with *Cyrtoma*.

Generally small (body length about 3 mm) dull black and distinctly bristled species (Fig. 261). Head almost circular in anterior view, semicircular when viewed from the side, with eyes contiguous on frons for a long distance in both sexes, and upper facets considerably enlarged. The incisions opposite base of antennae deeply triangular, face rather long, moderately broad or narrow (not depending on the sex). Antennae inserted at about middle of head in profile. Three ocelli on a distinct tubercle, usually 2 pairs of ocellar bristles, posterior pair (if present) always much smaller. Occiput densely bristled. Basal antennal segment indistinctly separated, 2nd with a circlet of preapical bristly hairs, one or two ventral bristles often much longer. 3rd segment with upper side almost straight, very convex on basal half beneath and narrowing towards tip, apical part narrow. Antennal style about as long as 3rd segment, slender and finely pubescent except for extreme tip, basal article distinct and square-shaped. Mouthparts (Figs. 262–264) small and pointing downwards, labrum strongly arched on basal half, apically narrowed to a 2-pointed tip, hypopharynx slender and as long as labrum. Palpi small ovate with one or several bristly hairs at tip and below, palpifer indistinctly developed and usually with 2 weak bristles beneath (only one bristle according to Bährmann (1961: 847)), stipites indistinct, lacinia absent. Labium with large membraneous labellae (with distinct pseudotracheae and sensory hairs) about as long as the soft membraneous basal part of labium.

Thorax often remarkably arched above, very "high", large pronotum (prothoracic collar) and proepisterna devoid of hairs or bristles, prosternum small, *Hybos*-like. Mesonotum well bristled, acr biserial and usually long, uniserial dc as long as acr, ending in 1 or 2 pairs of long prescutellars. Large bristles in full number, the prescutellar dc, postalars and 4–6 scutellars the longest.

Wings large, often uniformly darkened (particularly in males) and iridescent, with conspicuously modified venation. Costa ending at tip of R_{4+5}, Sc closely approximated to R_1 but turned up at middle of wing towards costa, often abbreviated at extreme tip. Stigma apparently not developed but the section of costa (and costal cell) between tips of Sc and R_1 darkened and somewhat stigma-like. Radial sector rather short, as long as the distance from its origin to humeral crossvein. R_{4+5} not forked, median section of vein M indistinct beyond anterior crossvein r-m, and also the base of fork M_1 and M_2. Discal cell not developed, the vein closing it (posterior crossvein m-m) absent. 2nd basal cell large, broader and larger than the equally long 1st basal and anal cells; the vein closing anal cell (vein Cu_{1b}) recurrent, more than is usual in the Ocydromiinae but still less than in *Empis* and its allies. Anal vein almost complete, usually reaching the wing-margin. Axillary lobe strongly developed, axillary excision almost rectangular, alula absent; a distinct costal bristle.

Fig. 261. Male of *Bicellaria spuria* (Fall.). Total length: 2.3–2.8 mm.

Legs rather long and slender, usually with remarkably long bristles especially on tibiae, hind tibiae more or less dilated apically, often also basal two segments of hind tarsi thickened. The tubular gland (sense organ) near base of fore tibiae very slightly developed, but tibiae slightly stouter there. All tibiae with small anteroventral spurs, tarsi not laterally compressed as in *Trichina*-complex.

Abdomen rather long and slender, cylindrical, indistinctly narrowing posteriorly but not remarkably downcurved; covered with numerous bristly hairs and stronger hind marginal bristles. 8th segment in ♂ not very modified, at least half as long as 7th segment, well visible, tergum scarcely smaller than sternum. Genitalia almost symmetrical and very slightly turned towards right. Periandrial lamellae narrowly pointed at tip and very narrowly connected anteriorly at base, anterior edge of each lamella produced into a fold linking periandrium with hypandrial bridge. Cerci generally large. Hypandrium cornet-like below, apically produced into two conspicuous specifically distinct prongs of various shapes, which offer the best diagnostic features. Hypandrial bridge always strongly developed, its upper edge (dorsal in normal position of hypopygium) connected with periandrial folds and proctiger (? 10th tergum). Aedeagus uniform in shape, bent downwards apically, with the tip lying between hypandrial prongs, coaxial apodeme long, and laterally aedeagus covered with flat leaf-shaped to almost rectangular postgonites. Female abdomen very narrowed apically, telescopic, with small slender cerci.

Distribution. A Holarctic genus, represented by 15 species in Europe (at least 3 more new species from Central Europe await description), 3 species in Japan (Kato, 1971a), and 9 species in North America; of these, *B. spuria* is considered to be Holarctic in distribution (Melander, 1965). Smith (1965) recorded one species from Nepal (probably *B. vana*) and I have seen an undescribed species close to *spuria* from Kazakhstan (Alma-Ata). The only record outside the Holarctic region, given by Bezzi (1912) from Formosa (*B. spuria* with a doubt), needs verification. The identification of *Bicellaria* species is by no means easy, in particular the female sex, and the older published records of the distribution of European species cannot be accepted because of common misidentifications. Eleven *Bicellaria* species have been found in Scandinavia.

Biology. The presence of large holoptic eyes in both sexes as in the Hybotinae clearly indicates the same habits; both males and females are highly predaceous, sitting on grasses, twigs and tips of leaves waiting for nearby flying insects, and prey is captured in flight with the help of the elongated hind legs. Swarming activity is very rare and mating activity has obviously been transferred to a solid substrate: the rare large aerial male aggregations are thought to be so-called "relict swarms", without obvious recent significance (Chvála, 1980). Adults are very abundant in various types of biotopes and larvae apparently live in the soil. Nectar feeding or flower visiting has never been observed.

Note. Originally the genus *Bicellaria* was placed in the Platypezidae, particularly because of its reduced wing venation which closely resembles that of *Microsania* and *Opetia*, but there are now no doubts about its inclusion in the Hybotidae. Its closest relative is the Nearctic *Hoplocyrtoma,* which has the same wing venation, but the two known species have distinctly thickened and spinose (*Oedalea*-like) hind femora, and the ♂ genitalia are more asymmetrical.

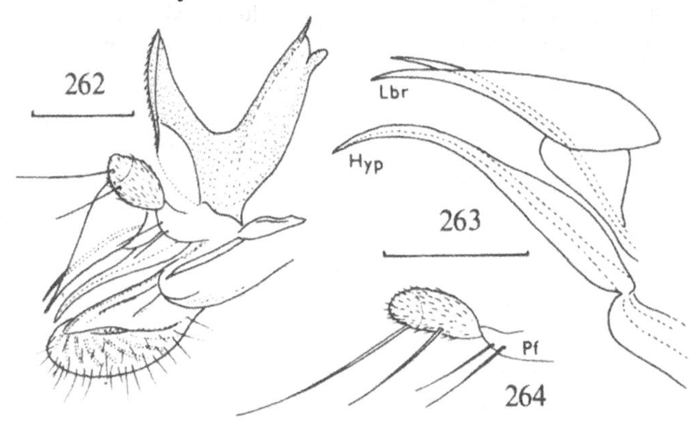

Figs. 262–264. Mouthparts of *Bicellaria spuria* (Fall.) ♂. – 262: lateral view; 263: labrum and hypopharynx; 264: palpus. Scale: 0.1 mm. For abbreviations see p. 13.

Key to species of *Bicellaria*

1 3rd ant.s. (Figs. 278, 279, 286) with 1 to 3 bristles above 2
– 3rd ant.s. without a bristle above 4
2 (1) Palpi with 3 or more strong black bristles. Face (Fig. 266) very broad, broader than 3rd ant.s. Fore tibiae with distinct pv bristly hairs longer than tibia is deep, and all tibiae with very long ad and pd bristles. The whole insect with long black bristles everywhere, halteres blackish in both sexes. Tibiae and femora rather short, particularly on hind legs, hind metatarsus (Fig. 290) not very long and the following segments becoming gradually shorter .... 13. *pilosa* Lundbeck
– Palpi with only 1 or 2 dark bristly hairs. Face not so broad, or narrowing towards mouth. Fore tibiae with pv hairs finer and not longer than tibia is deep. The whole insect more slender and with shorter hairs everywhere. All tibiae and femora long; hind metatarsi long, together with the next segment distinctly longer than the apical three tarsal segments combined 3
3 (2) Legs long and slender, mid femora uniformly slender and not very much deeper than tibia. Basal two segments of hind tarsi (Fig. 293) long and thickened, deeper than the

following tarsal segments. Wings faintly brown. ♂: halteres often brownish or almost pale; base of abdomen more or less pale pubescent; hypandrial prongs long and slender, apically with only fine hairs (Fig. 327). ♀: halteres pale; abdomen mostly with pale hairs .... 21. *intermedia* Lundbeck

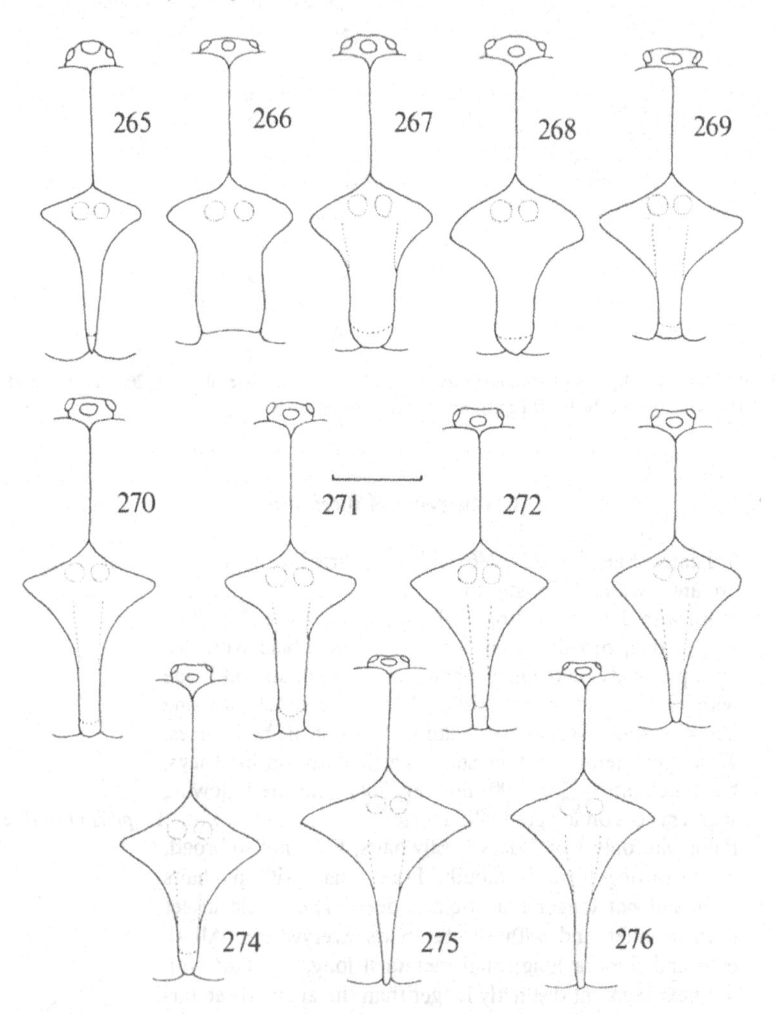

Figs. 265–276: Faces in frontal view of *Bicellaria*. 265: *simplicipes* (Zett.); 266: *pilosa* Lundb.; 267: *austriaca* Tuomik.; 268: *subpilosa* Coll.; 269: *bisetosa* Tuomik.; 270: *spuria* (Fall.); 271: *mera* Coll.; 272: *sulcata* (Zett.); 273: *vana* Coll.; 274: *intermedia* Lundb.; 275: *nigra* (Meig.); 276: *nigrita* Coll. Scale: 0.2 mm.

– Legs not so slender, mid femora deeper on basal half, about twice as deep as tibia. Basal two segments of hind tarsi (Fig. 291) long and rather slender, scarcely deeper than the following segments. Wings dark brown. ♂: halteres black; abdomen black pubescent, at most with several pale hairs at extreme base; hypandrial prongs shorter and stouter, apically with long bristly hairs (Fig. 323). ♀: halteres black-

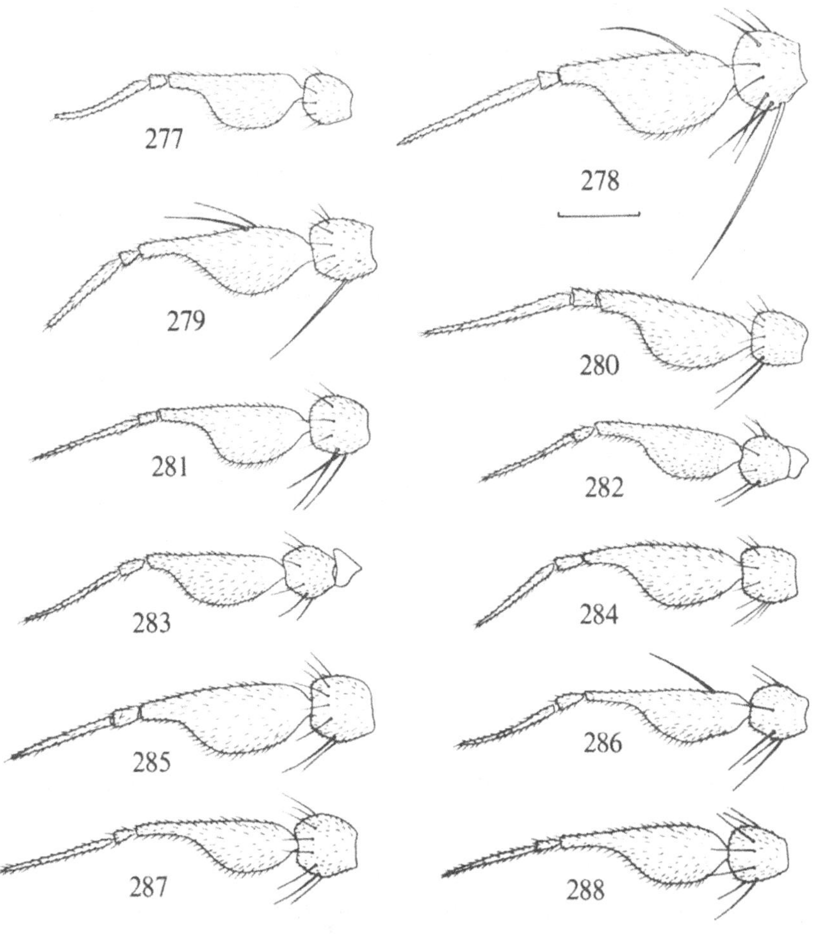

Figs. 277–288. Antennae of *Bicellaria*. – 277: *simplicipes* (Zett.); 278: *pilosa* Lundb.; 279: *austriaca* Tuomik.; 280: *subpilosa* Coll.; 281: *bisetosa* Tuomik.; 282: *spuria* (Fall.); 283: *mera* Coll.; 284: *sulcata* (Zett.); 285: *vana* Coll; 286: *intermedia* Lundb.; 287: *nigra* (Meig.); 288: *nigrita* Coll. Scale: 0.1 mm.

ish, or at least knobs very darkened; abdomen with black hairs, sometimes some pale hairs at base .... 14. *austriaca* Tuomikoski

4 (1) Hind tibiae very slender, not dilated towards tip (Fig. 289). All legs long and slender, covered with a few weak bristles. Vein M_1 usually complete. Abdomen mostly with pale hairs even in ♂, thorax very arched above. Halteres blackish in ♂, light brownish in ♀ .... 12. *simplicipes* (Zetterstedt)

– Hind tibiae dilated towards tip, stouter than mid tibiae,

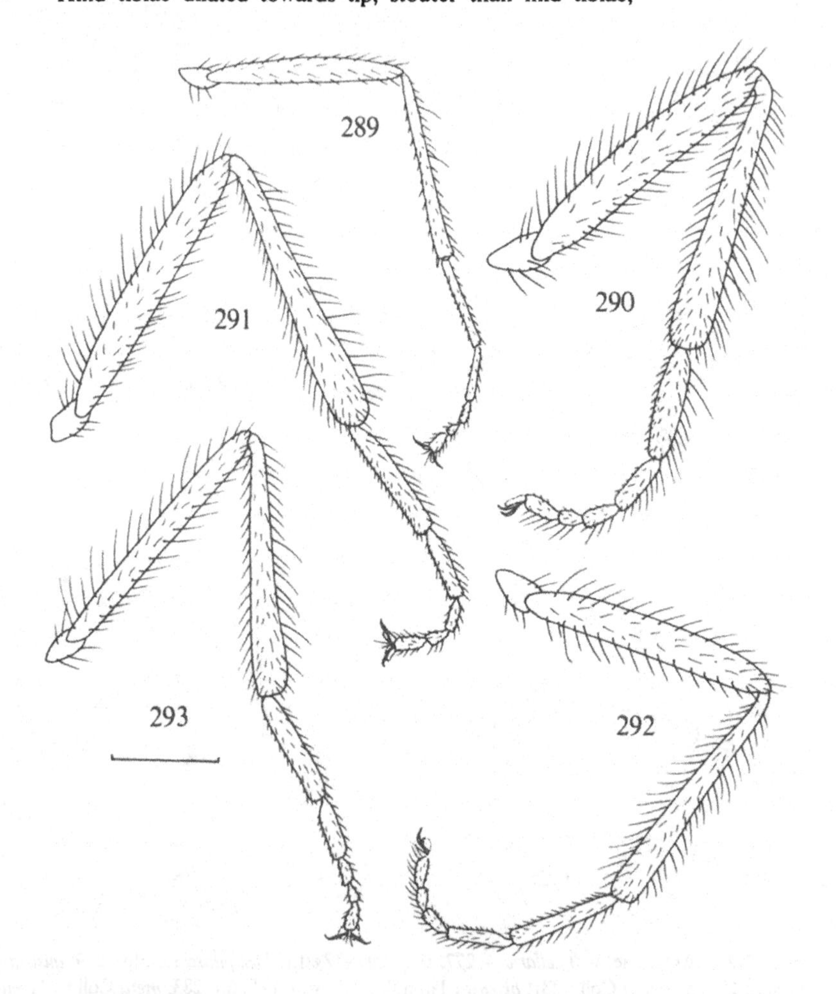

Figs. 289–293. Hind legs in anterior view of *Bicellaria*. – 289: *simplicipes* (Zett.); 290: *pilosa* Lundb.; 291: *austriaca* Tuomik.; 292: *subpilosa* Coll.; 293: *intermedia* Lundb. Scale: 0.5 mm.

and legs with stronger and longer bristles. Vein M_1 incomplete, abbreviated at base and not reaching M_2 5

5 (4) Basal two segments of hind tarsi distinctly dilated (Figs. 302, 304). Abdominal pubescence pale at least at base of abdomen, halteres in ♀ yellowish. Face uniformly very narrow 6

– Basal two segments of hind tarsi not dilated, all tarsal segments equally slender. Abdominal pubescence dark and halteres blackish in both sexes. Face broader, or narrowing towards mouth 8

6 (5) Large long-legged species; hind femora long and slender, hind tibiae 1.2–1.5 mm long and dilated only towards tip (Fig. 302). Mid tibia with 1–3 bristles above much longer than other pubescence. Halteres blackish in ♂, pale in ♀. ♂: hypandrial prongs rather short and stout (Fig. 366) .... 22. *nigra* (Meigen)

– Smaller and shorter-legged species; hind femora shorter and stouter, hind tibiae about 1.1 mm long and more evenly dilated (Fig. 304). Mid tibia with less differentiated bristles above 7

7 (6) Greyer species, body extensively pale bristled and pubescent. ♂: halteres with yellow knobs, hypandrial prongs rather short. ♀: abdominal pubescence almost all pale .... *halterata* Collin

– Blacker species, body mostly black bristled and pubescent. ♂: halteres blackish, hypandrial prongs long and slender (Fig. 368). ♀: abdomen with pale hairs only at sides about base .... *nigrita* Collin

8 (5) Face rather narrow, and distinctly narrowed towards mouth (Figs. 272, 273). Fore tibiae without pv hairs or these confined to several hairs near base, the hairs always shorter than tibia is deep; venter of tibia clothed with soft pile. Hind tibiae always distinctly dilated towards tip 9

– Face broader and parallel-sided (cf. Figs. 268–271). Fore tibiae with a complete pv ciliation, the hairs as long as or longer than tibia is deep 10

9 (8) ♂: thorax greyish black with brownish dust, not dull black from any angle. Fore tibiae more slender and with somewhat longer pv hairs especially on basal half. Hypandrial prongs rather short and stout (Fig. 353). ♀: thorax duller brownish black. A weak pv ciliation on fore tibiae, and these with rather long hairs above .... 19. *sulcata* (Zetterstedt)

– ♂: thorax mostly dull black. Fore tibiae somewhat stouter, pv cilia practically absent. Hypandrial prongs much longer and slender (Fig. 357). ♀: thorax more shining black. Practically no pv ciliae on fore tibiae, and these slightly stouter and with rather shorter hairs above .... 20. *vana* Collin

10 (8) Larger and everywhere longer bristled species, generally

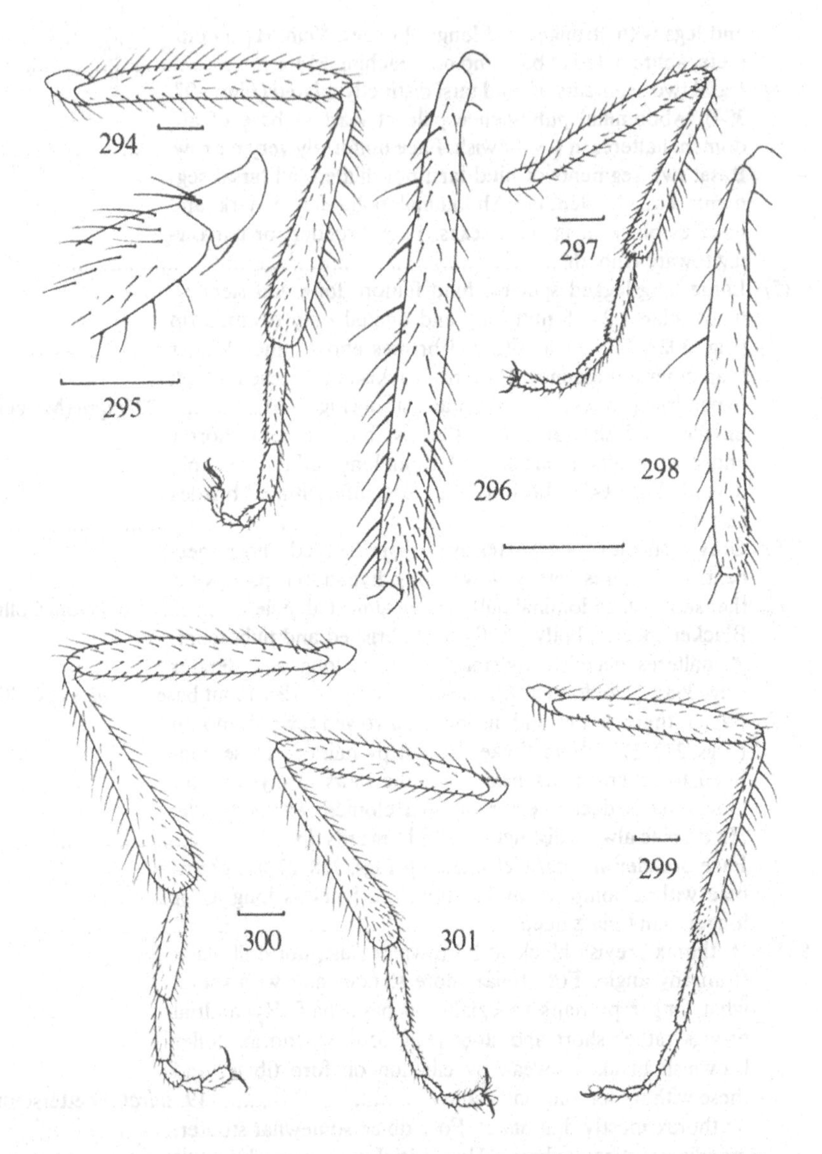

Figs. 294–301. Legs of *Bicellaria*. – 294: *bisetosa* Tuomik., hind leg in anterior view; 295: *bisetosa* Tuomik., fore trochanter and base of femur in posterior view; 296: *bisetosa* Tuomik., fore tibia in posterior view; 297: *spuria* (Fall.), hind leg in anterior view; 298: *spuria* (Fall.), fore tibia in posterior view; 299: *mera* Coll., hind leg in anterior view; 300: *sulcata* (Zett.), hind leg in anterior view; 301: *vana* Coll., hind leg in anterior view. Scale: 0.2 mm.

more than 3 mm in length. Usually 6 scutellar bristles. Hind tibiae distinctly dilated and with longer hairs beneath, the hairs on basal two-thirds in ♂ distinctly longer than tibia is deep. Thorax brownish black, mid femora with rather long av bristly hairs (only on basal third in ♀). Face very broad (Fig. 268). Slender apical part of 3rd ant.s. short .... 15. *subpilosa* Collin

– Smaller and everywhere shorter bristled species, about 2.5 mm in length. Usually 4 scutellar bristles. Hind tibiae rather slender and with shorter hairs beneath, the hairs more adpressed and shorter than tibia is deep. Thorax more greyish black, mid femora without distinct av bristly hairs (except *bisetosa*). Face somewhat narrower 11

11 (10) Fore tibiae with less differentiated long bristles dorsally. Fore trochanters with 2 bristly hairs dorsally and mid fe-

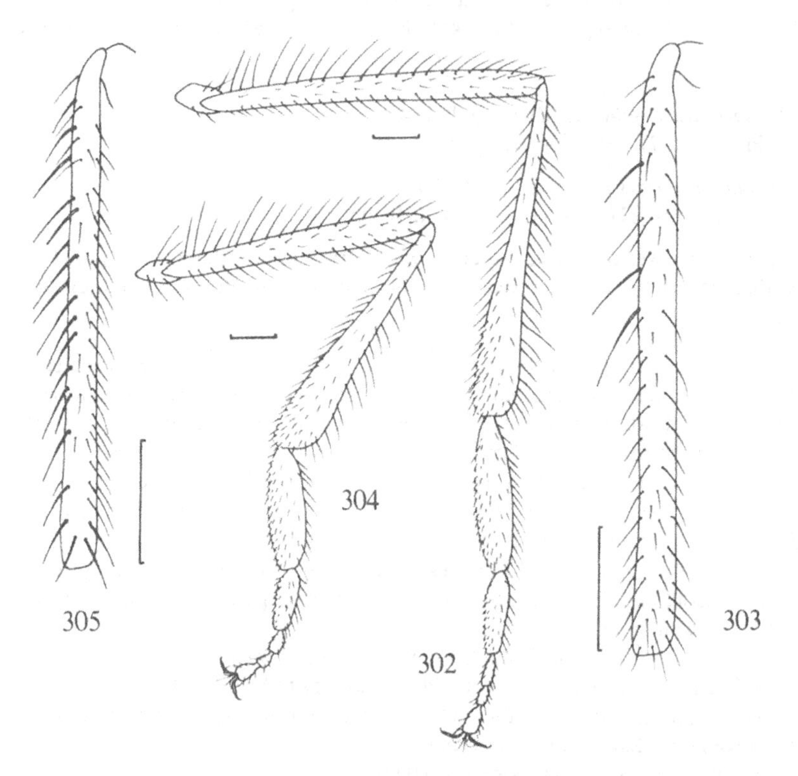

Figs. 302–305. Legs of *Bicellaria*. – 302: *nigra* (Meig.), hind leg in anterior view; 303: *nigra* (Meig.), mid tibia in posterior view; 304: *nigrita* Coll., hind leg in anterior view; 305: *nigrita* Coll., mid tibia in posterior view. Scale: 0.2 mm.

mora with av bristly hairs. Thorax and abdomen subshining black, thinly greyish brown dusted. ♂: hypandrial prongs (Fig. 337) very broad and stout .... 16. *bisetosa* Tuomikoski

– Fore tibiae with 1 or 2 bristles above much longer than other fine hairs. Fore trochanters with 1 bristly hair dorsally, and no av bristles on mid femora. Thorax and abdomen duller black 12

12 (11) Slender apical part of 3rd ant.s. longer, almost one-half as long as apical style (Fig. 282). Hind tibiae dilated towards tip, slightly stouter at end than the thickest part of hind femur (Fig. 297). ♂: hypandrial prongs long and slender (Fig. 344) .... 17. *spuria* (Fallén)

– Slender apical part of 3rd ant.s. short, much less than half length of style (Fig. 283). Hind tibiae rather slender, not as stout at end as the thickest part of hind femur (Fig. 299). ♂: hypandrial prongs (Fig. 349) shorter and stouter .... 18. *mera* Collin

12. ***Bicellaria simplicipes*** (Zetterstedt, 1842)
Figs. 265, 277, 289, 306–310.

Cyrtoma simplicipes Zetterstedt, 1842: 331.
Bicellaria melaena Haliday; Lundbeck, 1910: 22.

Sparsely and weakly bristled species with very humped thorax and rather distinct medial fork on wings. Hind tibiae extremely slender and all legs without distinct bristles.

♂. Antennae (Fig. 277) without dorsal bristle, face (Fig. 265) very narrowed below. Thorax remarkably arched above and mostly dull black, brownish dusted in side view. All thoracic bristles rather small and weak except for 1 notopleural, 1 postalar, 2 prescutellar dc and 4–6 scutellars.

Wings deep brown, often very darkened, medial fork almost complete, although fine. Axillary lobe prominent. Halteres blackish. Legs extremely slender and practically devoid of distinct bristles, all hairs rather short. Hind tibiae (Fig. 289) very slender, not dilated apically, dorsally with 3 bristles scarcely longer than tibia is deep. Anterior four tibiae at about middle with a dorsal bristle almost as long as tibiae are deep, pv ciliation on fore tibiae minute.

Abdomen brownish black dusted but dull black from some angles, venter usually yellowish. Pubescence rather short with a tendency to be paler, hind-marginal bristles rather long, black. Genitalia (Figs. 306–310) with conspicuously short and stout hypandrial prongs, median excision not very deep.

Length: body 2.4–2.8 mm, wing 2.6–3.0 mm.

♀. Wings much clearer than in male and halteres more yellowish brown. Thorax rather brownish dusted, abdomen dull brownish grey with pubescence shorter, denser and much paler. Legs obviously more bristled than in male, hind femora with av bristly hairs sometimes almost whitish and not very much shorter than femur is deep.

Length: body 2.8–3.3 mm, wing 2.9–3.0 mm.

Note. The sparse and weak bristling, the almost complete medial fork on wings, the slender hind tibiae and the distinctive hypandrium with only a shallow median cleft (with consequently short and broad hypandrial prongs) are plesiomorphous states within the genus. *B. simplicipes* is undoubtedly a very primitive member of the genus.

Lectotype designation. There are 6 specimens (3 ♂ 3 ♀) in the Dipt. Scand. Coll. at Lund under "*Cyrtoma simplicipes*" from Coll. Staeger (Dania), Gothem (Gottlandia) and Wadstena (Ostrogothia), all being syntypes and belonging to *B. simplicipes* Zett. The male from Denmark, labelled "Staeg. 18", was labelled in 1977 and is herewith designated as lectotype of *Bicellaria simplicipes* (Zett.).

Distribution and biology. Widespread throughout Fennoscandia and Denmark including the north, also from the Kola Peninsula, but everywhere uncommon; not yet recorded from Norway. – From Great Britain east to the north of the USSR (Archangelsk), south to Central Europe (Austria, Bavaria, Czechoslovakia, Hungary), rather common only in Great Britain. – Mid May to September, in temperate Europe until the first half of October.

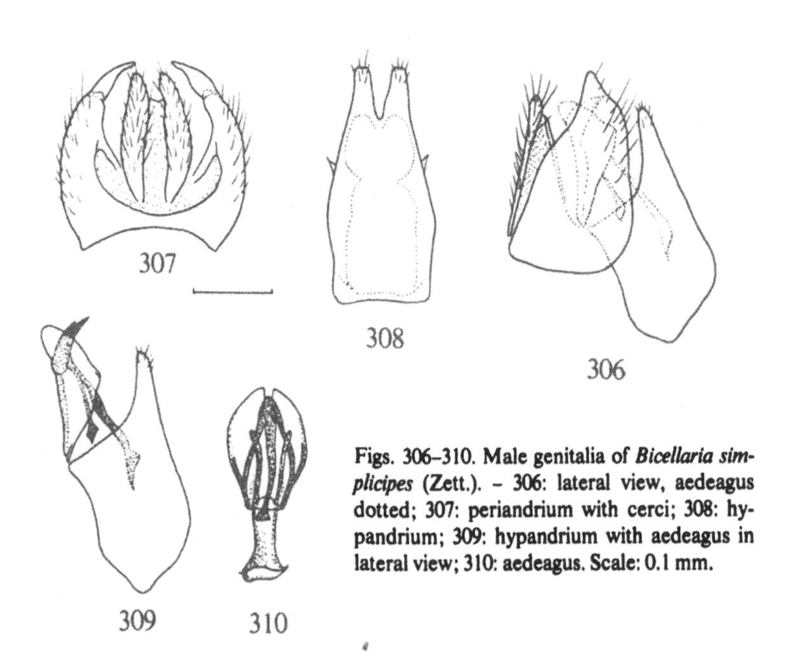

Figs. 306–310. Male genitalia of *Bicellaria simplicipes* (Zett.). – 306: lateral view, aedeagus dotted; 307: periandrium with cerci; 308: hypandrium; 309: hypandrium with aedeagus in lateral view; 310: aedeagus. Scale: 0.1 mm.

13. ***Bicellaria pilosa*** Lundbeck, 1910
Figs. 266, 278, 290, 311–315.

Cyrtoma spuria Fallén; Zetterstedt, 1842: 329.
Bicellaria pilosa Lundbeck, 1910: 27.

Conspicuously long black bristled species everywhere, 3rd ant.s. with 1 or 2 small bristles above and palpi with several strong black bristles. Face very broad.

♂. Face very broad and not narrowed below, broader than in any other species (Fig. 266), palpi with several (more than 3) strong black bristles. 3rd ant.s. (Fig. 278) with 1 or 2 small bristly hairs dorsally. Occiput clothed with densely-set, long black bristles. Mesonotum mostly dull black and with long black bristles, 4–6 scutellars.

Wings brownish black, squamae and halteres blackish. Legs not very long and slender, everywhere clothed with long black bristles. Hind tibiae dilated apically, hind metatarsus slightly thickened, following segments becoming shorter and narrower (Fig. 290). Anterior four femora with long pd and pv bristles, shorter av bristles on basal half, tibiae with distinct pv ciliation and several long, strong pd and ad bristles much longer than tibia is deep. Hind femur with densely-set long dorsal and av bristles, tibiae strongly bristled above.

Abdomen dull brownish black and densely clothed with long black bristles and hairs. Genitalia (Figs. 311–315) with hypandrial prongs rather slender but shorter than those of *intermedia*, short spinose apically.

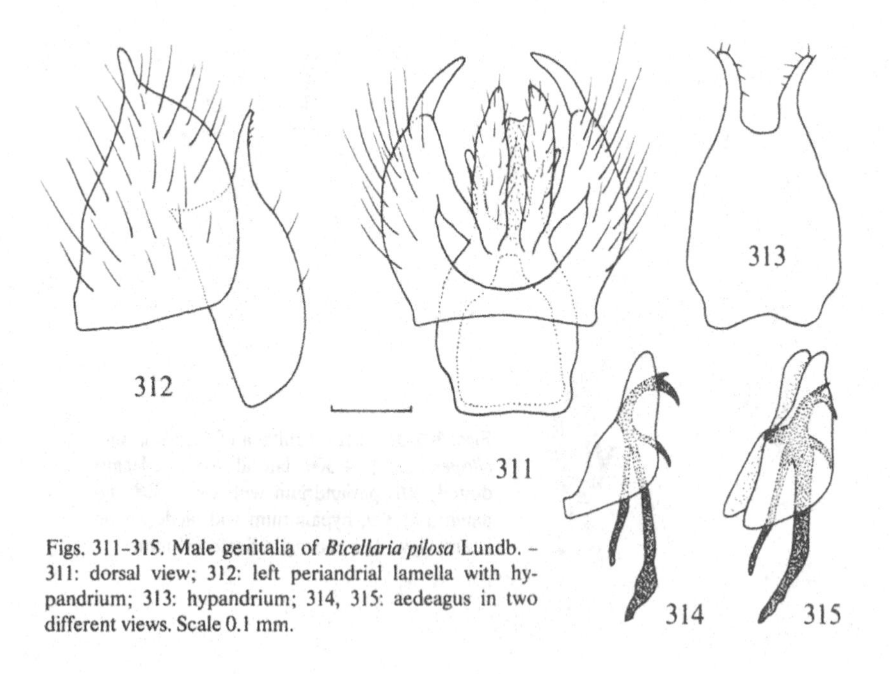

Figs. 311–315. Male genitalia of *Bicellaria pilosa* Lundb. – 311: dorsal view; 312: left periandrial lamella with hypandrium; 313: hypandrium; 314, 315: aedeagus in two different views. Scale 0.1 mm.

Length: body 2.9–3.3 mm, wing 2.8–3.3 mm.

♀. Resembling male very closely but generally larger, wings scarcely lighter brownish, and abdomen pointed with shorter pubescence.

Length: body 3.2–4.0 mm, wing 3.0–3.5 mm.

Distribution and biology. Not uncommon in Scandinavia but more abundant in the central and northern parts, including extreme north and the Kola Peninsula. - Great Britain, east to the north of the European USSR, rather common in mountains of Central Europe (Carpathians, Alps) and Romania (Mt. Rodnei). - From May to mid August (in the north) but mainly in June and first half of July.

Note. Lundbeck (1910) described this species from Danish specimens taken at Tisvilde (NEZ), Silkeborg (EJ) and Frederikshavn (NEJ) but all the syntypes have been destroyed by dermestids; the oldest material preserved in Lundbeck's collection dates from 1924 (Rebild, NEJ).

14. ***Bicellaria austriaca*** Tuomikoski, 1955
Figs. 267, 279, 291, 316–320, 321, 323–326.

Bicellaria austriaca Tuomikoski, 1955: 70.

Antennae with 1 or 2 small hairs on 3rd segment above. Resembling *pilosa* but a less bristled and longer-legged species, face narrower and palpi with 1 or 2 preapical bristles. Abdomen clothed with black bristly hairs.

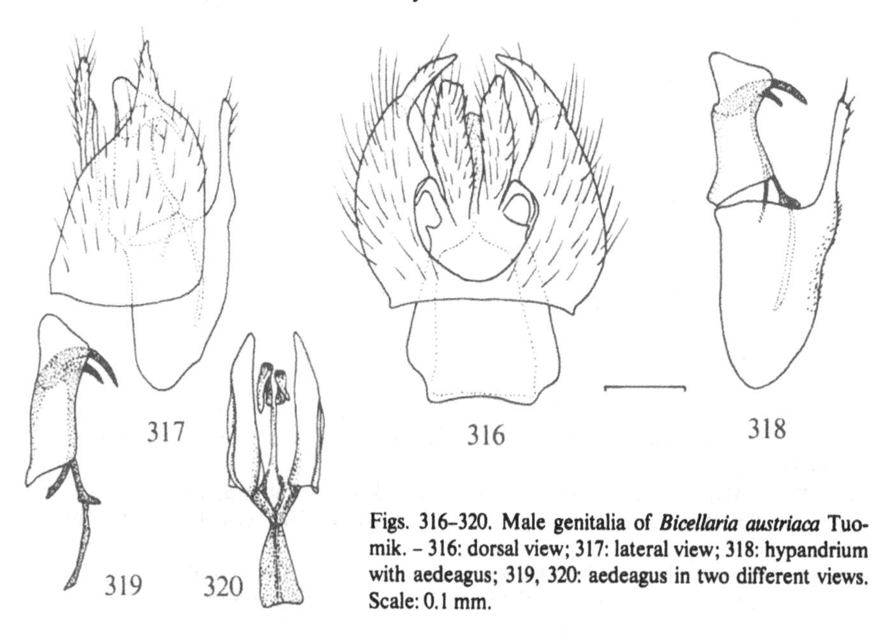

Figs. 316–320. Male genitalia of *Bicellaria austriaca* Tuomik. - 316: dorsal view; 317: lateral view; 318: hypandrium with aedeagus; 319, 320: aedeagus in two different views. Scale: 0.1 mm.

♂. Antennae (Fig. 279) as in *pilosa* or *intermedia* with 1 or 2 small hairs on 3rd segment above, but face (Fig. 267) moderately broad and almost parallel, scarcely as broad as 3rd ant.s. is deep. Palpi usually with 2 bristly hairs. Occiput with shorter and less numerous hairs than in *pilosa.* Mesonotum dull brownish black even on pleura, 4 scutellar bristles.

Wings dark brown, squamae and halteres blackish. Legs longer and less bristled than in *pilosa,* anterior four femora with long pv bristles (longer on mid tibia) but dorsally with only short adpressed bristly hairs; no av bristles except for 2 or 3 at base. Fore tibiae with fine pv ciliation shorter than tibia is deep, all tibiae with strong dorsal bristles, 2 or 3 pairs on anterior tibiae, numerous bristles on hind tibiae. Mid femora rather

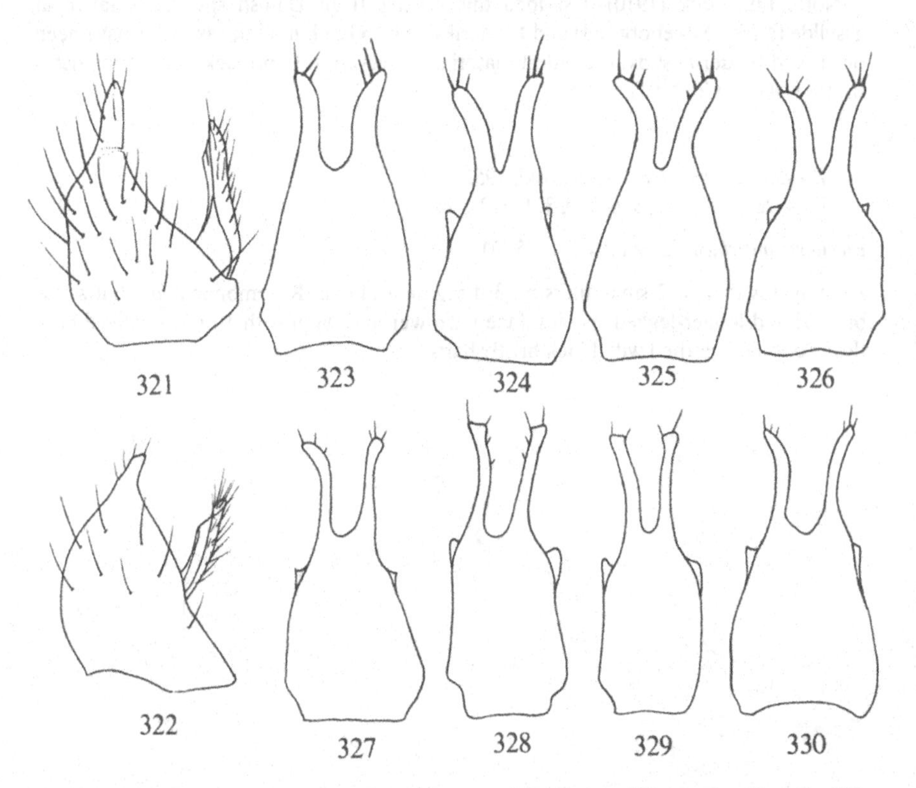

Figs. 321, 322. Right periandrial lamella with cercus of *Bicellaria.* – 321: *austriaca* Tuomik.; 322: *intermedia* Lundb.

Figs. 323–326. Hypandrium of *Bicellaria austriaca* Tuomik. – 323: Bohemia, Starkoc; 324: Dania, Bornholm; 325: Dania, Hald (syntype of *intermedia* Lundb.); 326: Dania, Tisvilde (syntype of *intermedia* Lundb.).

Figs. 327–330. Hypandrium of *Bicellaria intermedia* Lundb. – 327: Dania, Ermelund (lectotype); 328: Slovakia, Bel.Tatry; 329: Slovakia, Bel.Tatry; 330: Bohemia, Krkonose. Scale: 0.1 mm.

stouter on basal half, the tibiae half their length. Hind femora with long dorsal and almost as long av bristles. Hind tibiae (Fig. 291) evenly dilated apically, metatarsus and 2nd tarsal segment lengthened and scarcely stouter than following segments.

Abdomen subshining blackish brown, clothed with rather long but sparse bristly hairs, rarely with a few paler hairs at extreme base. Genitalia (Figs. 316–320) with hypandrial prongs somewhat shorter and stouter than in *intermedia*, armed apically with 3–4 long spine-like bristles (Figs. 323–326).

Length: body 2.6–3.7 mm, wing 3.0–3.6 mm.

♀. Resembling male very closely, wings scarcely lighter brown and halteres often rather blackish brown. Abdomen with sparse blackish pubescence, the hairs with a tendency to be paler near base.

Length: body 2.8–4.0 mm, wing 2.7–3.8 mm.

Holotype identification. The species was described from numerous specimens from Central European mountains. The holotype male, labelled "Typus 8202, Österreich, Obertilliach, R. Frey", is in Helsinki Museum and was studied by the author in 1980. There are 4 ♂ of *austriaca* from Tisvilde (NEZ) and 1 ♂ from Hald (EJ), Denmark (leg. Lundbeck) in Copenhagen Museum, which were identified by Lundbeck and included in the syntypic series of *B. intermedia* Lundb.

Distribution and biology. Uncommon southern species (35 specimens examined), known from Denmark and the south of Sweden, extending north to southern Norway (HOi) to approximately 60°N; not found in Finland. – A continental species known only from C. Europe, where it is very common especially at higher altitudes; clearly the commonest *Bicellaria* species in C. Europe, east to central parts of European USSR (Kovalev, in litt.), south to Bulgaria (Rila). – Mid June to August, in temperate Europe from early June to September, rarely to the beginning of October.

15. ***Bicellaria subpilosa*** Collin, 1926
Figs. 268, 280, 292, 331–334.

Bicellaria subpilosa Collin, 1926: 190.

Rather a robust and densely black bristled species, resembling *pilosa* but with the main differential features (antennae, palpi) of *spuria;* face rather broad, fore tibiae with distinct pv ciliation and mid femora with long av bristles.

♂. Face (Fig. 268) rather broad and parallel, scarcely narrower than in *austriaca*, but 3rd ant.s. (Fig. 280) without any bristly hair above. Palpi usually with 2 fine preapical hairs, occiput with long but rather sparse bristles. Mesonotum subshining brownish black, rather densely long bristled, usually 6 scutellar bristles.

Wings dark brownish grey, squamae and halteres blackish. Legs rather slender and distinctly bristled, although not as much as in *pilosa*. Fore femora with distinct but rather fine pv bristles and still smaller av bristly hairs, mid femora with very long pv bristles and, compared with *spuria*, with distinct shorter av bristles. Anterior tibiae with several strong dorsal bristles, fore tibiae with long pv ciliation, the hairs at least as long as tibia is deep. Hind femora with distinct recumbent bristles dorsally and much longer av bristles, pv bristles smaller and more numerous. Hind tibiae (Fig. 292) rather weakly dilated towards tip, ventrally with dense, rather recumbent bristles which are at least as long as tibia is deep, the hairs distinctly longer than those of *spuria* and even *pilosa.* Hind tarsi quite slender.

Abdomen rather subshining black with a slight coating of brownish dust, clothed with long black bristles especially at sides and below. Genitalia (Figs. 331–334) with hypandrial prongs rather long and slender.

Length: body 2.8–3.7 mm, wing 2.6–3.5 mm.

♀. Very much like male, the av bristles on mid femora less distinct and confined to base only, but hind tibiae with long bristles ventrally as in male.

Length: body 2.8–3.5 mm, wing 2.6–3.3 mm.

Distribution and biology. Very common throughout Scandinavia including the extreme north. – Widespread in Europe, Great Britain (mainly Scotland), C. Europe, south to Romania (Mt. Rodnei), Bulgaria (Leningrad, Repino) and Greece (Griechische Inseln, coll. Loew). Tuomikoski (1955) also recorded it from the Caucasus. – May to July, rather an early summer species, very rarely to mid August; one specimen taken in C. Europe (Czechoslovakia) on 13th November undoubtedly represents an accidental 2nd generation.

Note. In Coll. Loew in Berlin *B. subpilosa* is placed under Loew's MS name "*longipes* m.": there is a series of 22 specimens from Greece, the two Germanies and the European USSR including the Kola Peninsula.

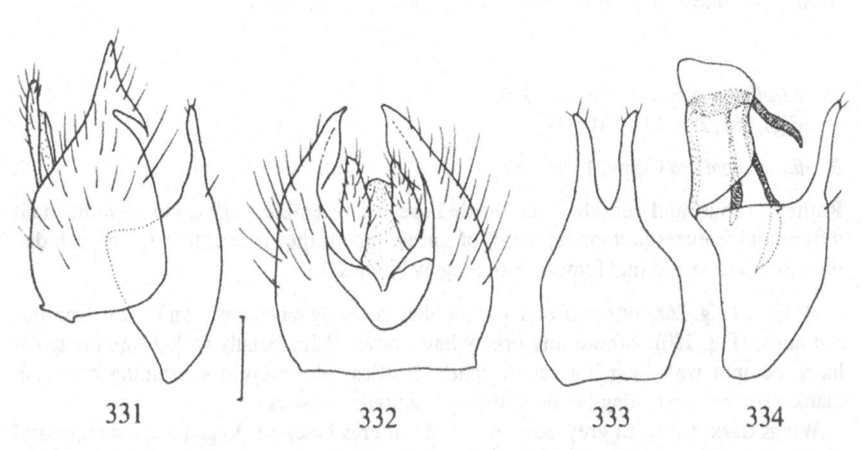

Figs. 331–334. Male genitalia of *Bicellaria subpilosa* Coll. – 331: lateral view, aedeagus omitted; 332: periandrium with cerci; 333: hypandrium; 334: hypandrium with aedeagus, lateral view. Scale: 0.1 mm.

16. ***Bicellaria bisetosa*** Tuomikoski, 1936
Figs. 269, 281, 294–296, 335–340.

Bicellaria bisetosa Tuomikoski, 1936: 84.

Subshining black, thinly greyish brown species with body and legs rather densely black bristled, 3rd ant.s. without a bristle above. Face rather broad, hind legs with tibiae dilated apically and basal two tarsal segments with a tendency to be slightly stouter than the following segments. Fore tibiae with a distinct pv ciliation, fore trochanters with 2 bristles above.

♂. Face rather broad and parallel-sided (Fig. 269), although not as broad as in *pilosa* or *subpilosa*. Antennae (Fig. 281) with 3rd segment bare above, rather broad at base and very narrowed apically. Palpi with one long, and usually another small, bristly hair at tip. Mesonotum subshining black but densely greyish to greyish brown dusted, not so densely and long bristled as in *subpilosa;* 4–6, very rarely 8 scutellar bristles.

Wings deep brown clouded but paler on lower half, squamae and halteres blackish. Hind tibiae (Fig. 294) considerably dilated apically (more slender in *subpilosa* and *spuria*), basal two tarsal segments long and apparently stouter than apical three segments. Fore femora (Fig. 295) with only fine small av and pv bristly hairs, pd bristles stouter but very adpressed, 2 distinct diverging bristles on fore trochanters above. Mid femora with long pv bristles, longer than femur is deep, and with shorter av bristles.

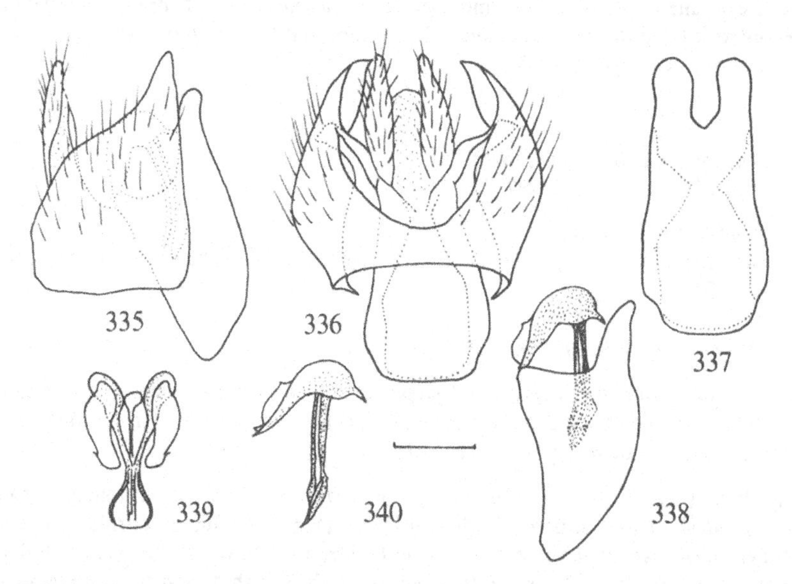

Figs. 335–340. Male genitalia of *Bicellaria bisetosa* Tuomik. – 335: lateral view, aedeagus dotted; 336: dorsal view; 337: hypandrium; 338: hypandrium with aedeagus in lateral view; 339, 340: aedeagus in two different views. Scale: 0.1 mm.

Fore tibiae (Fig. 296) with pv hairs about as long as tibia is deep, dorsally with numerous long bristly hairs, the few strong dorsal bristles hardly differentiated. Mid tibiae shorter pubescent above, 2 or 3 pairs of strong dorsal bristles much longer; ventral ciliation as long as tibia is deep. Hind femora with long ad bristles, those on basal half longer than femur is deep, about equally long av bristles, but pv bristles very inconspicuous, practically absent. Hind tibiae with numerous strong bristles dorsally, ventral ciliation distinct, although not as long as in *subpilosa*.

Abdomen subshining like thorax and also greyish brown dusted above, all bristles black, more numerous and longer at sides and on venter. Genitalia (Figs. 335–340) with remarkably short and stout hypandrial prongs closely resembling those of *simplicipes*.

Length: body 2.8–3.3 mm, wing 2.6–3.1 mm.

♀. Entirely resembling male except for sexual characters.

Length: body 3.0–3.7 mm, wing 3.0–3.5 mm.

Lectotype designation. The syntypic series in the Helsinki Museum includes 11 ♂ and 3 ♀ from Kuusamo (Ks) and Suistamo (Kr); one male labelled "Suistamo, Tuomikoski" was selected in 1980 and is herewith designated as lectotype of *Bicellaria bisetosa* Tuomik.

Distribution and biology. A northern species known only from N. Sweden (Nb., Lu. Lpm. and T. Lpm.), N. Finland (Ks, Li), Russian Karelia; and 1 ♀ taken in S. Finland, in Helsinki (N). Rather a rare species in Fennoscandia, and only 18 ♂ and 9 ♀ have been examined. – Unknown to me outside Fennoscandia but Tuomikoski (1955) also recorded it from the Ussuri region. – In swamps and bogs in July (6–31), in Ussuri as early as mid June (Tuomikoski, 1955).

17. ***Bicellaria spuria*** (Fallén, 1816)
Figs. 261–264, 270, 282, 297, 298, 341–347.

Empis spuria Fallén, 1816: 33.
Bicellaria nigra Macquart, 1823: 155 (nec Meigen, 1824); for the synonymy see Collin, 1961: 254.
Cyrtoma atra Meigen, 1824: 2.
Cyrtoma melaena Haliday, 1833: 158.

Resembling a small and less bristled *subpilosa*, with similar pv ciliation on fore tibiae. 3rd ant.s. very constricted on apical half, hind tibiae rather slender, although more dilated apically than in *simplicipes* and *mera*.

♂. Face (Fig. 270) not very broad, not as narrow as in *sulcata* but much narrower than in *subpilosa*, almost parallel-sided. 3rd ant.s. (Fig. 282) without a bristle above, the slender apical part long, quite half length of style, and the ventral bristles on 2nd segment moderately long. Thorax rather dull black, brownish dusted from some angles, mesonotal bristles rather long and strong, although less developed than in *subpilosa;* usually 4 scutellar bristles, seldom 6 bristles.

Wings brownish with medial fork (M_1 and M_2) very incomplete, halteres and squamae blackish. Legs rather slender and less bristled than in *subpilosa* or *bisetosa*, hind tibiae (Fig. 297) slightly dilated apically although at least as deep at tip as the thickest part of hind femur. Fore tibiae (Fig. 298) with a complete pv ciliation, the hairs as long as tibia is deep, dorsally with 1 or 2 strong bristles much longer than other dorsal bristly hairs. The double row of ventral bristles on anterior four femora rather fine and small, av bristles on mid femora practically absent (distinct in *subpilosa*) except for 2 or 3 hairs at base. Mid tibia which several long bristles above, and the ventral adpressed bristles on hind tibiae much shorter than in *subpilosa*. Hind femora (Fig. 297) distinctly bristled above and below, av bristles longest, almost as long as femur is deep. Tarsi quite slender.

Abdomen dull brownish black, covered with black bristles, hind marginal and lateral bristles long. Genitalia as in Figs. 341–347, with hypandrial prongs conspicuously long and slender.

Length: body 2.3–2.8 mm, wing 2.3–2.6 mm.

♀. Head, wings and legs as in male but thorax and abdomen more greyish, rather subshining greyish black, abdomen very pointed and clothed with shorter black bristly hairs.

Length: body 2.3–3.1 mm, wing 2.3–2.9 mm.

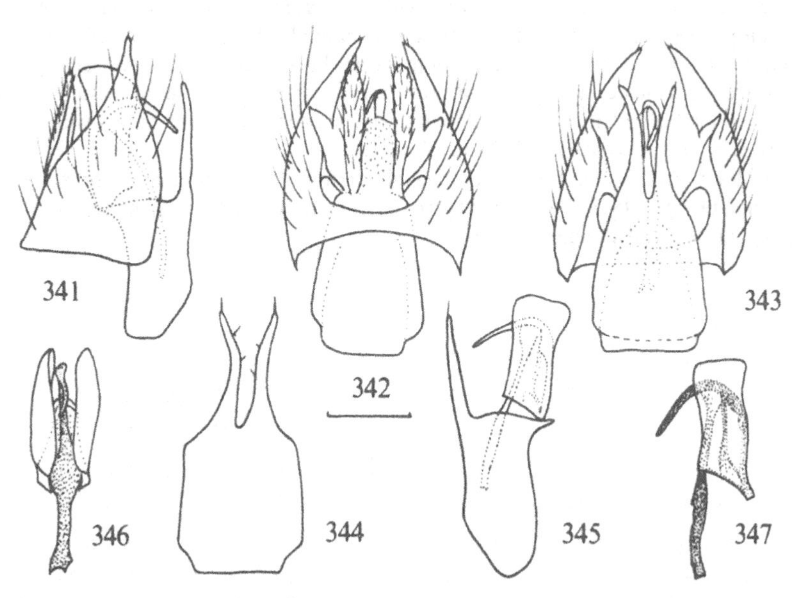

Figs. 341–347. Male genitalia of *Bicellaria spuria* (Fall.). – 341: lateral view; 342: dorsal view; 343: ventral view; 344: hypandrium; 345: hypandrium with aedeagus in lateral view; 346, 347: aedeagus in two different views. Scale: 0.1 mm.

Distribution and biology. Widespread and very common throughout Scandinavia, including the extreme north. – Common in temperate Europe, south to Bulgaria (Vitoscha, Strandja); Strobl (1899) recorded it from Spain, but according to Collin (1961) it is rather a northern species in Great Britain; also N. America, Ontario and Mass. (Melander, 1965). – Mid May to August, rarely September, 1 ♂ from Denmark from 13.x. (Tyvekrog, NEZ); mostly June and July.

18. ***Bicellaria mera*** Collin, 1961
Figs. 271, 283, 299, 348–350.

Bicellaria mera Collin, 1961: 258.

Hind tibiae rather slender but not as much as in *simplicipes*, other main characters as in *spuria*. Antennae with slender apical part of 3rd segment short, hypandrial prongs in ♂ rather short and stout, very much as in *sulcata*.

♂. Head including face (Fig. 271) as in *spuria* but 3rd ant. s. (Fig. 283) less constricted before tip, its slender apical part much shorter than half length of style; the bristles beneath 2nd segment obviously shorter. Thorax very dull black, brownish dusted when viewed from in front, thoracic bristles shorter than in *spuria*; 4 scutellars.

Wings rather brownish grey clouded, not at all deep brown. Squamae and halteres black. Legs with pv ciliation on fore tibiae as in *spuria*, but hind tibiae (Fig. 299) more slender, not so deep at tip as the thickest part of hind femur. Legs obviously with fewer and weaker bristles than in *spuria*, large strong bristles confined to dorsal bristles on posterior four tibiae, those on hind tibiae more numerous, and hind femora with distinct pv bristles not much shorter than femur is deep. Fore trochanters with a single bristle above as in *spuria* (2 bristles in *bisetosa*).

Abdomen rather dull black with a slight coating of brownish pile, all bristles black. Genitalia (Figs. 348–350) with hypandrial prongs decidedly shorter and stouter than in *spuria*, armed with a single seta at tip.

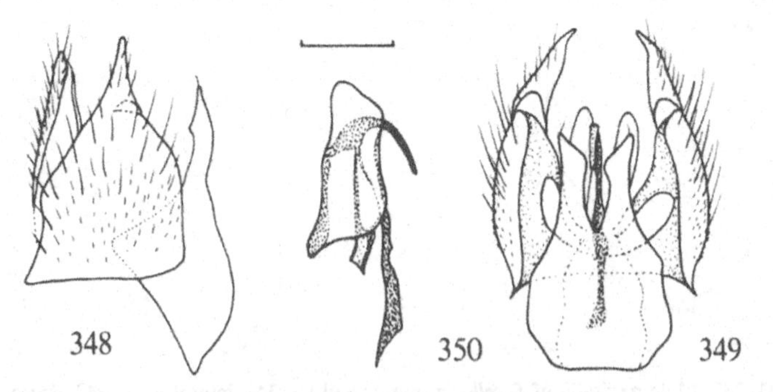

Figs. 348–350. Male genitalia of *Bicellaria mera* Coll. – 348: lateral view, aedeagus omitted; 349: ventral view, 350: aedeagus, lateral view. Scale: 0.1 mm.

Length: body 2.3–2.8 mm, wing 2.3–2.7 mm.

♀. Resembling male very closely but thorax and abdomen more subshining black, only very thinly brownish grey dusted.

Length: body 2.4–3.0 mm, wing 2.4–2.9 mm.

Distribution and biology. Very rare in Denmark; 1 pair from Saxkøbing, Lolland (20.viii.1964, N. M. Andersen) and a ♀ from "Sjælland" (vi.1819, Mus. Westerm.). – England (not Scotland) and C. Europe (Czechoslovakia, Hungary). – Outside Scandinavia from end May to August.

Note. The rather slender hind tibiae only resemble those of *B. simplicipes;* however, in *simplicipes* the tibiae are not at all dilated, the face is very narrowed below, the medial fork on wings is almost complete, and abdomen is clothed with mostly pale hairs.

19. ***Bicellaria sulcata*** (Zetterstedt, 1842)
Figs. 272, 284, 300, 351–354.

Cyrtoma sulcata Zetterstedt, 1842: 331.
Bicellaria sulcata ssp. *vanella* Tuomikoski, 1936: 82 – **syn.n.**

3rd ant.s. without a bristle above, face very narrowed below and hind tibiae considerably thickened apically. Fore tibiae with microscopic pile beneath, pv ciliation absent.

♂. Head with a large incision opposite base of antennae, which is gradually narrowed below to a very narrow face (Fig. 272) resembling that of *nigra.* Apical slender part of 3rd ant.s. rather short (Fig. 284). Mesonotum black but distinctly brown dusted, not dull black from any point of view, and with 2 narrow greyish stripes between acr and dc when viewed from behind. All thoracic bristles black and rather weak, humeral and posthumeral bristles not differentiated from small weak bristles, 4–6, rarely 8 scutellars.

Wings brownish to brown-grey, squamae and halteres blackish. Legs not very long but rather slender, hind tibia (Fig. 300) distinctly dilated apically, wider at tip than the widest part of hind femur, basal two hind tarsal segments slightly stouter than following segments. Fore tibiae clothed with a soft pile beneath as in *nigra,* the pv hairs fine, pale and very inconspicuous, dorsally with longer bristly hairs about as long as tibia is deep. Anterior four femora with rather short bristly hairs beneath, pv bristles on mid femur the longest, av bristles very inconspicuous.

Abdomen subshining black from some angles, thinly brown dusted and with rather long black bristles. Genitalia (Figs. 351–354) with hypandrial prongs rather short and stout.

Length: body 2.7–3.3 mm, wing 2.6–3.1 mm.

♀. Resembling male in all essential characters, thorax dulled by brownish dust, not subshining black as in *vana.*

Length: body 2.8–3.5 mm, wing 2.7–3.1 mm.

Lectotype designations. *B. sulcata* was described from 3 specimens from S. Sweden and Denmark, from 1 ♂ from Arrendala (Scania, Dahlbom) and a pair sent by Stæger from Denmark. All specimens are in the Dipt. Scand. Coll. at Lund under *Cyrtoma sulcata*. The ♂ labelled "Arrendala, Dahlb. Jun. 38 Scania" is *B. spuria* ♂, but the Danish pair labelled "Staeg." represents *B. sulcata* as described above. The male sent by Stæger (No. 116) was labelled in 1977 and is herewith designated as lectotype of *B. sulcata* (Zett.).

B. sulcata ssp. *vanella* was described by Tuomikoski from males (number not stated) taken in Helsinki and Nilsiä, Finland. In the Palaearctic collection in Helsinki Museum there is 1 ♂ of *sulcata* labelled "*Bicellaria vanella* Tuomik. ♂, Helsinki, Tuomikoski, Spec. typ. No. 8221", which was labelled in 1980 and is herewith designated as lectotype of *B. sulcata* ssp. *vanella* Tuomik. In the Finnish Coll. in Helsinki there is a box with 4 ♀(Suistamo, Tuomik. 3 ♀, Helsinki, Tuomik. 1 ♀) labelled "*vanella* n. sp." – evidently not types as the ♀ sex was not described. The 3 ♀ from Suistamo, in rather

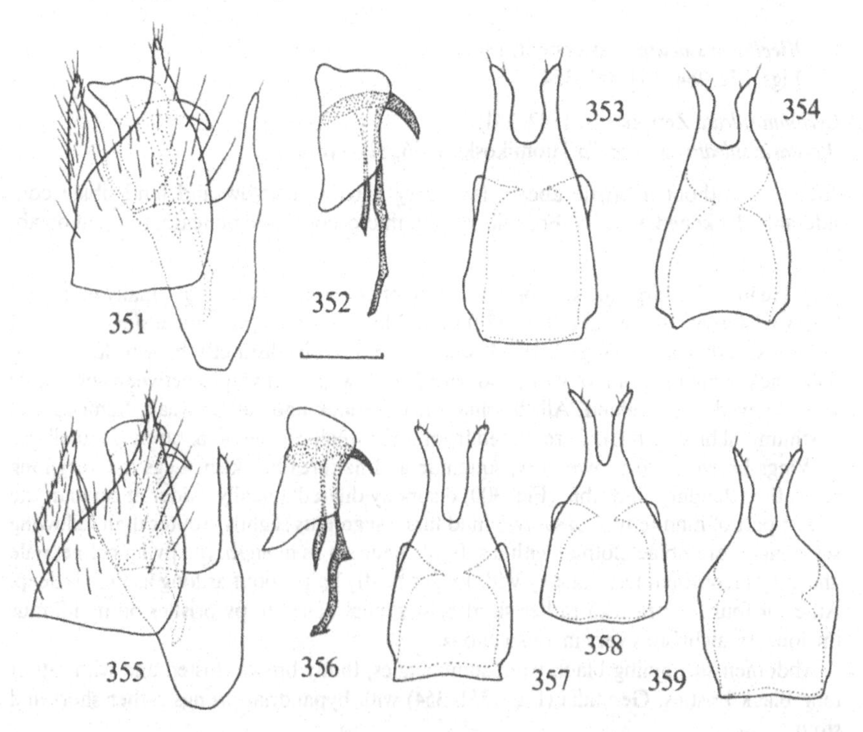

Figs. 351–354. Male genitalia of *Bicellaria sulcata* (Zett.). – 351: lateral view; 352: aedeagus; 353: hypandrium (Bohemia, Vysoké Tatry); 354: the same (Dania, Frerslev hegn).
Figs. 355–359. Male genitalia of *Bicellaria vana* Coll. – 355: lateral view; 356: aedeagus; 357: hypandrium (Bohemia, Novohradské hory); 358: the same (Dania, Bornholm); 359: the same (England, Hunts). Scale: 0.1 mm.

bad condition, are females of *spuria*, and the single ♀ from Helsinki (fore legs and right antenna broken) is very probably a ♀ of *sulcata*.

Distribution and biology. Throughout Fennoscandia and Denmark, including the extreme north and the Kola Peninsula, but much less common than *spuria* or *subpilosa*. – Scotland, absent in England according to Collin (1961); rather uncommon but widespread in Europe, known to me from C. Europe, south to Spain, Bulgaria and the Caucasus; also Siberia (Yakutsk) according to Tuomikoski (1955). – Flight period long in Scandinavia, from early May to late October.

20. ***Bicellaria vana*** Collin, 1926
Figs. 273, 285, 301, 355–359.

Bicellaria vana Collin, 1926: 190.

Very much like *sulcata*, legs somewhat shorter and stouter, with shorter bristling. Mesonotum in ♂ mostly dull black and genitalia very different, hypandrial prongs longer and slender.

♂. Head (Fig. 273) as in *sulcata*, mesonotum almost dull black when viewed from above, uniformly thinly blackish brown dusted when viewed from behind. Abdomen also deeper black, and general coloration of *vana* much blacker than in *sulcata;* thoracic pleura thinly blackish grey dusted; scutellum and prescutellar depression more brownish grey dusted. Thorax obviously more humped than in *sulcata* but with the same bristling.

Wings often lighter greyish, squamae and halteres black. Legs perhaps slightly shorter and stouter than in *sulcata*, hind tibiae (Fig. 301) considerably dilated apically as in *sulcata* but corresponding tarsi indistinctly shorter and more slender (2nd tarsal segment in particular). Fore tibiae generally stouter than in *sulcata*, with similar soft pile beneath but the pv pale ciliation confined to only a few minute hairs near base; however, dorsal bristly hairs in some specimens as long as in *sulcata*, scarcely shorter.

Genitalia (Figs. 355–359) very distinctive, hypandrial prongs much longer and more slender, a feature which is quite constant among different populations.

Length: body 2.6–3.3 mm, wing 2.5–3.0 mm.

♀. The female sex is very difficult to differentiate from that of *sulcata:* apart from the slight differences in the legs, as in ♂, the thorax is more subshining black and the abdomen is also blacker.

Length: body 2.5–3.5 mm, wing 2.8–3.3 mm.

Distribution and biology: Uncommon in Denmark and S. Sweden, north to Norway (On), approximately 61°N. Southern species, not found in Finland. – England, where it is the commonest *Bicellaria* species according to Collin (1961), very common also in C. Europe (Czechoslovakia, Hungary, Austria, Switzerland) and much commoner than *sulcata;* Kovalev (in litt.) has recorded it from central parts of European USSR. – June to September, in England and C. Europe from mid May to the end of October.

21. ***Bicellaria intermedia*** Lundbeck, 1910
Figs. 274, 286, 293, 322, 327–330, 360–363.

Bicellaria intermedia Lundbeck, 1910: 25.

Legs rather long and slender (not as long as in *nigra*), 3rd ant.s. with a bristle above. Fore tibiae with a short pv ciliation, basal two segments of hind tarsi dilated. Base of abdomen with pale hairs.

♂. Face (Fig. 274) rather broader but distinctly narrowing below, lower occipital bristles pale. 3rd ant.s. (Fig. 286) with a fine bristle above at base, 2nd segment with two equally long bristles beneath; apical slender part of 3rd segment long, half length of style. Thorax brownish grey dusted, mesonotum almost dull black when viewed from above, thoracic bristles black, rather long and strong, 4 scutellars.

Wings brownish to brownish grey, squamae and halteres blackish, latter sometimes almost light brownish. Legs not very long but very slender, particularly anterior four femora, mid femora not very much stouter than corresponding tibiae. Hind tibiae (Fig. 293) slightly dilated apically, about as deep at tip as hind femur, basal two tarsal segments slightly swollen, although not as much as in *nigra* and its allies. Fore femora with fine av (on basal half), pv and pd bristles, the pv the longest, not much shorter than femur is deep. Mid femora with long pv bristles, but practically no av bristles except for a few hairs at base. Hind femora with long ad and av bristles longer than femur is deep. Anterior tibiae with several strong bristles dorsally, hind tibiae densely long bristled above.

Abdomen clothed with rather weak but long, mostly pale bristles, thinly brownish grey dusted, subshining from some angles. Genitalia (Figs. 322, 327–330, 360–363) with hypandrial prongs long and slender, aedeagus apically bifid.

Length: body 2.5–3.3 mm, wing 2.5–3.2 mm.

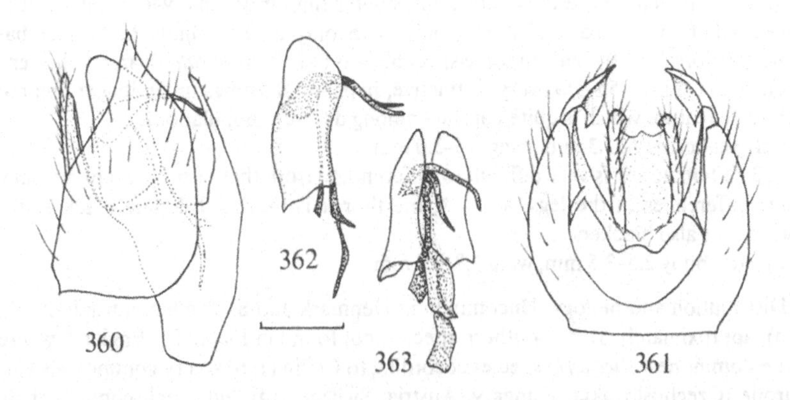

Figs. 360–363. Male genitalia of *Bicellaria intermedia* Lundb. – 360: lateral view; 361: periandrium with cerci; 362, 363: aedeagus in lateral and anterolateral views. Scale: 0.1 mm.

♀. Resembling male very closely except for sexual differences, halteres yellowish and abdominal pubescence shorter and mostly pale.
Length: body 2.7–3.5 mm, wing 2.9–3.2 mm.

Lectotype designation. The syntypic series in the Copenhagen Museum includes 9 ♂ and 29 ♀ of *intermedia* and 5 ♂ of *austriaca*, collected at many Danish localities. One male labelled "Ermelund, 20.viii.1909, Lundbeck" was selected in 1977 and is herewith designated as lectotype of *B. intermedia* Lundb.

Distribution and biology. Common species in southern Scandinavia, particularly abundant in Denmark and S. Finland, extending north to central Scandinavia at approximately 62°N, in Sweden to Jmt., in Finland to Tb and Sb, very rarely to Ks (4 specimens). – Great Britain, east to NW and central European USSR, rather common in C. Europe, south to Bulgaria (Boianski vodopad, 18.vii.1972, Beschovski).– Mid June to September.

Note. *B. intermedia* closely resembles the following species, *nigra;* however, the latter has 3rd ant.s. without a bristle above, face much narrower, fore tibiae clothed with only soft pile beneath (no pv bristly hairs), hind tibiae longer and much more slender on basal half, and basal two tarsal segments of hind legs more swollen, almost as deep as tip of tibia. *B. austriaca,* having 3rd ant.s. with a dorsal bristle as in *intermedia,* has broader parallel-sided face, hind tarsal segments only indistinctly dilated, and is generally a more densely black bristled species with abdominal pubescence extensively black.

22. ***Bicellaria nigra*** (Meigen, 1824)
Figs. 275, 287, 302, 303, 364–366.

Cyrtoma nigra Meigen, 1824: 3.
Cyrtoma rufa Meigen, 1824: 3.

Large, long-legged, and with extensive pale bristles on base and venter of abdomen. Basal two segments of hind tarsi dilated, hind tibiae very elongate and dilated on apical half only. Antennae without a dorsal bristle on 3rd segment, face very narrow.

♂. Face (Fig. 275) very narrow, gradually narrowing towards mouth, occipital bristles pale below neck. Antennae (Fig. 287) very much as in *intermedia,* 3rd segment slender apically but without a dorsal bristle near base, and 2nd segment with only short bristles beneath. Upper eye-facets very enlarged, more than usual in this genus. Thorax brownish dusted but somewhat shining when viewed from above, all bristles black, 4 scutellars.

Wings large, brownish grey clouded, deeper brown along costa, squamae dark with pale fringes, halteres blackish. Legs conspicuously long and slender, hind pair in particular; hind tibiae (Fig. 302) very slender on basal half, dilated apically and basal two tarsal segments distinctly swollen. Anterior four femora with rather fine short bristly hairs both below and above, tibiae (Fig. 303) black bristled above and 1 or 2 bristles very strong and longer than other bristles. Fore tibiae with soft pale pile beneath, no pv ciliation. Hind femora long and slender, with rows of long black bristles beneath and above, dorsal bristles on basal half the longest, more than twice as long as femur is deep.

Abdomen shining brownish black with brownish bronze pollen, dorsum mostly black bristled but extreme base, sides on basal half and practically whole venter clothed with long pale bristly hairs. Genitalia (Figs. 364–366) with hypandrial prongs rather slender but short and diverging, yellowish towards tip; aedeagus very long and slender, not bifurcated.

Length: body 3.3–3.8 mm, wing 3.2–3.6 mm.

♀. Very much like male but squamae and halteres yellowish, abdomen with predominantly pale hairs and fore tibiae slightly spindle-shaped.

Length: body 3.2–4.0 mm, wing 3.2–3.6 mm.

Distribution and biology. Rather common throughout Scandinavia including the ex-

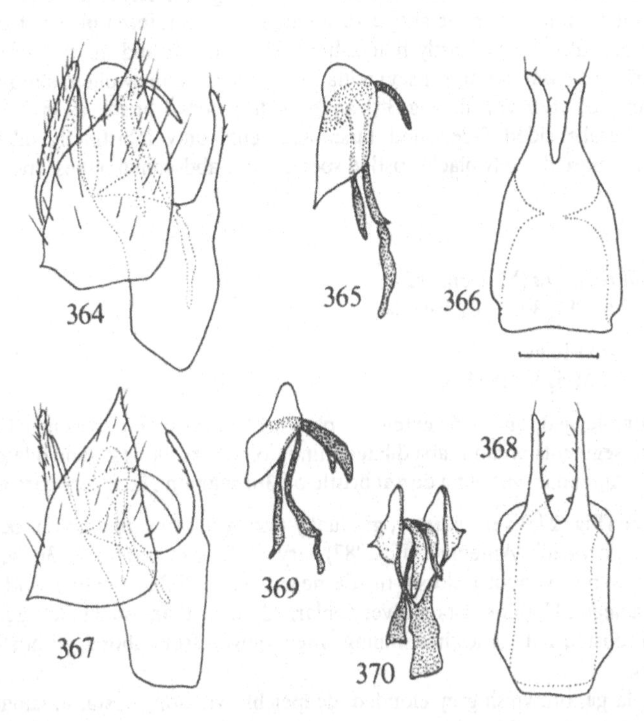

Figs. 364–366. Male genitalia of *Bicellaria nigra* (Meig.). – 364: lateral view; 365: aedeagus in lateral view; 366: hypandrium.

Figs. 367–370. Male genitalia of *Bicellaria nigrita* Coll. – 367: lateral view; 368: hypandrium; 369, 370: aedeagus in lateral and anterolateral views. Scale: 0.1 mm.

treme north of Norway and the Kola Peninsula. - Widespread and not uncommon in Europe from Great Britain through the central parts south to Spain and the Caucasus. - May to mid September, in Great Britain (Collin, 1961) to early October.

Bicellaria nigrita Collin, 1926
Figs. 276, 288, 304, 305, 367–370.

Bicellaria nigrita Collin, 1926: 190.

Very much like *nigra* but generally smaller, legs not as long and slender, hind tibiae more evenly swollen and all bristles on legs and body darker and longer.

♂. Head (Fig. 276) as in *nigra* but 2nd ant.s. (Fig. 288) beneath with longer bristles (not as long as in *intermedia*) and lower part of occiput with darker bristly hairs. Thorax not so subshining, duller black when viewed from above, all bristles somewhat longer and stronger.

Wings as in *nigra* but distinctly shorter and squamae with black fringes. Legs generally shorter but not very much stouter, although hind tibiae (Fig. 304) decidedly stouter on basal half, more evenly dilated towards tip; hind basal tarsal segments swollen as in *nigra*. Legs more bristled, hind femora with longer av and dorsal bristles, latter unusually long on basal half of femur, and anterior four tibiae (Fig. 305) dorsally with longer black bristles, so that the few strong bristles (prominent in *nigra*) are only slightly differentiated. Fore tibiae with similar soft pile beneath, compared with *nigra* also with short pale pv ciliation.

Abdomen more densely brownish bronze dusted, mostly black bristled, the long bristly hairs at base of abdomen and on venter distinctly darker. Genitalia (Figs. 367–370) with hypandrial prongs very long and slender, at most slightly diverging, usually parallel and covered with numerous minute spines on the inner side; aedeagus much shorter and stouter than in *nigra*, and bifurcated.

Length: body 2.6–3.5 mm, wing 2.6–3.0 mm.

♀. Resembling male but halteres yellowish, squamae blackish with pale fringes, abdomen with paler bristly hairs at base and on venter, and fore tibiae slightly dilated.

Length: body 2.8–3.2 mm, wing 2.5–3.0 mm.

Distribution and biology. Not yet found in Scandinavia, although it may well be found at least in Denmark. - England, C. Europe (Czechoslovakia, Austria, Hungary) and Bulgaria (Grgau, 1250 m). In temperate central Europe *nigrita* is at least as common as *nigra*, if not commoner. According to Kovalev (in litt.) also in central parts of European USSR. - Mid May to the beginning of July, a spring and early summer species.

Note. There is another closely related species, *B. halterata* Coll., still only known from England; I have not found it in Scandinavia, nor in temperate central Europe. *B. halterata* belongs to the *B. nigra*-complex (dilated hind tarsi, narrow face, 3rd ant.s. without a bristle above), with the main characters (smaller size, shorter legs) of *nigrita*; it is generally a greyer species with all hairs and bristles on head, thorax and abdomen paler than in both *nigra* and *nigrita*, hind tibiae and tarsi somewhat less dilated, knobs of halteres yellowish in both sexes and male genitalia with hypandrial prongs shorter, resembling those of *nigra*.

Tribe Oedaleini

Males holoptic (with a tendency to be dichoptic in *Euthyneura*, broadly dichoptic in *Allanthalia*), females always broadly dichoptic, face broad in both sexes. Antennae with 3rd segment very long or broadly ovate, style always extremely short, sometimes 1-articulated, or quite absent. Proboscis usually rather long, very long in *Euthyneura*, ex-

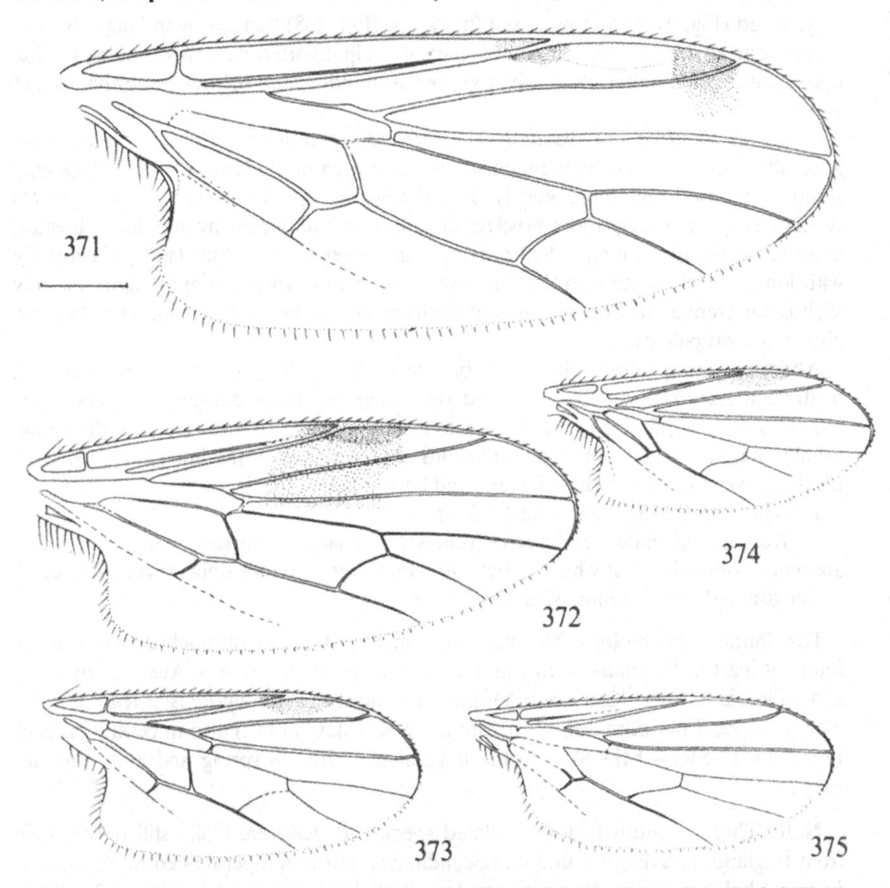

Figs. 371–375: Wings of Oedaleini. – 371: *Oedalea hybotina* (Fall.); 372: *Euthyneura gyllenhali* (Zett.); 373: *Euthyneura myrtilli* Macq.; 374: *Anthalia schoenherri* Zett.; 375: *Allanthalia pallida* (Zett.). Scale: 0.5 mm.

tremely short in *Allanthalia*, projecting forwards or vertical. Small mesonotal hairs multiserial, acr at least 4-serial, large thoracic bristles distinct but humeral bristle often small or quite absent *(Oedalea)*. Discal cell very large, usually also 2nd basal cell, discal cell emitting 3 veins to wing-margin with a tendency for vein M_2 to be abbreviated before tip. Legs simple, hind femora thickened in *Oedalea*, spinose or with bristly hairs beneath towards tip. Male genitalia asymmetrical, rotated towards right and generally upturned, hypandrium long and very asymmetrical. Aedeagus short and broad, coaxial apodeme short, rarely with unpaired postgonite *(Oedalea)*. Abdomen in female with a prolonged ovipositor-like 8th segment. Adults are nectar feeders (*Euthyneura*-complex) or predators *(Oedalea)*, and larvae often develop in rotten wood.

The species of this tribe, and of *Euthyneura* in particular, superficially resemble, and are often misidentified as, the species of *Anthepiscopus* (Empididae, Oreogetoninae). However, *Anthepiscopus* species have biserial acr, fore tibiae without the tubular gland, basal antennal segments prolonged, mouthparts (also long and forwardly projecting) with distinct maxillae (laciniae) and very long maxillary palps, wings with costa continued as an ambient vein around posterior wing-margin, and the vein closing anal cell very recurrent, *Empis*-like. Moreover, the male genitalia in *Anthepiscopus* are very different, symmetrical and unrotated, with more or less discrete gonopods, and the female abdomen is not ovipositor-like.

Genus ***Oedalea*** Meigen, 1820

Oedalea Meigen, 1820: 355.

Type-species: *Empis hybotina* Fallén, 1816 (des. Westwood, 1840).

Xiphidicera Macquart, 1834: 356.

Type-species: *Xiphidicera rufipes* Macquart, 1834 (mon.) = n. dubium.

Rather larger (body length 2.5–4.5 mm), polished black species (Fig. 376) with conspicuously long antennae and incrassate hind femora, black spinose beneath towards tip. Head small, eyes holoptic in ♂ with upper facets often considerably enlarged, broadly dichoptic in ♀ with all facets equally small, separated by polished black frons. Face short and broad in both sexes, polished black and almost square-shaped. Occiput clothed with fine bristly hairs, several upper postocular bristles about as long as a pair of anterior ocellar bristles, posterior ocellars minute. Antennae black, basal segment short but slender, fairly well separated from the globular 2nd segment, which has a circlet of short preapical hairs; 3rd segment conspicuously long and laterally compressed, 2-articulated terminal style short. Mouthparts (Figs. 377, 378) rather short, in normal position about one-third as long as head height, pointing downwards or very slightly obliquely forwards. Labrum strongly chitinised, rather broad and with a complicated structure at tip, hypopharynx very slender and as long as labrum. Maxillae reduced to slender stipites, and small narrow palpi situated on bristled palpifers. Labium as long as labrum, membraneous, labellae large and soft, with distinct sensory hairs and pseudotracheae.

Thorax arched above, humeri and postalar calli distinct. Mesonotum uniformly clothed with numerous short hairs, multiserial acr and dc, which are only indistinctly separated anteriorly by very narrow bare lines, all hairs directed upwards towards median line. Large thoracic bristles confined to 1–2 notopleurals, 1 postalar, 1 pair of widely spaced prescutellar dc and 6–8 scutellars.

Wings (Fig. 371) more or less infuscated, rarely almost clear, stigma elongate but not reaching tip of vein R_{2+3}; wings entirely covered with microtrichia. Costa running a short distance beyond vein M_1, vein Sc closely approximated to R_1 and disappearing

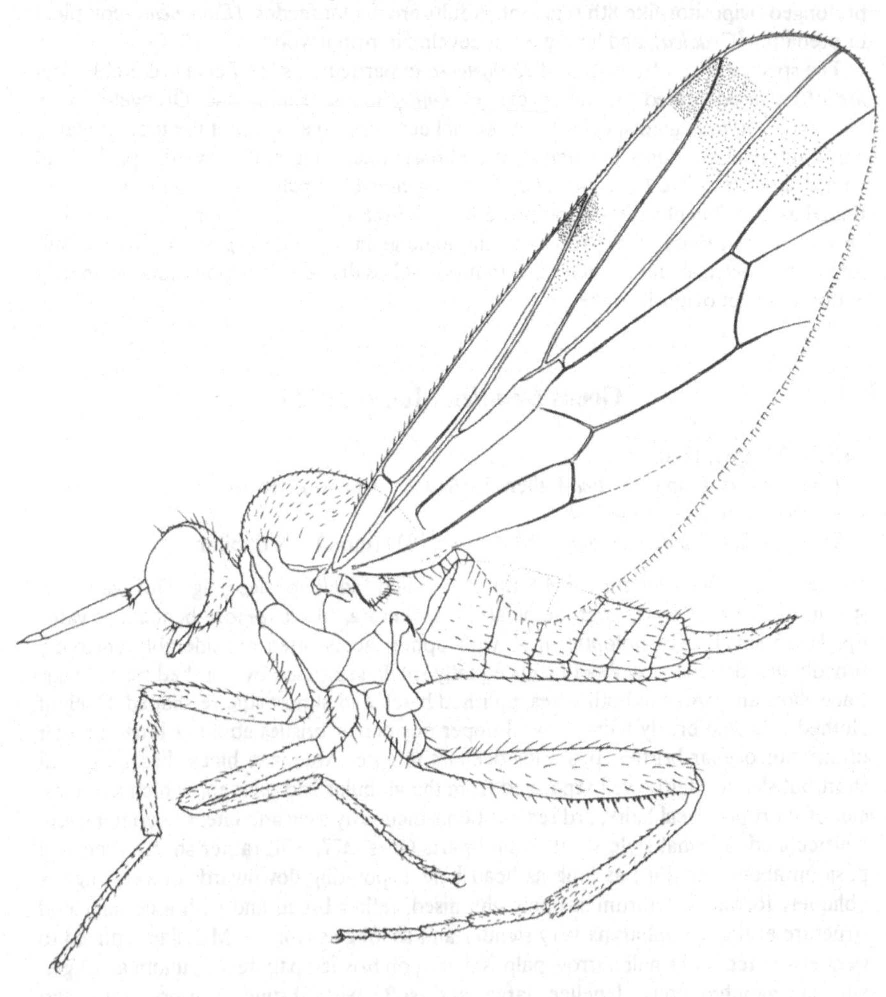

Fig. 376. Female of *Oedalea hybotina* (Fall.). Total length: 3.6 mm.

before reaching costa. Discal cell very large, emitting 3 veins to wing-margin; basal cells equally long but 2nd basal cell almost twice as broad, anal cell shorter than basal cells, the vein closing it slightly recurrent. Anal vein almost complete, axillary lobe well-developed but axillary angle obtuse.

Legs with hind femora very thickened and armed with 4 rows of black spines on apical half beneath; the spines in outer rows longer and less numerous (4–6), inner spines very small and numerous. Hind tibiae apically slightly dilated and with bent "knees". Anterior legs rather short and not very slender, fore tibiae slightly dilated, spindle-shaped, with tubular gland indistinct. Legs covered with only fine hairs, without distinct bristles.

Abdomen clothed with fine hairs, marginal bristles scarcely differentiated. Abdomen in ♂ with seven visible segments, 7th segment rather small, 8th quite concealed within the 7th. Genitalia (Figs. 379, 443–447) small, more or less upturned and rotated towards right, with clearly visible, small, slender, remarkably erect cerci. Periandrial lamellae rather uniform in shape and not substantially asymmetrical. Hypandrium very asymmetrical, apically with an ovate process which is firmly connected to aedeagus. Hypandrial bridge usually incomplete, but on the right-hand side there is a firm connection to the inner posterior fold of right periandrial lamella. Aedeagus rather short and membraneous (and consequently pale), tubular, with a short coaxial apodeme and another

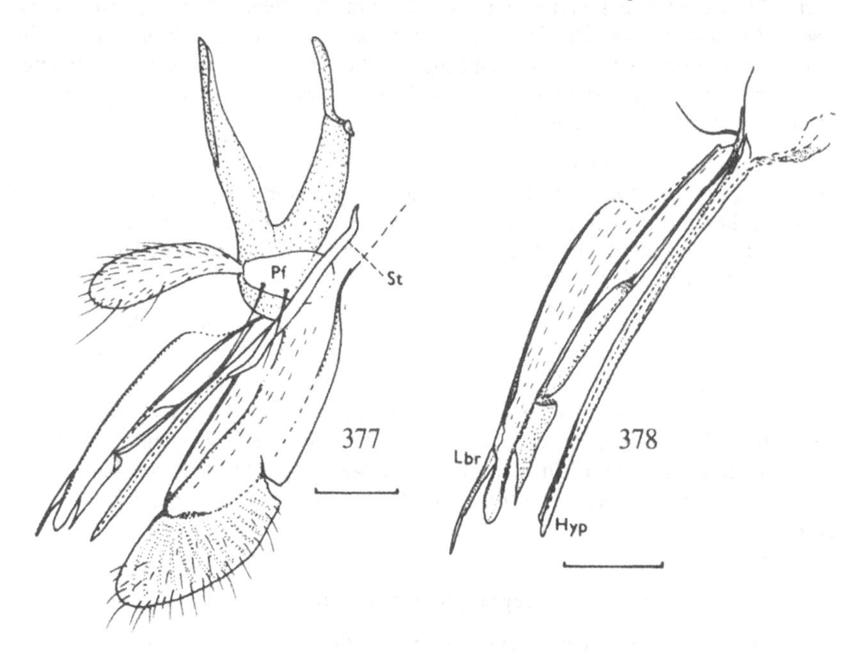

Figs. 377, 378. Mouthparts of *Oedalea zetterstedti* Coll. ♂. – 377: lateral view; 378: details of labrum and hypopharynx. Scale: 0.1 mm. For abbreviations see p. 13.

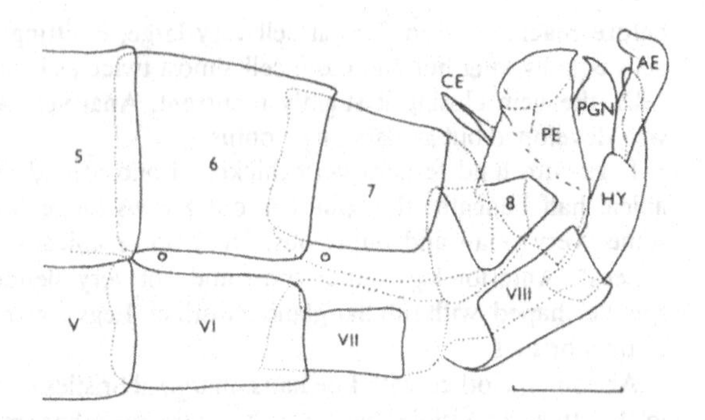

Fig. 379. Male postabdomen of *Oedalea stigmatella* Zett. (macerated). Scale: 0.2 mm. For abbreviations see p. 13.

still shorter horizontal one, which is connected with a sclerotised (and consequently dark) apically pointed postgonite. Abdomen in ♀ with seven segments unmodified, 8th and 9th segments very long and slender, forming a long terminal "ovipositor".

Some twenty species of *Oedalea* are known and, as they are very uniform, their identification is far from easy and is mainly based on coloration. Many characters of considerable significance in related genera, for instance the structure of raptorial legs, the form of frons in females, and even the structure of male genitalia, are of practically no use in *Oedalea* species. This indicates rather recent speciation, although two groups within the Palaearctic fauna may be distinguished on the basis of the structure of antennal style: these groups are represented by *O. zetterstedti* and *O. hybotina*, with the continental *O. tristis* being somewhat intermediate.

Distribution. A Holarctic genus, although one species described from the mountains of Burma, *Oedalea longicornis* Frey, 1953, is very probably congeneric and the genus may well be more widely distributed in eastern Asia. Four species are recorded from North America south to Georgia; 14 species are known with certainty from Europe, of which 9 have been found in Scandinavia.

Biology. Adults are often swept from low herbage in deciduous forest, or slowly fly above the ground vegetation. Larvae very probably develop in rotten wood, as they have been repeatedly bred from it (Zetterstedt, 1852; Lundbeck, 1910; Collin, 1961). Adults are predators of small insects. Hovering specimens in swarms have been observed, either of many individuals (Collin, 1961) or of individual specimens at the tips of tree-branches after sunset (Chvála, 1981b), but swarming activity is probably not connected with mating. Nectar feeding, or flower visiting in general, has not been recorded.

Key to species of *Oedalea*

1 Antennal style slender, bristle-like (cf. Fig. 380) 2
– Antennal style stout, as deep at base as tip of 3rd segment (cf. Fig. 388) 8

2 (1) Large thoracic bristles pale, brownish yellow 3
– Large thoracic bristles black 4
3 (2) Hind femora yellow on basal half, blackish apically (Fig. 392). Last prescutellar pair of dc minute. ♂: eye-facets equally small .... 23. *stigmatella* Zetterstedt
– Hind femora wholly yellow (Fig. 393). Last pair of prescu-

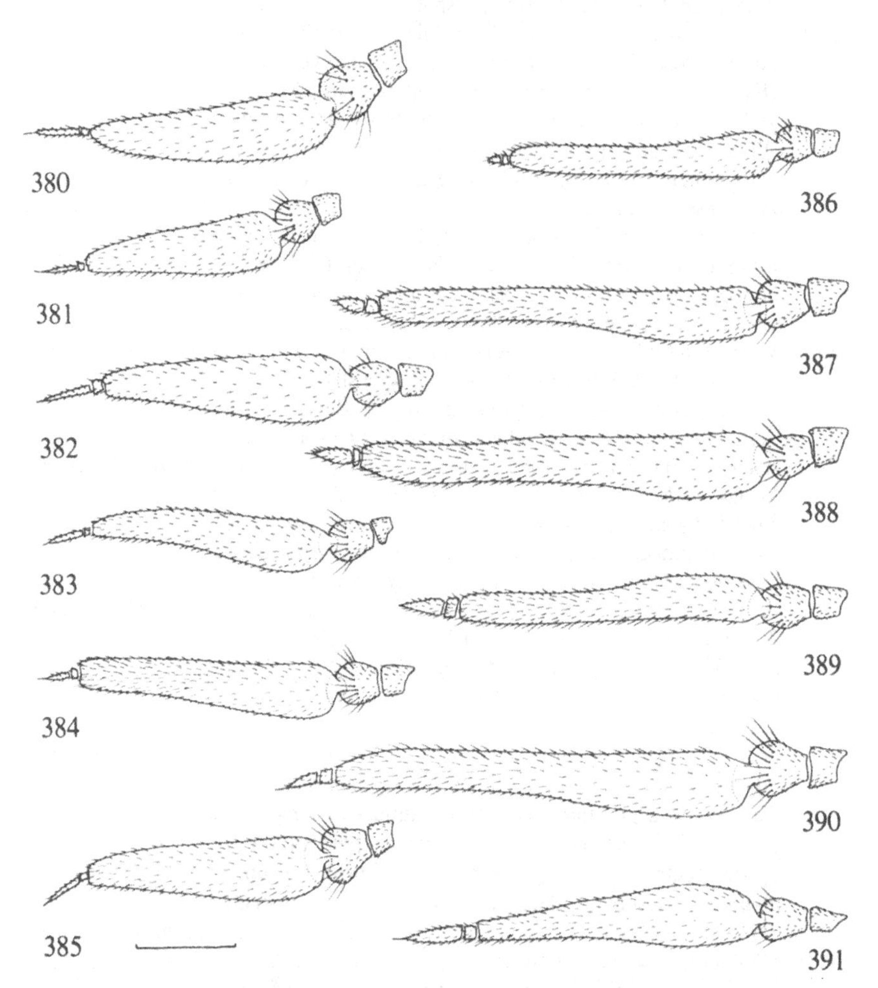

Figs. 380–391. Antennae of *Oedalea*. – 380: *stigmatella* Zett. ♂; 381: *freyi* sp.n. ♂; 382: *freyi* sp.n. ♀; 383: *zetterstedti* Coll. ♂; 384: *holmgreni* Zett, ♂; 385: *ringdahli* sp.n. ♂; 386: *tibialis* Macq. ♂; 387: *flavipes* Zett. ♂; 388: *flavipes* Zett. ♀; 389: *kowarzi* Chv. ♂; 390: *oriunda* Coll. ♂; 391: *hybotina* (Fall.) ♀. Scale: 0.2 mm.

tellar dc longer, about half length of the longest inner scutellars. ♂: upper eye-facets considerably enlarged .... 24. *freyi* sp.n.

4 (2) Hind femora yellow, at most slightly brownish at extreme tip. Last prescutellar pair of dc small and fine 5

– Legs extensively darkened including hind coxae and femora. A pair of very strong prescutellar dc about as long as large scutellars. Halteres blackish in ♂ 7

5 (4) Halteres pale, yellowish. Wings faintly greyish clouded, almost clear. 3rd ant.s. rather slender, narrowed apically (Fig. 383). Notopleural bristle long, abdomen more or less yellowish at base. ♂: upper facets slightly enlarged (about 1.3 times), dorsum of abdomen subshining and covered with rather long hairs. Hind coxae and trochanters yellow in both sexes .... 25. *zetterstedti* Collin

– Halteres darkened, blackish in ♂, brownish in ♀. Wings more or less brownish clouded. Notopleural bristle fine, abdomen blackish even at base. ♂: upper facets considerably enlarged, about twice as large 6

6 (5) Wings faintly brownish clouded, less so in ♀. 3rd ant.s. almost equally stout towards a blunt tip, style short (Fig. 384). ♂: hind coxae and trochanters darkened, dorsum of abdomen brownish dusted and almost bare on a broad median stripe. ♀: hind femora entirely yellow .... 26. *holmgreni* Zetterstedt

– Wings darker brown clouded. 3rd ant.s. distinctly narrowed apically, style longer. ♂: hind coxae and trochanters pale, dorsum of abdomen subshining, clothed with a distinct pale pubescence. ♀: hind femora faintly brownish towards tip .... *austroholmgreni* Chvála

7 (4) 3rd ant.s. rather long, about 4.5 times as long as style (Fig. 385). A notopleural and a postalar bristle small and fine. Squamae yellowish brown. ♂: upper facets considerably large (about twice as large), dorsum of abdomen brownish dusted and almost bare. Larger, 3.3 mm in length .. 27. *ringdahli* sp.n.

– 3rd ant.s. shorter, about 3.5 times as long as style. A notopleural and a postalar bristle large, almost as long as the strong pair of prescutellar dc. Squamae pale yellow. ♂: upper facets very enlarged (about 2.3 times), dorsum of abdomen subshining black with long pale hairs. Smaller, 2.6–2.9 mm in length .... *montana* Chvála

8 (1) Hind femora wholly yellow 9

– Hind femora distinctly darkened at least on apical fifth; in doubtful cases (*tristis* ♀) halteres black 12

9 (8) Scutellar bristles pale. 3rd ant.s. very long and slender, at least as long as head is high 10

– Scutellar bristles black. 3rd ant.s. not so long, shorter than head is high (Fig. 390) .... 30. *oriunda* Collin

10 (9) Larger species, over 3 mm in length. Tibiae and tarsi more or less darkened 11

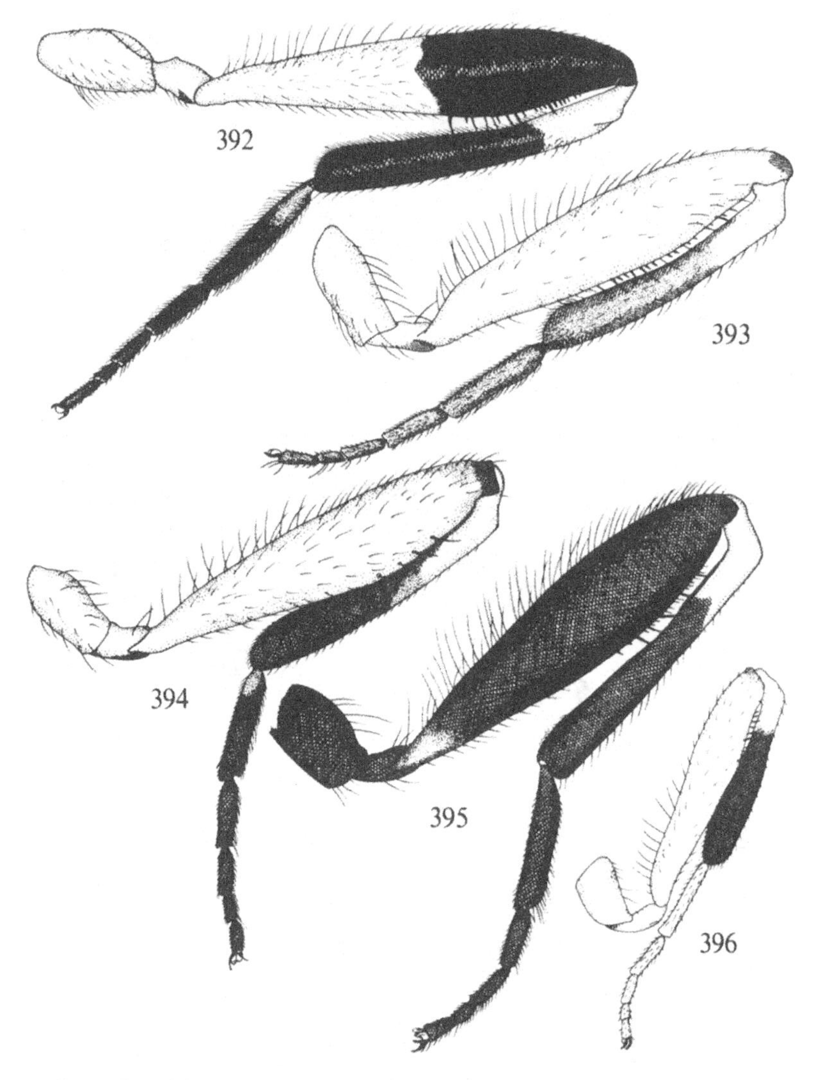

Figs. 392–396. Hind legs in anterior view of *Oedalea*. – 392: *stigmatella* Zett. ♂; 393: *freyi* sp.n. ♂ holotype; 394: *zetterstedti* Coll. ♂; 395: *ringdahli* sp.n. ♂ holotype; 396: *tibialis* Macq. ♂.

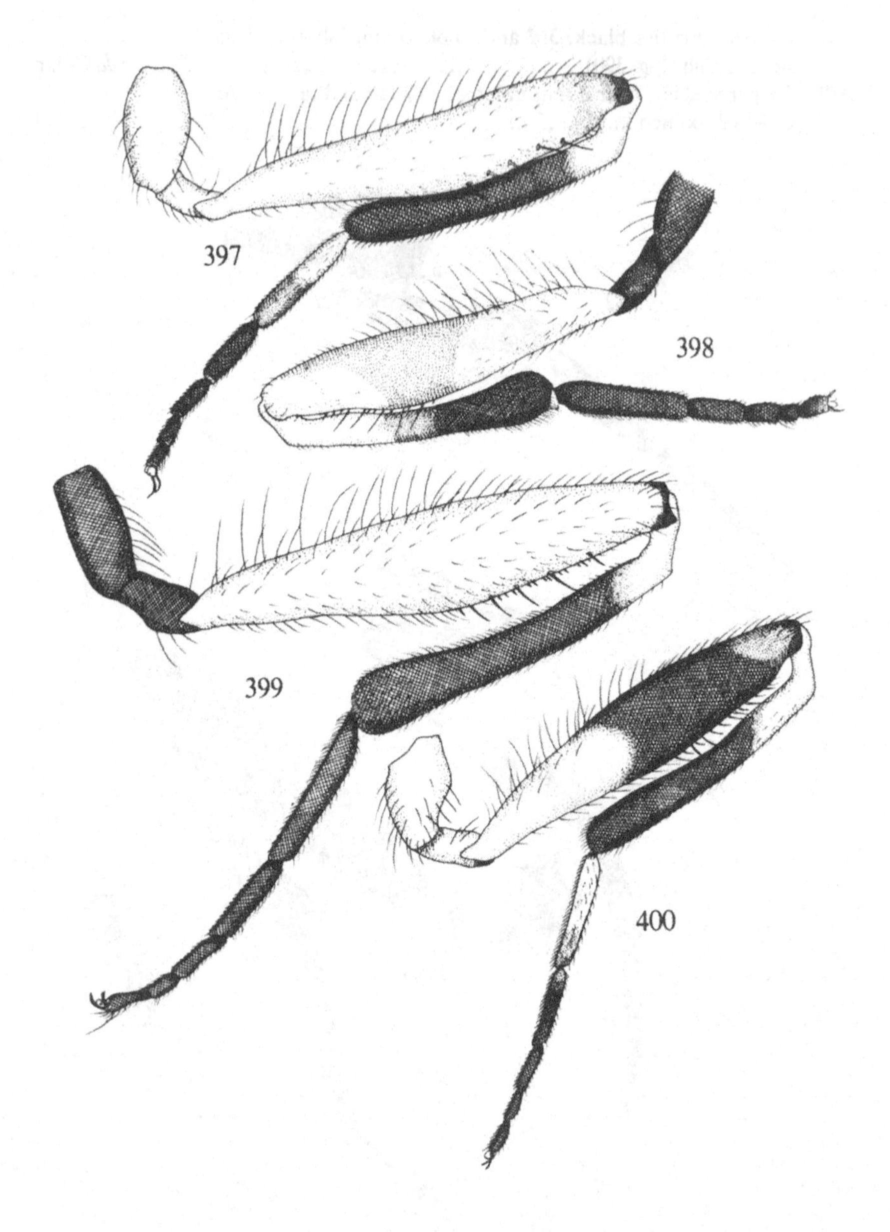

Figs. 397–400. Hind legs in anterior view of *Oedalea*. – 397: *flavipes* Zett. ♂; 398: *kowarzi* Chv. ♂; 399: *oriunda* Coll. ♂; 400: *hybotina* (Fall.) ♀.

- Smaller species, 2.5–3 mm in length. Anterior four tibiae and all tarsi pale yellow, hind tibiae contrasting dark. Halteres brownish to dark yellowish brown, paler in ♀. ♂: upper facets only indistinctly enlarged, dorsum of abdomen extensively polished .... 28. *tibialis* Macquart

11 (10) Larger species, generally over 3.5 mm in length. Notopleural bristle black, legs long and slender with pale parts clear yellow and metatarsi extensively pale. ♂: upper facets considerably enlarged, halteres brownish yellow, hind coxae and trochanters pale, dorsum of abdomen brownish dusted .... 29. *flavipes* Zetterstedt

- Generally smaller, less than 3.5 mm in length. Notopleural bristle pale, legs shorter and stouter with pale parts more yellowish brown and metatarsi dark. ♂: all facets almost equally small, halteres clear yellowish, hind coxae and trochanters dark, abdomen polished on dorsum .... *kowarzi* Chvála

12 (8) Wings blackish in ♂, paler in ♀, with a distinct dark stigma. Large thoracic bristles black, halteres blackish .... *tristis* Scholtz

- Wings very slightly infuscated, almost clear; stigma whitish with a dark patch at base and another at tip (Fig. 371). Scutellar bristles brownish yellow, halteres pale 13

13 (12) Wing-tip clear; wings smaller, with less space between faint stigma and end of vein R_{2+3} (Fig. 371) .... 31. *hybotina* (Fallén)

- The dark patch at tip of faint stigma enlarged, occupying the whole apex of wing. Wings larger, with more space between stigma and end of vein R_{2+3} .... *apicalis* Loew

23. ***Oedalea stigmatella*** Zetterstedt, 1842
Figs. 73, 379, 380, 392, 401–404.

Oedalea stigmatella Zetterstedt, 1842: 246.
Oedalea stigmatica; Boheman, 1852: 190 (lapsus calami).

Antennae with a slender bristle-like style, large thoracic bristles yellowish. Legs extensively yellow with hind femora blackish on apical half.

♂. Upper facets very slightly enlarged, about 1.2 times as large as those below; a pair of anterior ocellar bristles rather short, scarcely as long as postocular ciliation. Antennae (Fig. 380) black, 3rd segment narrowed apically, style long and slender, almost 1/3 as long as 3rd segment, the bare terminal part rather long. Thorax polished black, notopleural depression greyish pollinose only on a small part below and behind notopleural bristle. Small thoracic hairs pale, fine and rather sparse, acr and dc more broadly separated by bare stripes right up to posterior half of mesonotum. Large bristles yellowish to yellowish brown: 1 long but fine notopleural with 2 small hairs anteriorly, a pair of prescutellar dc and 1 postalar small, 6–8 scutellars the longest.

Wings very faintly greyish brown infuscated or almost clear, with long ovate brownish stigma. Squamae and halteres pale yellow. Legs yellow on anterior two pairs, tarsi brownish apically; hind leg (Fig. 392): coxae, trochanters and basal half of femora yellow, apical half of femora almost blackish, tibiae whitish on basal third, otherwise very darkened, like the tarsi except for pale base of metatarsus.

Abdomen subshining black even on dorsum, covered with long pale hairs, dorsum with shorter hairs. Extreme base of abdomen and venter often brownish. Genitalia as in Figs. 401–404.

Length: body 3.5–4.4 mm, wing 3.6–4.3 mm.

♀. Eyes broadly separated by a polished frons, all facets equally small. Thorax more brownish on pleura, abdomen almost polished black with a long "ovipositor" (Fig. 73).

Length: body 3.6–4.5 mm, wing 3.9–4.8 mm.

Lectotype ♂ from Ostrogothia, Sweden, was designated by Chvála (1981b) and is in Zetterstedt's Dipt. Scand. Coll. at Lund.

Distribution and biology. The commonest species of the genus in northern Europe and Scandinavia (195 specimens examined), widely distributed along the Baltic coast and particularly common in S. Finland, extending north to about 62°N, in Sweden to Jmt., in Finland to Sb, isolated records up to Ks (Kuolajärvi, 30.vii.1934, 4 ♂ 1 ♀, Krogerus); also in Norway ("Norvegia", J. Sahlberg). – Great Britain, east to NW and central European USSR, C. Europe, south to Romania and Yugoslavia, rather uncommon in temperate Europe. – Mid May to August, but mainly in June and first half of July.

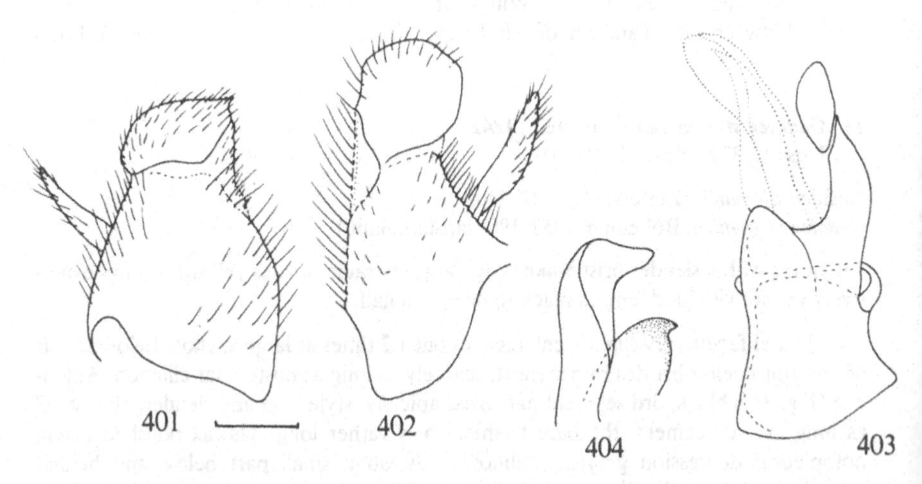

Figs. 401–404. Male genitalia of *Oedalea stigmatella* Zett. – 401: left periandrial lamella with cercus; 402: right periandrial lamella with cercus; 403: hypandrium, aedeagal complex dotted; 404: tip of aedeagal complex. Scale: 0.1 mm.

24. ***Oedalea freyi*** sp.n.
Figs. 381, 382, 393, 405–409.

Antennae with 3rd segment evenly conical, style slender and bristle-like. Large thoracic bristles yellowish brown and legs extensively yellowish with hind femora wholly yellow.

♂. Eyes meeting on frons with upper facets distinctly enlarged, about 1.6 times as large as those below antennae. Face broad, polished black. Anterior pair of ocellar bristles rather long, black, posterior pair minute. Occiput black, thinly greyish dusted, clothed with black bristly hairs above, those below neck longer and pale. Antennae (Fig. 381) small, brownish black, 3rd segment broad at base and evenly narrowed towards tip; style bristle-like, rather short (3rd ant.s. almost 5 times as long as style) and its bare terminal part very indistinct. Proboscis polished black, palpi dark with 2 rather brown preapical bristly hairs.

Thorax polished black to brownish black, notopleural depression only narrowly polished anteriorly, covered with greyish pile on more than posterior half. Mesonotal pubescence (multiserial acr and dc, which are narrowly separated anteriorly) silvery grey, large thoracic bristles pale to yellowish brown: 1 strong dark notopleural with 2 smaller whitish bristles below, 1 rather small postalar, a pair of prescutellar dc not very much shorter than scutellars; usually 8 scutellars, inner pair the strongest and extensively darkened.

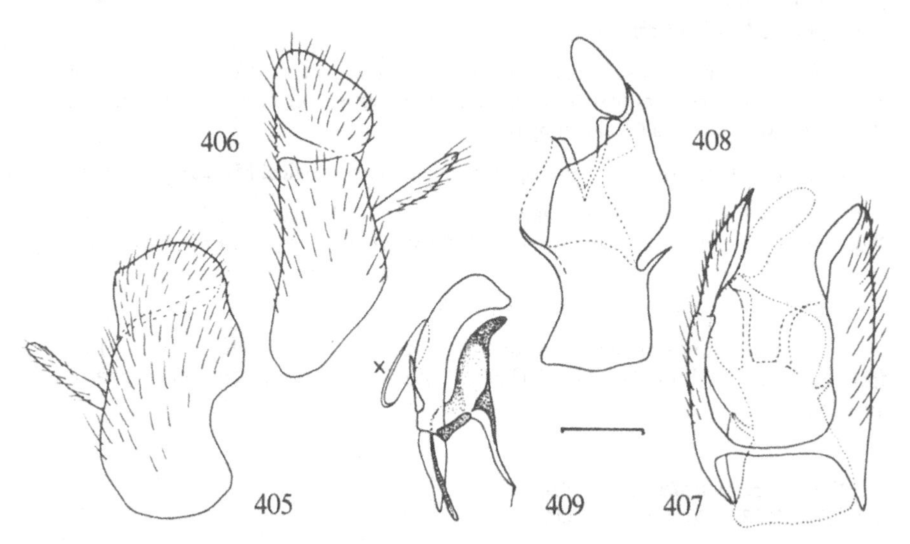

Figs. 405–409. Male genitalia of *Oedalea freyi* sp.n. (paratype, Lapponia ross.). – 405: left periandrial lamella with cercus; 406: right periandrial lamella with cercus; 407: periandrium (hypandrium dotted); 408: hypandrium; 409: aedeagal complex (x – apical process of hypandrium). Scale: 0.1 mm.

Wings faintly brownish clouded especially on costal half, paler posteriorly, stigma light brownish, rather indistinct. Squamae pale yellow with whitish fringes, halteres yellow.

Legs extensively yellow, all coxae, trochanters, femora and anterior four tibiae pale yellow, tarsi and hind tibiae more or less darkened, rarely fore tibiae slightly brownish. Compared with *stigmatella* hind femora (Fig. 393) wholly yellow, at most somewhat brownish at extreme tip. All pubescence pale, anterior four femora with fine pale hairs above and below, tibiae with short hairs, hind femora with long pale bristly hairs dorsally, those near base about as long as femur is deep; apical half of hind femora with black spines beneath, outer rows with 5–6 long black spines, the two inner rows with short stubby spines.

Abdomen clothed with fine pale hairs, sides with longer pubescence, dorsum thinly brassy brown dusted, subshining, sides and venter quite polished black to blackish brown. Genitalia as in Figs. 405–409.

Length: body 3.0–3.3 mm, wing 3.3–3.6 mm (holotype body 3.1 mm, wing 3.5 mm).

♀. Frons polished brownish black and very broad, widening above, eyes with all facets equally small. Antennae (Fig. 382) with 3rd segment considerably longer than in ♂, at least 6 times as long as style. Notopleural depression more extensively polished, at least in anterior half. Wings paler, almost clear, with stigma very indistinct. Legs generally paler, tarsi slightly brownish, or legs entirely yellow including tarsi. Abdomen with sparse shorter pale pubescence, polished even on dorsum. "Ovipositor" long and slender, brownish.

Length: body 3.0–3.6 mm, wing 3.5–4.2 mm.

In view of the bristle-like slender antennal style and pale thoracic bristles, *O. freyi* sp.n. needs comparison only with *stigmatella;* the latter has bicolored hind femora with the apical two-fifth to half contrasting black.

Holotype ♂. Finland: Le, Malla, leg. R. Frey; in the Zoological Museum, Helsinki.

Paratypes. Finland: Le, Enontekis, 1 ♂, R. Frey, Malla, 2 ♀, R. Frey; LkW, Pallastunturi, 4 ♀, R. Frey, Muonio, 1 ♀, R. Frey; Ks, Salla, 1 ♀, R. Frey, Oulankaj., 1 ♀, R. Frey. USSR: Lapponia rossica, 1 ♂ 2 ♀, J. Sahlberg. In the Zoological Museum, Helsinki; 1 ♀ from Malla in author's collection.

Distribution. Northern species, known only from N. Finland (Ks, LkW, Le) and from the north of European USSR (Lr). – Dates of occurrence and other data not available.

25. ***Oedalea zetterstedti*** Collin, 1926
Figs. 377, 378, 383, 394, 410–416.

Oedalea holmgreni Zetterstedt; Lundbeck, 1910: 199.
Oedalea zetterstedti Collin, 1926: 213.

Large thoracic bristles black, antennal style bristle-like and 3rd ant.s. tapering apically.

Hind femora yellow, halteres yellow in both sexes. ♂: dorsum of abdomen subshining with distinct pubescence, hind coxae and trochanters yellow.

♂. Upper facets slightly larger, about 1.3 times as large as lower facets. Antennae (Fig. 383) black with 3rd segment distinctly narrowed towards tip, style slender and rather long, 3rd segment about 4–5 times as long as style. Thorax polished black, postalar calli and humeri at tip almost yellowish. Notopleural depression greyish dusted posteriorly, at least apical quarter polished. Short thoracic hairs pale, large bristles black: 1 very strong notopleural with 2 fine pale bristly hairs anteriorly, 1 shorter postalar and equally long pair of prescutellar dc, 6–8 scutellars, at least 2 inner pairs very strong.

Wings very faintly greyish brown clouded, stigma brown. Discal cell large, squamae whitish yellow with pale fringes, halteres clear to dirty yellow. Legs yellow except for brown fore and hind tibiae (Fig. 394) (almost whitish yellow on at least basal fifth) and dark brown tarsi, at least mid metatarsus broadly pale at base. Hind femora (Fig. 394)

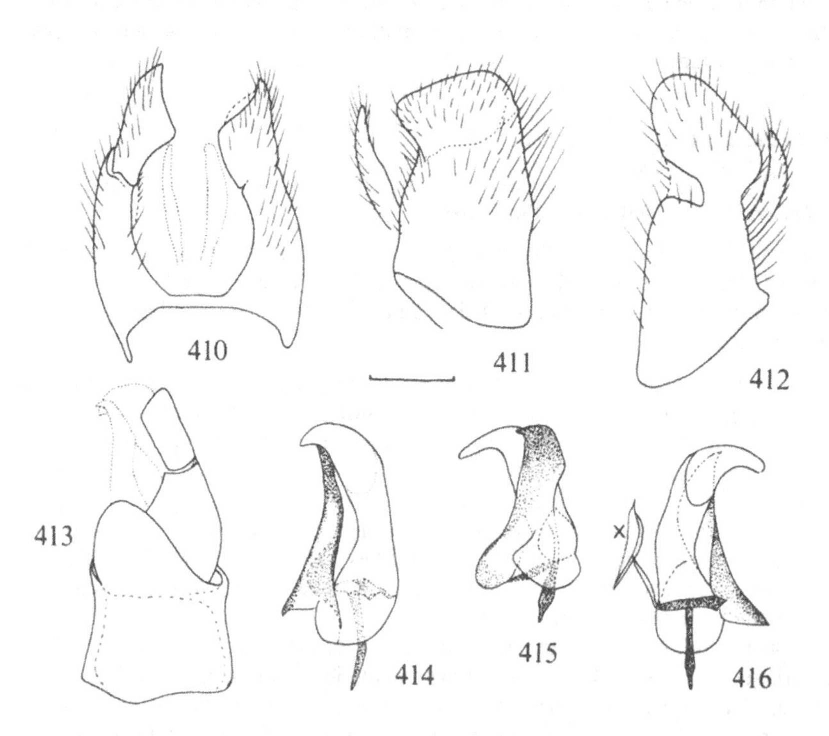

Figs. 410–416. Male genitalia of *Oedalea zetterstedti* Coll. – 410: periandrium, cerci dotted; 411: left periandrial lamella with cercus; 412: right periandrial lamella with cercus; 413: hypandrium, tips of aedeagus and postgonite dotted; 414–416: aedeagal complex in three views (x – apical process of hypandrium). Scale: 0.1 mm.

often slightly darkened at extreme tip, pubescence pale except for the usual black spines on apical half beneath.

Abdomen polished black, subshining on dorsum and covered all over with rather long pale hairs, basal segments often translucent yellowish, more extensively so on venter. Genitalia as in Figs. 410–416.

Length: body 2.7–3.5 mm, wing 3.0–4.0 mm.

♀. Head with a broad polished black frons and equally small eye-facets. Notopleural depression largely polished, dusted in posterior quarter only. The brownish parts of legs more darkened than in ♂, abdomen wholly polished, "ovipositor" long, black.

Length: body 2.7–3.9 mm, wing 3.2–4.1 mm.

Distribution and biology. Widely distributed in Scandinavia but not at all common. There are only scattered records from central and northern areas, including the extreme north of Norway (Fø), Sweden (T. Lpm.), Finland (Le) and the north of the Kola Peninsula; more frequent in Denmark and S. Sweden (Sk.). – Great Britain (Scotland), common in temperate C. Europe, south to Romania, Yugoslavia and the Caucasus. – May to July, dates ranging from 19.v. (Denmark, NEZ) to 27.vii. (Sweden, T. Lpm.), but mainly in June.

26. ***Oedalea holmgreni*** Zetterstedt, 1852
Figs. 384, 417–420.

Oedalea Holmgreni Zetterstedt, 1852: 4267.

Antennae with a short bristle-like style, 3rd segment rather evenly broad and apically rounded. Hind femora wholly yellow, large thoracic bristles black. ♂: dorsum of abdomen dull brown and with short hairs, hind coxae, trochanters and halteres extensively darkened.

♂. Upper facets very enlarged, about twice as large as those below, anterior ocellar bristles long, black. Occiput subshining black, sides brown tomentose, clothed with black hairs becoming longer and almost whitish below. Antennae (Fig. 384) black, 3rd segment long (about 7 times as long as style) and almost evenly broad towards rounded tip, style slender and rather short. Thorax black, postalar calli and anterior margin of scutellum often brownish. Notopleural depression greyish dusted right to its anterior corner. Small thoracic hairs pale, large bristles rather small and fine, black: 1 notopleural with 2 smaller bristles anteriorly, 1 postalar, a pair of widely separated prescutellar dc and 6–12 scutellars, 3 pairs longer.

Wings faintly brownish clouded, stigma darker brown. Squamae dusky with long pale fringes, halteres blackish. Legs yellow on anterior four coxae, trochanters and all femora, tibiae brown, mid tibiae paler, hind tibiae almost blackish brown except for base, tarsi very darkened. Pubescence pale, fore femora with long av bristly hairs as long as femur is deep, hind femora with a double row of long outstanding hairs dorsally at least on basal half.

Abdomen black, basal sterna scarcely paler. Sterna and side margins of terga polished and covered with rather long brownish hairs, disc of terga very dulled by brownish dust and covered with minute hairs only. Genitalia as in Figs. 417–420.

Length: body 2.9–3.6 mm, wing 3.0–3.8 mm.

♀. Antennae with style more slender and longer than in ♂, notopleural depression with greyish dusting confined to posterior half, wings not so clouded, squamae pale yellow and halteres paler, yellowish brown. Legs more yellowish on mid tibia, hind coxae and trochanters, the dark coloration (darker than in ♂) confined to fore and hind tibiae except for the whitish basal part, and to all tarsi; pubescence shorter. Abdomen uniformly polished black and more evenly pale pubescent even on dorsum. "Ovipositor" very long and strongly curved, polished black.

Length: body 3.3–3.9 mm, wing 3.8–4.0 mm.

Lectotype designation. Zetterstedt (1852) described this species from 2 ♀ taken by Holmgren in Ostrogothia, Sweden; one ♀ (monte Omberg ad Stocklycke, 15. Jul. 1850) was donated to Zetterstedt, the second (Swartswald 8. Jul. 1851) to the Museum in Stockholm. Collin (1961: 293) saw only the latter ♀ in Stockholm, labelled "OG" and "Hgn" but the first ♀, not found by Collin in Zetterstedt's Collection, is actually present in the Göteborgsamlingen at Lund; it is labelled "*Oed. Holmgreni* Zett. ♀" and "Omberg. Ostr.", and it is evidently the missing syntype; the 3rd segment is missing on both antennae but otherwise the specimen is in perfect condition; it was labelled in 1977 and is herewith designated as lectotype of *Oedalea holmgreni* Zett.

Distribution and biology. Rare but apparently widespread in Scandinavia; in Finland northwards to LkW, also in Alandia; in Sweden in Sk. and Ög. Not yet recorded from Denmark and Norway. – Outside Scandinavia only from Great Britain, common in England according to Collin (1961), western Germany (Caspers, 1982), France, and Kovalev (in litt.) has recorded it from the NW of European USSR. – June and the beginning of July, in England as early as May.

Note. The Central European *O. austroholmgreni* Chv. is compared with *holmgreni* in

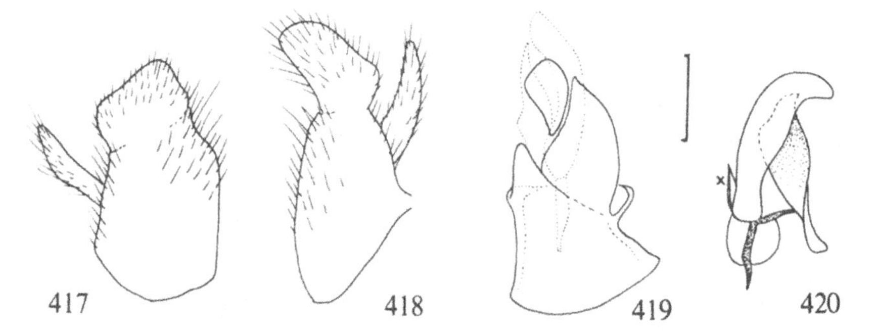

Figs. 417–420. Male genitalia of *Oedalea holmgreni* Zett. – 417: left periandrial lamella with cercus; 418: right periandrial lamella with cercus; 419: hypandriun, aedeagal complex dotted; 420: aedeagal complex (x – apical process of hypandrium). Scale: 0.1 mm.

the key; in addition to the characters given, it also has the large thoracic bristles stronger and notopleural depression polished on anterior third.

27. ***Oedalea ringdahli*** sp.n.
Figs. 385, 395, 421–424.

Legs including hind femora brownish black, knees whitish. Antennal style slender, 3rd segment broad at base and narrowed apically. ♂: halteres blackish, dorsum of abdomen densely brown dusted with minute hairs. A pair of strong black prescutellar bristles.

♂. Eyes meeting on frons, upper facets very enlarged, about twice as large as those below. Occiput subshining black with a slight greyish tomentum, occipital hairs black, those on postocular margins and above mouth paler. A pair of strong black anterior ocellar bristles, posterior pair minute. Face short, almost square-shaped, polished dark brown. Antennae (Fig. 385) black, 3rd segment about 3.7 times as long as deep, very broad at base and evenly narrowing towards somewhat blunt tip; style slender and rather short, 3rd segment 4.5 times as long as style. Palpi very small, black, covered with grey pile and a single black dorsal preapical bristle. Proboscis black, pointing obliquely forwards.

Thorax polished black, notopleural depression polished on anterior half, greyish dusted posteriorly. Multiserial acr and dc (narrowly separated by bare stripes anteriorly) small and numerous, greyish, large bristles black: 1 small fine notopleural, 1 perhaps still smaller postalar, and a pair of strong black prescutellars almost as long as the inner two pairs of the six strong marginal scutellars.

Wings faintly brownish clouded, veins dark brown, large ovate stigma deeper brown. Discal cell large, apical section of vein M_1 distinctly shorter than the length of discal

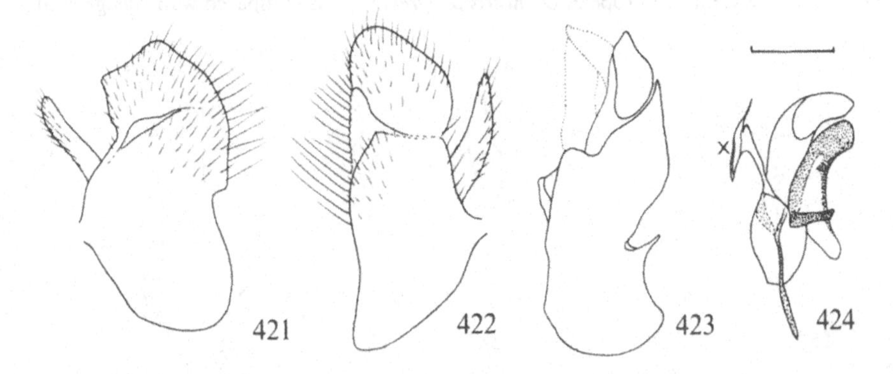

Figs. 421–424. Male genitalia of *Oedalea ringdahli* sp.n. (holotype). – 421: left periandrial lamella with cercus; 422: right periandrial lamella with cercus; 423: hypandrium, tips of aedeagus and postgonite dotted; 424: aedeagal complex (x – apical process of hypandrium). Scale: 0.1 mm.

cell, a very small black costal bristle. Squamae yellowish brown with densely-set pale fringes, halteres blackish brown.

Legs uniformly blackish brown leaving only "knees" (extreme tips of femora and extreme base of tibiae) narrowly whitish yellow, mid tibiae perhaps lighter brownish on basal half, as is extreme base of all femora. Pubescence mostly short and pale, long (almost whitish) prominent hairs on posterior four coxae anteriorly and on at least basal half of hind femora dorsally. Hind legs (Fig. 395) stout, hind metatarsus about as long as following three tarsal segments.

Abdomen black to blackish brown, basal two sterna paler. Venter and sides of terga subshining and covered with long pale hairs, discs of terga densely brown dusted and covered with minute hairs, very much as in *holmgreni*. Genitalia small (Figs. 421–424).

Length: body 3.3 mm, wing 3.6 mm.

♀. Unknown.

In view of the extensively darkened legs and the shape of the antennae, this species cannot be confused with any other Scandinavian species. *O. ringdahli* is closely related to the Central European *O. montana* Chv., but the latter has 3rd ant.s. shorter and style longer, very strong notopleural and postalar bristles, pale squamae, and abdomen polished on dorsum, covered with long pale hairs as in *zetterstedti*.

Holotype ♂. Sweden: Ly. Lpm., Umfors, 20.vi.1937, O. Ringdahl; in Coll. Ringdahl at Lund.

This species is known to me only from the holotype ♂, but the two dark-legged specimens of *O. holmgreni* Zett. mentioned by Collin (1961: 293) from England may well belong to this new species.

Distribution and biology. Probably a northern rather mountainous species; the holotype was taken in the mountains about 10 km N of Tärna (Ly. Lpm.), Sweden. – June.

28. ***Oedalea tibialis*** Macquart, 1827
Figs. 386, 396, 425–430.

Oedalea tibialis Macquart, 1827: 142.

A tiny species with very long slender antennae and a plump style as in *flavipes*, large thoracic bristles pale. Legs wholly yellow including tarsi, only hind tibiae contrasting blackish brown.

♂. Upper eye-facets only very slightly enlarged, all facets almost equally small. Anterior pair of ocellar bristles black and almost as long as a row of distinct upper postocular bristles. Antennae (Fig. 386) black, 3rd segment very long and evenly slender, apical style thickened and very short, not at least slender, its bristle-like apex short. Thorax shining black, mesonotum with rather sparse greyish multiserial acr and dc, notopleural depression largely polished. Large thoracic bristles pale, light yellowish brown, except for somewhat blackish but not very strong notopleural bristle; a very small postalar, a pair of prescutellar dc longer, 3 pairs of scutellars the longest.

Wings indistinctly brownish, almost clear, with a short ovate darker brown stigma which occupies only the smaller basal half of costal section between R_1 and R_{2+3}. Squamae whitish with pale fringes, halteres yellowish brown to brownish. Legs pale yellow including tarsi, hind tibiae (Fig. 396) conspicuously blackish brown, whitish at extreme base.

Abdomen polished black even on dorsum, pubescence pale, longer at sides, minute on tergal discs. Genitalia as in Figs. 425–430, with distinctly convex hypandrium.

Length: body 2.6–2.8 mm, wing 2.5–2.8 mm.

♀. Frons broad and polished black, wings less infuscated than in ♂ and halteres paler, sometimes wholly yellow. "Ovipositor" long and slender, polished black.

Length: body 2.9–3.2 mm, wing 2.8–3.0 mm.

Distribution and biology. Rare in Fennoscandia: 1 ♀ from S. Sweden (Sk., Fulltofta, 26.vii.1967, H. Andersson) and 1 pair from the USSR (Vib, Viborg, R. Frey). – Only a few records available from C. and SW. Europe, although according to Collin (1961) it is not uncommon in England from May to August.

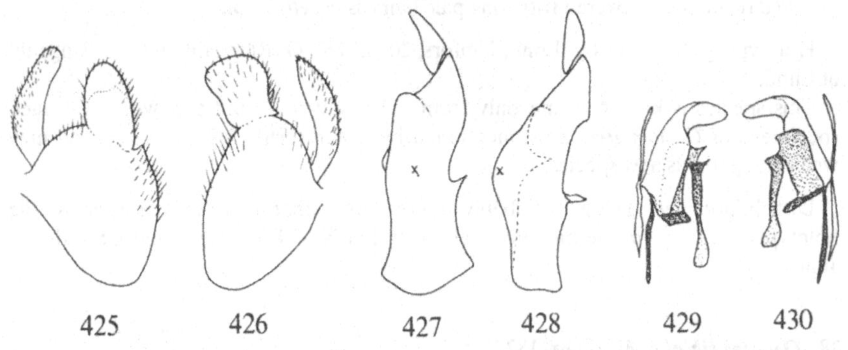

Figs. 425–430. Male genitalia of *Oedalea tibialis* Macq. – 425: left periandrial lamella with cercus; 426: right periandrial lamella with cercus; 427, 428: hypandrium in dorsal and side view; 429, 430: aedeagal complex in two different views. Scale: 0.1 mm.

29. ***Oedalea flavipes*** Zetterstedt, 1842
Figs. 387, 388, 397, 431–434.

Oedalea flavipes Zetterstedt, 1842: 247.
Oedalea infuscata Loew, 1859: 48.

Antennae with very long and slender 3rd segment, style short and stout. Large thoracic bristles pale but notopleural bristle black. Legs with all coxae, femora and mid tibiae pale yellow. ♂: upper facets considerably enlarged.

♂. Upper facets about 1.8 times as large as those below. Antennae (Fig. 387) with basal segments often brownish, 3rd segment conspicuously long and evenly slender, style short and plump, its terminal bristle-like part practically absent; antennae longer than head is high. Thorax polished black, postalar calli and humeri at tip often yellowish, notopleural depression with grey pile on posterior half, polished anteriorly. Notopleural bristle strong, blackish, a postalar and a pair of prescutellar dc small and fine, yellowish like the 6–8 large scutellar bristles.

Wings brownish clouded, more intensely so on costal half, base almost clear, the long ovate stigma deep brown. Squamae whitish, halteres yellowish brown with stalk darkened. Legs conspicuously clear pale yellow except for blackish brown fore and hind tibiae (Fig. 397) and fore tarsi; posterior four tarsi brownish, with metatarsi mostly yellowish, darkened at tip.

Abdomen polished black, discs of 1st to 4th (or 5th) terga very dulled by brownish dust, pubescence long and almost whitish at sides of abdomen and on venter, dorsum with sparse minute hairs. Genitalia (Figs. 431–434) rather large and clothed with long dark hairs.

Length: body 3.2–3.8 mm, wing 3.6–4.0 mm.

♀. Frons broadly polished black and all facets small, halteres yellow, legs even paler yellow on pale parts and abdomen polished black even on dorsum towards base.

Length: body 3.3–4.1 mm, wing 3.3–4.3 mm.

Lectotype, a ♀ from Esperöd (Sk., Sweden), was designated by Chvála (1981b) and is in the Dipt. Scand. Coll. at Lund.

Distribution and biology. A southern species in Scandinavia; in Denmark in S. Jutland and Zealand (9 localities) and in S. Sweden in Skåne (6 localities), the northern

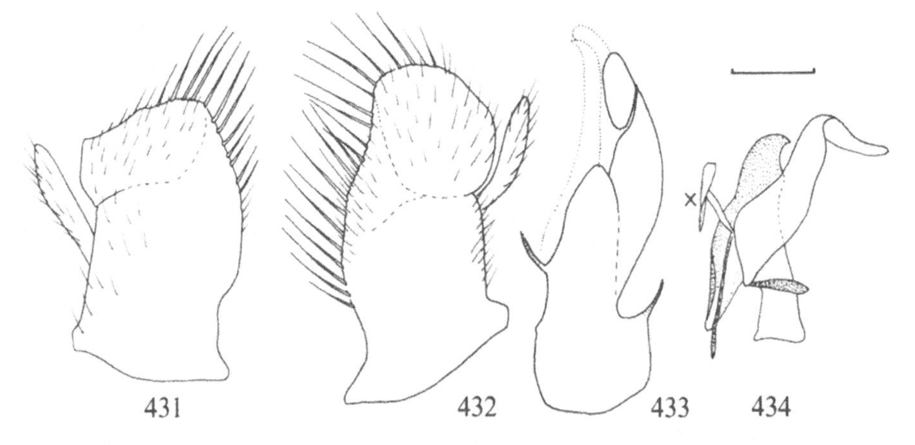

Figs. 431–434. Male genitalia of *Oedalea flavipes* Zett. – 431: left periandrial lamella with cercus; 432: right periandrial lamella with cercus; 433: hypandrium, tips of aedeagus and postgonite dotted; 434: aedeagal complex (x – apical process of hypandrium). Scale: 0.1 mm.

border of its distribution being approximately 56°N. – Great Britain and C. Europe, including central areas of European USSR; rather a common species in lowlands. – May to August.

Oedalea kowarzi Chvála, 1981

Figs. 389, 398, 435–438.

Oedalea kowarzi Chvála, 1981: 133.

Resembling *flavipes* but smaller, with all large thoracic bristles pale, yellow halteres and legs rather yellowish brown on pale parts. ♂: all facets equally small, hind coxae darkened.

♂. Eye-facets scarcely differentiated, upper facets at most 1.2 times as large, antennae (Fig. 389) and mouthparts as in *flavipes*. Notopleural depression with the grey dusting extending more anteriorly right up to anterior third, small mesonotal hairs greyish yellow and notopleural bristle pale, rather small and fine.

Wings very faintly brownish infuscated, almost clear, and halteres yellow, much paler than in *flavipes*. Legs not so long and slender, with pale coloured parts more dirty yellow to yellowish brown, hind coxae dusky and hind femora (Fig. 398) translucent brownish towards tips. Tarsi extensively darkened and posterior metatarsi scarcely paler at base.

Abdomen more extensively polished black even on dorsum, at most 1st tergum and base of the 2nd slightly brownish dusted, abdominal pubescence pale and rather long, dorsum of abdomen almost bare. Genitalia (Figs. 435–438) with shorter pubescence than in *flavipes* and shorter and stouter aedeagus.

Length: body 3.2 mm, wing 3.3 mm.

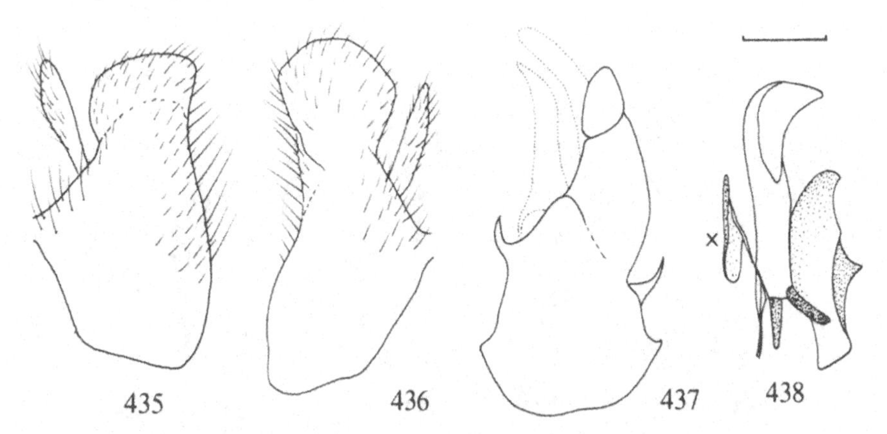

Figs. 435–438. Male genitalia of *Oedalea kowarzi* Chv. – 435: left periandrial lamella with cercus; 436: right periandrial lamella with cercus; 437: hypandrium, tips of aedeagus and postgonite dotted; 438: aedeagal complex (x – apical process of hypandrium). Scale: 0.1 mm.

♀. Resembling male but notopleural depression polished on anterior half, hind coxae yellow and eyes separated by a broad frons. Abdomen polished with shorter pubescence, "ovipositor" slender, blackish brown.
Length: body 3.2 mm, wing 3.5 mm.

Distribution and biology. A continental species known up to now only from C. Europe, but there was 1 ♂ in the Helsinki Museum taken by Frey in NW of European USSR at Archangelsk (identified as *O. infuscata* Loew). Its occurrence in Russian Karelia and the adjacent parts of Finland is very probable. – In C. Europe in June.

30. ***Oedalea oriunda*** Collin, 1961
Figs. 390, 399, 439–442.

Oedalea oriunda Collin, 1961: 297.

Antennae as in *flavipes* but style apparently more slender, and large thoracic bristles mostly black. Hind femora yellow, wings with a distinct dark stigma.

♂. Eyes with upper facets considerably enlarged, about twice as large as those below (according to Collin "only slightly enlarged"). Antennae (Fig. 390) black, 3rd segment long and slender, although not as conspicuously long as in *flavipes*, style short and plump, apically pointed. Thorax polished black, notopleural depression largely polished, silvery dusted posteriorly behind notopleural bristle. Small thoracic hairs silvery grey, a strong black notopleural bristle with 2–3 small whitish hairs anteriorly, a postalar and a pair of prescutellar dc pale and very fine, 8–10 scutellar bristles, inner 3 pairs always strong and black, outer pairs fine and pale.

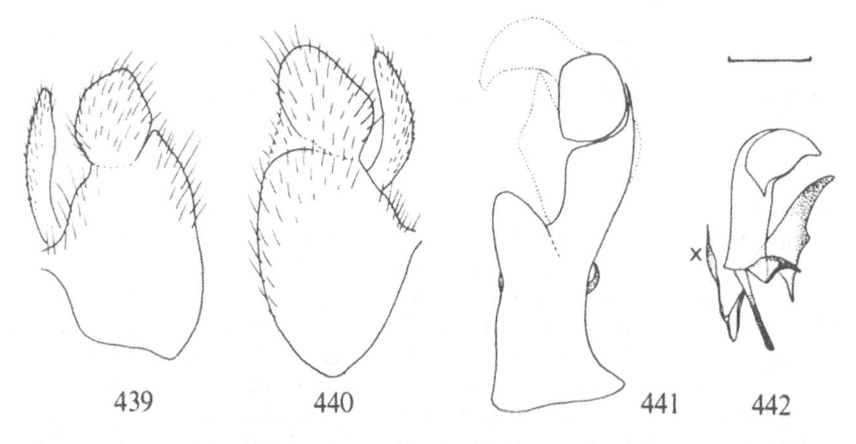

Figs. 439–442. Male genitalia of *Oedalea oriunda* Coll. – 439: left periandrial lamella with cercus; 440: right periandrial lamella with cercus; 441: hypandrium, tips of aedeagus and postgonite dotted; 442: aedeagal complex (x – apical process of hypandrium). Scale: 0.1 mm.

Wings considerably brownish infuscated, base paler, stigma deeper brown. Squamae pale with concolorous fringes, halteres blackish brown. Legs yellow on anterior four coxae and all femora; fore coxae often brownish at base, hind coxae and trochanters (Fig. 399) blackish brown. Tibiae brown with pale base, tarsi evenly dark. Fore and hind femora (Fig. 399) with a double row of distinct pale hairs beneath.

Abdomen polished black, anterior two terga and base of 3rd tergum with brown tomentum dorsally. Pubescence whitish, very long at sides, dorsum with only short hairs. Genitalia (Figs. 439–442) rather large, covered with long pale and dark hairs.

Length: body 3.3–4.0 mm, wing 3.6–4.0 mm.

♀. Unknown.

Distribution and biology. Only 2 ♂ from Denmark: F, Bukke Skov, 14.vi.1879, Schlick and SJ, Sønderborg, 1.vi.1904, Wüstnei. – Known up to now from 3 ♂ swept from conifers in Suffolk, England (Collin, 1961). – June, in England from May.

Note. *O. oriunda* especially needs comparison with the common central European *O. tristis* Scholtz, particularly in the female sex, as females of *tristis* have hind femora often only indistinctly darkened on apical half. Distinguishing characters are given in the Key.

31. ***Oedalea hybotina*** (Fallén, 1816)
Figs. 371, 376, 391, 400, 443–447.

Empis hybotina Fallén, 1816: 31.

Wings almost clear with whitish stigma and 2 small brown patches at tip of veins R_1 and R_{2+3}. Hind femora blackish on apical half, antennae long with a plump style. Large thoracic bristles and halteres pale.

♂. Eyes with all facets equally small, upper facets not differentiated. Anterior pair of ocellar bristles and upper postocular bristles blackish. Antennae (Fig. 391) black, often brownish on basal segments, 3rd segment very long and slender, slightly narrowed apically; style very stout, its bare bristle-like terminal part very short. Thorax polished black but notopleural depression almost entirely greyish pollinose. Small mesonotal hairs almost whitish and very numerous, large bristles, including 6–8 scutellars, yellowish.

Wings (Fig. 371) practically clear with distinct brown veins, stigma whitish with a small brown patch on both tips (at tips of veins R_1 and R_{2+3}), apical patch occupying at most apical third of costal section between R_1 and R_{2+3}; wing-tip clear. Squamae whitish with pale fringes, halteres whitish yellow with brownish base to stalk. Anterior four legs pale yellow except for brownish tarsi; hind legs (Fig. 400) yellow on coxae, trochanters and basal half of femur, apical half of femora blackish, tibiae blackish except for whitish base, and tarsus dark brown; metatarsus pale at base.

Abdomen polished black even on dorsum, discs of terga almost bare, sides of abdomen and venter with rather long pale hairs. Genitalia as in Figs. 443–447.

Length: body 3.6 mm, wing 3.7 mm.

♀. Resembling male but eyes separated by a broad polished frons and notopleural depression polished on anterior half. "Ovipositor" long and slender, mostly yellowish, darkened at tip.

Length: body 3.3–3.7 mm, wing 3.8–4.2 mm.

Distribution and biology. Widespread but uncommon in Scandinavia; males are rare (3 ♂ and 21 ♀ examined), even outside Scandinavia. Denmark, in Sweden north to Jmt., in Finland to Ks and ObN, also Russian Karelia and the Kola Peninsula up to 69°N; not yet found in Norway. – Rather rare in C. Europe (Poland, GFR, Czechoslovakia and Austria), also in central parts of European USSR (Kovalev, in litt.), but not found in Great Britain and in the south. – June and July, in C. Europe as early as May.

Note. In view of the distinctive wing-pattern, this species only needs comparison with *O. apicalis* Loew, known from England and C. Europe. For distinguishing features, see the Key.

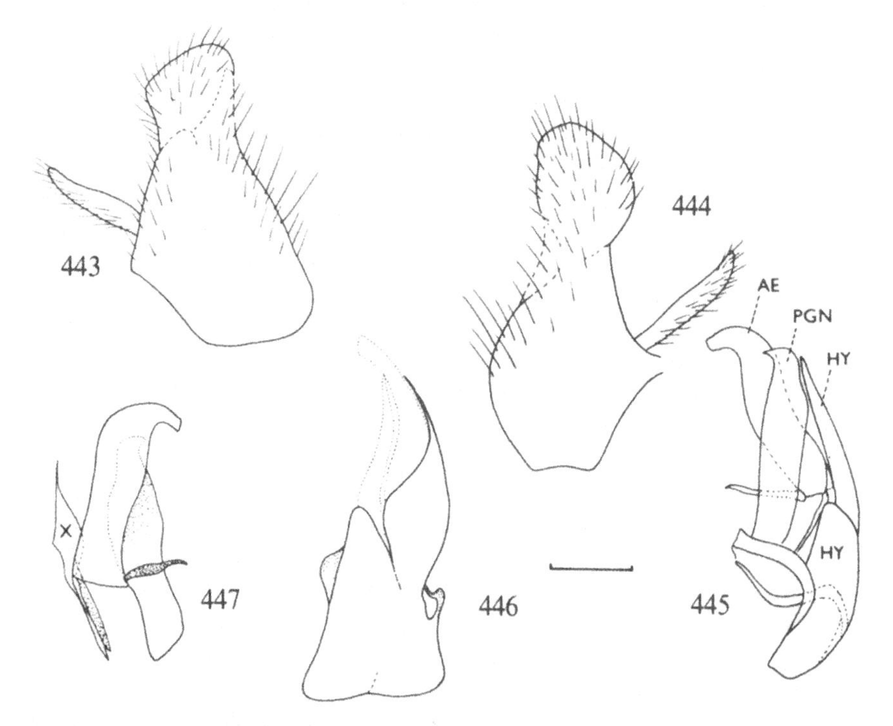

Figs. 443–447. Male genitalia of *Oedalea hybotina* (Fall.). – 443: left periandrial lamella with cercus; 444: right periandrial lamella with cercus; 445: hypandrium with aedeagal complex in lateral view; 446: hypandrium, tips of aedeagus and postgonite dotted; 447: aedeagal complex (x – apical process of hypandrium). Scale: 0.1 mm. For abbreviations see p. 13.

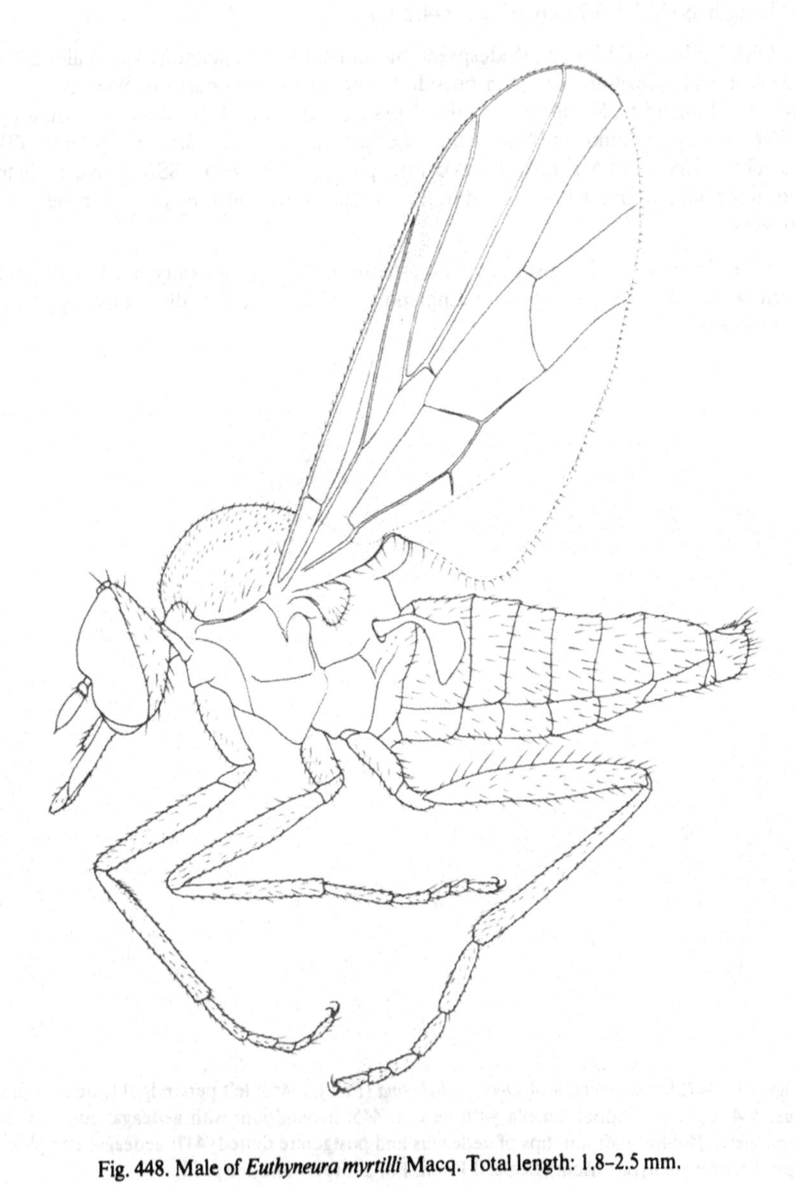

Fig. 448. Male of *Euthyneura myrtilli* Macq. Total length: 1.8–2.5 mm.

Genus *Euthyneura* Macquart, 1836

Euthyneura Macquart, 1836: 518.
Type-species: *Euthyneura myrtilli* Macquart, 1836 (mon.).

Small, body length about 2–3 mm, polished black species (Fig. 448) with generally pale simple legs devoid of distinct spines or bristles, and with a long proboscis directed forwards.

Head almost globular or slightly deeper than long, eyes touching on frons in ♂ or very narrowly separated, upper facets usually enlarged; eyes in ♀ separated by a broad triangular polished black frons, all facets always equally small. Occiput clothed with fine bristly hairs, those on postocular margin above prominent, as long as anterior pair of ocellar bristles; posterior pair minute. Antennae black, 1st segment slender and distinctly longer than deep, 2nd segment globular and armed with a circlet of small preapical bristles; 3rd segment short ovate to conical, at most 3 times as long as deep, distinctly shorter than in *Oedalea* species; terminal style 1- or 2-articulated, slender and very short, extreme tip bristle-like, basal article (if present) inconspicuous. Mouthparts (Figs. 449–453) long, labrum about as long as head is high, or at least half as long, pointing forwards or obliquely forwards. Labrum heavily sclerotised and prominent, apically pointed; hypopharynx closely attached to labrum, equally long, slender and very chitinised. Maxillae reduced to stylet-like stipites, palpifer large, finely sclerotised

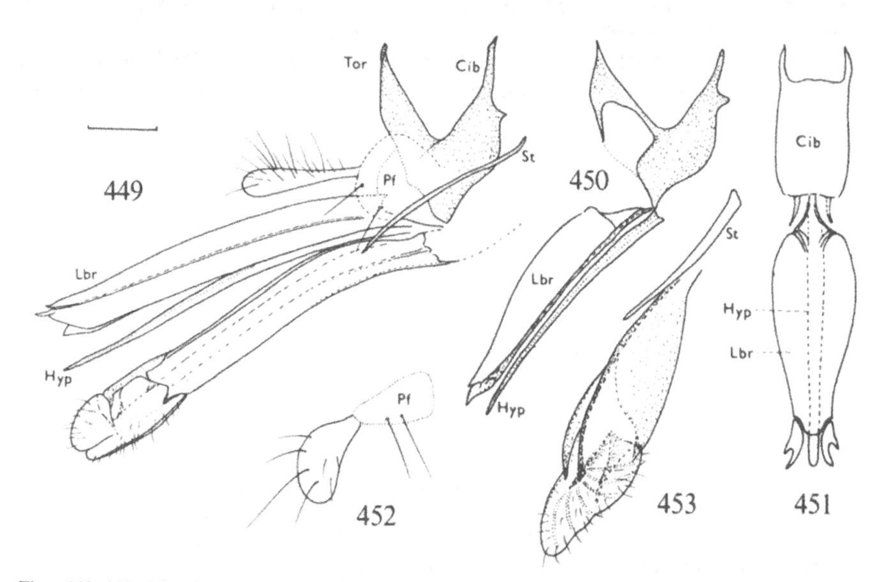

Figs. 449–453. Mouthparts of *Euthyneura*. – 449: *myrtilli* Macq. ♂, lateral view. 450–453: *gyllenhali* (Zett.) ♂ – 450: labrum and hypopharynx in lateral view; 451: the same in dorsal view; 452: palpus; 453: labium and maxilla in lateral view. Scale. 0.1 mm. For abbreviations see p. 13.

and bristled (bearing two hair-like bristles beneath); palpi long and slender, about 1/3 as long as labrum, clothed with several bristle-like hairs above. Labium rather broad and finely sclerotised except for membraneous labellae with pseudotracheae (Fig. 453).

Thorax slightly elongated, anterior part of mesonotum convex, posterior half almost flat on prescutellar depression. Small thoracic hairs numerous, acr 4- to 6-serial, rather narrowly separated from multiserial dc, which are, however, often narrowly 2-serial or even 1-serial at about middle; last prescutellar pair (rarely two) of dc strong. Large bristles confined to a small humeral, 1 or 2 large notopleurals (with 2 smaller hairs below), 1 postalar, and 4–6 marginal scutellars, inner pair the strongest.

Wings (Fig. 373) rather large, clear, sometimes almost milky-white or with a distinct brown pattern, entirely covered with fine microtrichia. Axillary lobe large, axillary excision obtuse or almost right-angled, a small costal bristle. Vein C reaching tip of vein M_1, Sc closely approximated to R_1 and disappearing before reaching costa, veins R_{4+5} and M_1 indistinctly diverging. Three veins from large discal cell, veins M_1 and M_2 distinctly separated or often almost petiolate. Both basal cells equally long but 2nd basal cell very broad, the vein closing it (basal crossvein m-cu) almost perpendicular, and the discal cell looking like a continuation of 2nd basal cell. Anal vein fine but complete, anal cell distinctly shorter than basal cells, the vein closing it almost at right angles.

Legs rather short and slender, without bristles and mostly short pubescent, the longest hairs on hind femur dorsally near base. Hind femora not spinose as in *Oedalea*, at most with a few (2 or 3) spine-like hairs at tip beneath. Fore tibiae slightly spindle-shaped with traces of a tibial gland near base beneath, tarsi slender.

Abdomen cylindrical, more or less laterally compressed in ♂, clothed with fine hairs (longer at sides) and more distinct hind marginal bristly hairs. 8th segment unmodified

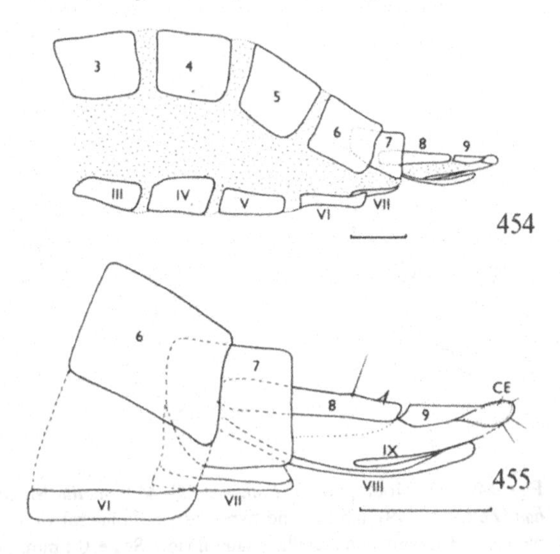

Figs. 454, 455. Female abdomen of *Euthyneura myrtilli* Macq. Scale: 0.2 mm. For abbreviations see p. 13.

in ♂ but smaller than the 7th, usually clearly visible. Male genitalia very small and distinctly upturned, only very slightly rotated towards right and very uniformly built without conspicuous specific characters; in general structure very much as in *Oedalea* species but aedeagus without unpaired postgonite. Periandrial lamellae very narrowly connected anteriorly, not very asymmetrical except for the terminal processes, inner walls firmly connected by narrow folds with the hypandrial bridge. Cerci rather small, ovate and finely pubescent. Hypandrium slightly convex, apically with two sclerites which are firmly connected to the base of aedeagus, forming part of the hypandrial ring complex. Aedeagus cylindrical to cornet-like, short, apically with a large ovate opening, at central part rather loosely attached to the inner wall of hypandrium. A simple coaxial apodeme of varying (specific) length. Female abdomen (Figs. 454, 455) more circular in cross-section and telescopic, 7th segment narrowed, 8th and 9th segments forming an "ovipositor" which is very distinct if exserted (in living specimens), in dried specimens abdomen often truncate at tip, with the "ovipositor" more or less withdrawn. Cerci small and slender.

Distribution. A Holarctic genus; 6 species are known from North America, 6 species are West Palaearctic and one, *E. aerea* Frey, was described from Japan. Three species have been found in Scandinavia.

Biology. The adults are exclusively nectar or pollen feeders, frequently visiting flowers; they are often swept from flowers and low herbage, but also from conifers. They are frequently found in peat-bogs on *Ledum palustre*, and in moist deciduous forests on blossoming *Frangula* and *Salix*. Hovering males of *E. gyllenhali* were observed in a swarm above bracken by Collin (1961) in England, but no other data on swarming or mating habits are available. Larvae very probably develop in rotten wood, as the adults have repeatedly been bred from it (Collin, 1961; Cole, 1964).

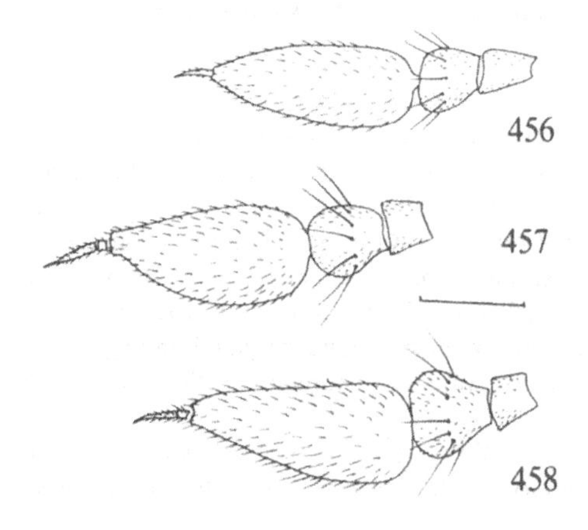

Figs. 456–458. Antennae of *Euthyneura*. - 456: *myrtilli* Macq. ♂; 457: *gyllenhali* (Zett.) ♂; 458: *albipennis* (Zett.) ♂. Scale: 0.1 mm.

Key to species of *Euthyneura*

1 Wings (Fig. 372) with a blackish stigma and a brown cross-band. Legs yellow, mesonotal hairs and bristles pale. ♂: eyes narrowly separated on frons, with all facets equally small .... 34. *gyllenhali* (Zetterstedt)

– Wings (Fig. 373) clear, stigma at most yellowish. Legs brownish or extensively darkened. ♂: eyes meeting on frons (very rarely separated), with upper facets enlarged 2

2 (1) Mesonotal hairs brown, large bristles black. Labrum about as long as head is high, legs brownish, wings clear with yellowish stigma. ♂: dorsum of abdomen subshining black .... 32. *myrtilli* Macquart

– Mesonotal hairs and bristles pale. Labrum distinctly shorter than head is high, legs brown to dark brown, wings milky-white with stigma whitish. ♂: dorsum of abdomen silvery dusted .... 33. *albipennis* (Zetterstedt)

32. ***Euthyneura myrtilli*** Macquart, 1836
Figs. 373, 448, 449, 454–456, 459–461.

Euthyneura myrtilli Macquart, 1836: 519.
Rhamphomyia consobrina Zetterstedt, 1838: 571.
Hemerodromia brevipes Loew, 1840: 22 – **syn.n.**
Anthalia rostrata Zetterstedt, 1842: 250.
Euthyneura simillima Strobl, 1893: 97 – **syn.n.**
Euthyneura myrtilli var. *incompleta* Strobl, 1898: 208.
Euthyneura myricae Haliday; Lundbeck, 1910: 211.

Large thoracic bristles black, wings clear with indistinct veins and stigma. Proboscis about as long as head is high, legs almost uniformly yellowish brown to brownish.

♂. Eyes meeting on frons with upper facets enlarged, about 1.7 times as large as those below, very rarely eyes separated on frons (see under 'Variation'). A pair of ocellar bristles, and somewhat longer postocular occipital bristles, black. Antennae (Fig. 456) brownish black, 3rd segment almost ovate but pointed at tip, about 2.5 times as long as deep, terminal style very short. Labrum almost as long as head is high, directed forwards, dark palpi elongate (Fig. 449). Mesonotum polished black on central part; anteriorly (including humeri), at sides and on prescutellar depression clothed with fine silver-grey pile. Acr and dc rather small, dark brownish, former broadly 4-serial, latter almost uniserial at about middle, more numerous anteriorly and ending in one (occasionally two) large prescutellar pair. All large thoracic bristles black, 6–8 scutellars.

Wings (Fig. 373) clear with yellowish to light brownish veins, stigma pale yellowish, rather indistinct. Squamae pale, halteres dirty greyish. Legs almost uniformly yellowish brown to brownish but fore coxae always paler. Pubescence short and pale, fore femora with longer pv hairs, hind femora with almost as long av bristly hairs towards tip, the longest being the erect bristly hairs on basal half of hind femur dorsally.

Abdomen shining black, clothed with long almost whitish hairs at sides, dorsum subshining and with much shorter hairs. Genitalia as in Figs. 459–461.

Length: body 1.8–2.5 mm, wing 2.1–2.5 mm.

♀. Eyes broadly separated by a polished black frons, all facets equally small. Labrum slightly longer than head is high, acr more conspicuous than in ♂ and irregularly 6-serial, halteres pale. Hind femora with 2 darker av bristles before tip and abdomen with shorter hairs.

Length: body 2.0–2.8 mm, wing 2.2–2.7 mm.

Variation. Rather a variable species in coloration of legs and wing-venation; crossveins may be absent (posterior crossvein m-m in var. *incompleta*) or duplicated, veins M_1 and M_2 either widely separated at base or almost petiolate. Eyes holoptic in ♂, but I have examined 2 ♂ with eyes separated by a polished frons about as deep as anterior ocellus: 1 ♂ from Denmark (NEZ, Tyvekrog, 7.vi.1908, W. Lundbeck), which perhaps led Lundbeck (1910) to describe this as a distinct species (*myricae*) with eyes separated in ♂ by a narrow black line (see also Collin, 1961: 303), and another ♂ from Russian Lapland (Lr (LPS), Kurvernööri, Hellén). There is a third ♂ in the Finnish Collection in Helsinki from N. Finland (Le, Saana, R. Frey) with a very broad frons (much deeper than anterior ocellus), but in view of other characters, including the structure of the ♂ genitalia, all three specimens are undoubtedly conspecific with *E. myrtilli*.

Holotype identifications, lectotype designation, synonymies. Through the kindness of Prof. G. Morge, Eberswalde, I studied the holotype ♀ of *E. simillima* Str. (Austria, Seitenstetten, 9.vi.1891, G. Strobl) and *E. myrtilli* var. *incompleta* Str. (Austria, Styriae alp., Kalbling, 28.viii.1896, G. Strobl) from the Coll. Strobl at Admont in 1980, and found them both to be identical with *E. myrtilli*. In the Dipt. Scand. Coll. at Lund, there are 2 ♂ and 5 ♀ of *myrtilli* under the name *Anthalia rostrata* Zett. from Sweden (Jmt., Åreskutan and Mullfjellet, and Ostrogothia) and Norway (TRy, Giebostad and NTi, Suul); the male from Åreskutan is herewith designated as lectotype of *A. rostrata* Zett.

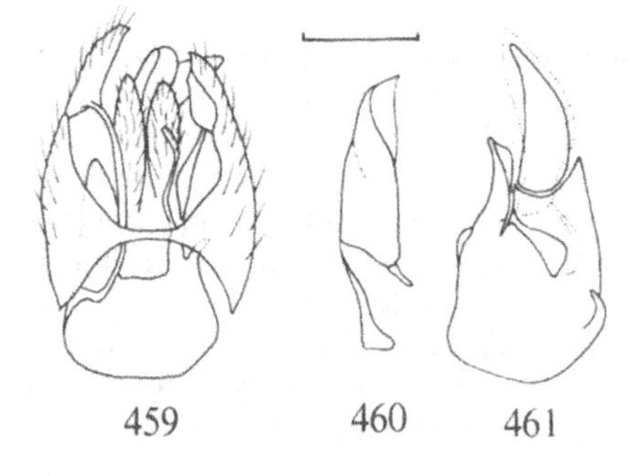

Figs. 459–461. Male genitalia of *Euthyneura myrtilli* Macq. - 459: dorsal view; 460: aedeagus; 461: hypandrium, aedeagus dotted. Scale: 0.1 mm.

and was labelled accordingly in 1977. Finally, in the Zoologisches Museum, Berlin, there is a single ♀ in Coll. Loew under *Anthepiscopus consobrinus* Zett., labelled "*Hemerodr. brevipes* Lw." in Loew's handwriting, which bears a small violet label (= Coll. Loew). This specimen, undoubtedly the holotype ♀ of *Hemerodromia brevipes* Loew, 1840, is also only a female of *E. myrtilli* Macq.

Distribution and biology. Common species in Denmark and Fennoscandia including the extreme north, also the Kola Peninsula. – The commonest, and the only *Euthyneura* species that is common in cold and temperate Europe; the southernmost occurrence known to me is from northern Romania (Mt. Rodnei); probably absent in the south. – Mid May to the beginning of August, but mainly in June and the first half of July. In C. Europe sporadically also in August.

Note. There are two further British *Euthyneura* species closely resembling *myrtilli* and not found in Scandinavia, *E. halidayi* Coll. and *E. myricae* Halid.; both may be easily distinguished from *myrtilli* by the pale mesonotal hairs and bristles, and by the presence of small spines beneath hind femora at tip. *E. myricae* is still known only from the old holotype ♀ (antennal style absent) and seems to be a rather problematic species. All the records in Lundbeck (1910), Frey (1950), Ringdahl (1951) and Lyneborg (1965) of *E. myricae* from Scandinavia, and that of Frey (1956) of *simillima*, actually refer to *E. myrtilli*. The S. European (Corsica) *E. inermis* (Beck.) (described as *Oedalea*), of which I have studied the holotype ♂, is another species closely related to *myrtilli* and *halidayi* with rather long whitish 6-serial acr and 2-serial dc, and with large mesonotal bristles (including 6 scutellars) pale, not "black" as stated by Becker (1910).

33. ***Euthyneura albipennis*** (Zetterstedt, 1842)
Figs. 458, 462–466.

Anthalia albipennis Zetterstedt, 1842: 250.
Euthyneura myrtilli Macquart; Frey, 1956: 597.

Small dark-legged species, wings milky-white with whitish veins. Thoracic hairs and bristles pale. ♂: eyes meeting on frons with upper facets enlarged, dorsum of abdomen silvery.

♂. Upper eye-facets about 1.7 times as large as those below. A pair of ocellar bristles and upper occipital bristly hairs blackish. Antennae (Fig. 458) black, 3rd segment more than twice as long as deep, conical and very pointed, almost straight above and below. Proboscis distinctly shorter than head is high. Thorax polished black, notopleural depression and scutellum largely greyish pollinose. 4-serial acr rather broadly separated from irregularly 2- to 3-serial dc, a pair of prescutellar dc very long and bristle-like, about as long as 2 notopleurals, a postalar and 4–6 scutellar bristles, all pale.

Wings almost milky-white with whitish veins, stigma invisible. Squamae with pale fringes, halteres whitish, stalk often brownish. Legs almost uniformly brown to dark brown, "knees" and hind metatarsi paler. All pubescence pale and rather short, hind femora before tip with av hair-like bristles not much shorter than those on dorsum.

Abdomen black, clothed with fine pale hairs which are especially long at sides, venter and sides of abdomen polished, dorsum densely almost silvery dusted and practically bare. Genitalia as in Figs. 462–466.

Length: body 1.6–1.9 mm, wing 2.2–2.5 mm.

♀. Resembling male except for sexual differences, abdomen not so silvery on dorsum.

Length: body 1.8–2.0 mm, wing 2.1–2.5 mm.

Lectotype designation. Described from Swedish Jämtland (Mullfjellet, Skalstugan) and Norway (Garnæs). There are 3 specimens of *E. albipennis* under *Anthalia albipennis* in the Dipt. Scand. Coll. at Lund, all syntypes: 1 ♂ from Garnæs, 1 ♀ labelled "Jemtl. Boh.", and another ♂ from Jämtland, labelled "Skalstugan, Jug. Alp. Jemtl. 16.–21. 7.40"; the ♂ from Skalstugan is in perfect condition and is herewith designated as lectotype of *Euthyneura albipennis* (Zett.), as labelled by me in 1977. Frey (1956) was unable to see Zetterstedt's types when sinking *albipennis* as a synonym of *myrtilli*. One ♀ from Gällivare (Sweden, Lu. Lpm.) in Coll. Becker in Berlin was identified by Becker as *Allanthalia pallida* "Lw.".

Distribution and biology. Rather a rare northern species (12 ♂ and 12 ♀ examined), known from N. Finland south to Ta, Sweden (Lu. Lpm., Jmt.) and Norway (NTi); also in the north of European USSR (Lr, Petsamo and Bjäloguba). – There is a single unconfirmed record outside Fennoscandia, by Cole (1964) who reported a male (possibly *albipennis*) bred from rotten wood on 29th April in Oxfordshire, England. – End June, July.

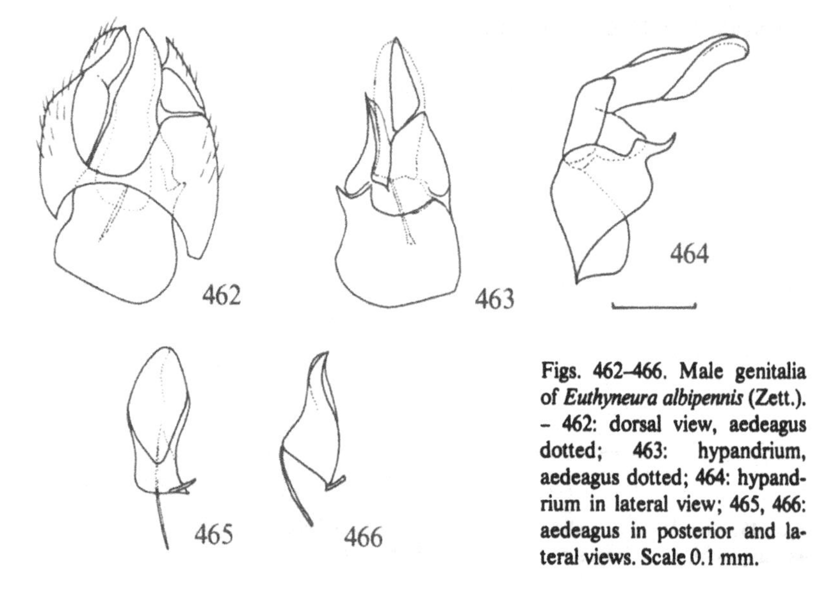

Figs. 462–466. Male genitalia of *Euthyneura albipennis* (Zett.). – 462: dorsal view, aedeagus dotted; 463: hypandrium, aedeagus dotted; 464: hypandrium in lateral view; 465, 466: aedeagus in posterior and lateral views. Scale 0.1 mm.

34. ***Euthyneura gyllenhali*** (Zetterstedt, 1838)
Figs. 372, 450–453, 457, 467–469.
Anthalia Gyllenhali Zetterstedt, 1838: 538.

Wings with a dark stigma and an incomplete brown cross-band. Legs yellowish, large thoracic bristles pale. ♂: eyes narrowly separated on frons with all facets equally small.

♂. Frons very linear, upper eye-facets not enlarged. Occiput thinly brownish dusted with all hairs and postocular ciliation dark. Antennae (Fig. 457) black, 3rd segment short ovate and very pointed apically, style slender and rather long, almost half as long as 3rd segment. Proboscis half as long as head is high, palpi dark with several long black hairs. Thorax polished black, greyish dusted on notopleural depression, all hairs and bristles pale; acr 4-serial anteriorly, irregularly 5-serial behind, broadly separated from equally small dc which are nearly uniserial at about middle but more numerous anteriorly and posteriorly, 1 prescutellar pair long and bristle-like, as long as 1 notopleural; 1 shorter postalar and 4 scutellar bristles, outer pair smaller.

Wings (Fig. 372) with a dark stigma and a brown middle cross-band, usually in the form of a brownish streak reaching from costa to vein R_{4+5}. Costa and radial veins dark brown. Squamae and halteres pale yellow. Legs wholly yellow or at most tarsi towards tip and hind femora on apical half more or less brownish. Pubescence pale and fine, the longest hairs on hind femora on basal half above, ventrally before tip several short dark bristly hairs.

Abdomen almost polished black, dorsum indistinctly greyish pollinose. Pubescence pale and rather dense, hind marginal hairs longer and more bristle-like. Genitalia as in Figs. 467–469.

Length: body 2.0–2.8 mm, wing 2.5–3.1 mm.

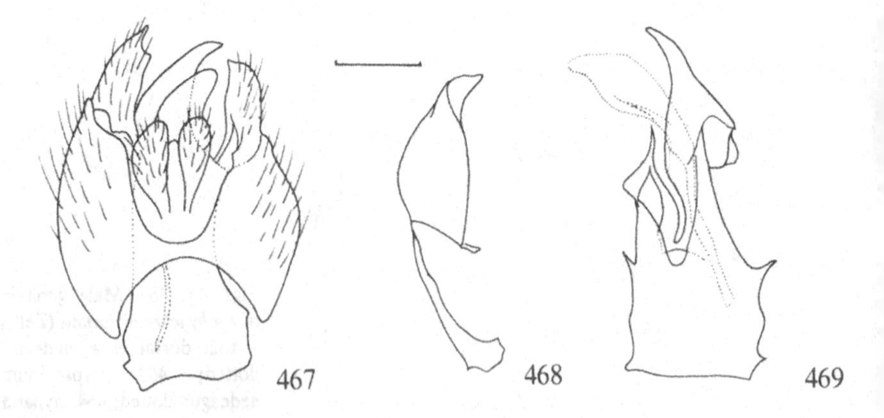

Figs. 467–469. Male genitalia of *Euthyneura gyllenhali* (Zett.). – 467: dorsal view; 468: aedeagus; 469: hypandrium, aedeagus dotted. Scale: 0.1 mm.

♀. Resembling male but frons broad, polished black, and abdomen very telescopic, apical two segments "ovipositor"-like.
Length: body 2.4–3.3 mm, wing 2.6–3.3 mm.

Lectotype designation. There are 3 ♀ under *Anthalia Gyllenhali* in the Ins. Lapp. Coll. at Lund from Umenäs, Wilhelmina and Åreskutan; the ♀ from Åreskutan (Jmt.) is not a syntype as it is dated 31.7. - 1–4.8.40. The ♀ labelled "Umenäs" (Ly. Lpm.) is herewith designated as lectotype of *Euthyneura gyllenhali* (Zett.) and was labelled accordingly in 1977. There are a further 5 specimens of *gyllenhali* in the Dipt. Scand. Coll., a pair from Mullfjellet and 3 ♀ from Umenäs.

Distribution and biology. Widespread but uncommon; Denmark, in Sweden north to Ly. Lpm., in Finland to Ks and LkW, approximately to 67°N; also in Lapponia rossica (USSR), but not yet found in Norway, although it should occur there. - Great Britain, in C. Europe preferring higher altitudes but not at all common, south to Bulgaria (Rila 1500–1850 m, 23.viii.1972, 1 ♀, A. Merta), and Kovalev (in litt.) has recorded it from central parts of European USSR. - Mainly in June and July, the dates ranging from 24 May to 4 August.

Genus *Anthalia* Zetterstedt, 1838

Anthalia Zetterstedt, 1838: 538.
Type-species: *Anthalia schoenherri* Zetterstedt, 1838 (des. Melander, 1928)*.

Very small (body length about 1.5–2.0 mm) black or yellow species (Fig. 470) with long clear wings. Eyes meeting on frons in ♂, with upper facets more or less enlarged, separated by a very broad frons in ♀, with all facets equally small. Face broad and rather short in both sexes. A long pair of anterior ocellar bristles, posterior pair only minute, and several distinct upper postocular occipital bristles. Antennae inserted at about middle of head in profile in ♀, head larger in ♂ (Fig. 471) and antennae inserted further below. Basal antennal segment very small, indistinct, 2nd segment globular and armed with a circlet of small preapical bristly hairs, 3rd segment rather large, very short ovate and laterally compressed; terminal style very small, 1-articulated, rarely (in Nearctic species) almost as long as 3rd segment and then 2-articulated (Melander, 1928). Mouthparts (Figs. 474–476) short, more or less pointing forwards, rarely surpassing the length of head (in Nearctic species). Palpi conical, usually with several preapical bristly hairs, situated on large bristly palpifers but maxillae (stipites) apparently firmly fused with labial paraphyses. Labrum heavily sclerotised and very convex,

*) Zetterstedt (1838) included 3 new species in the genus *Anthalia*, and Coquillett (1903) designated the first of these, *gyllenhali*, as type of the genus. By doing this Coquillett synonymised *Anthalia* with the originally monotypic *Euthyneura* Macquart, 1836. To preserve the generic name *Anthalia*, Melander (1928) designated the second originally included species, *schoenherri*, as type of the genus. The I. C. Z. N. has recently been requested (Chvála & Smith, 1982) to set aside Coquillett's original type-species designation and to validate Melander's.

hypopharynx stylet-like, as long as labrum. Labium with large membraneous labellae armed with pseudotracheae.

Thorax distinctly arched above, with well developed humeral and postalar calli, distinct scutellum and numerous large bristles (humerals, posthumerals, notopleurals, a postalar, usually 2 pairs of prescutellar dc and 4–10 scutellars); small mesonotal hairs (numerous acr and dc) fine and short, acr at least 4-serial and rather narrowly separated from multiserial dc.

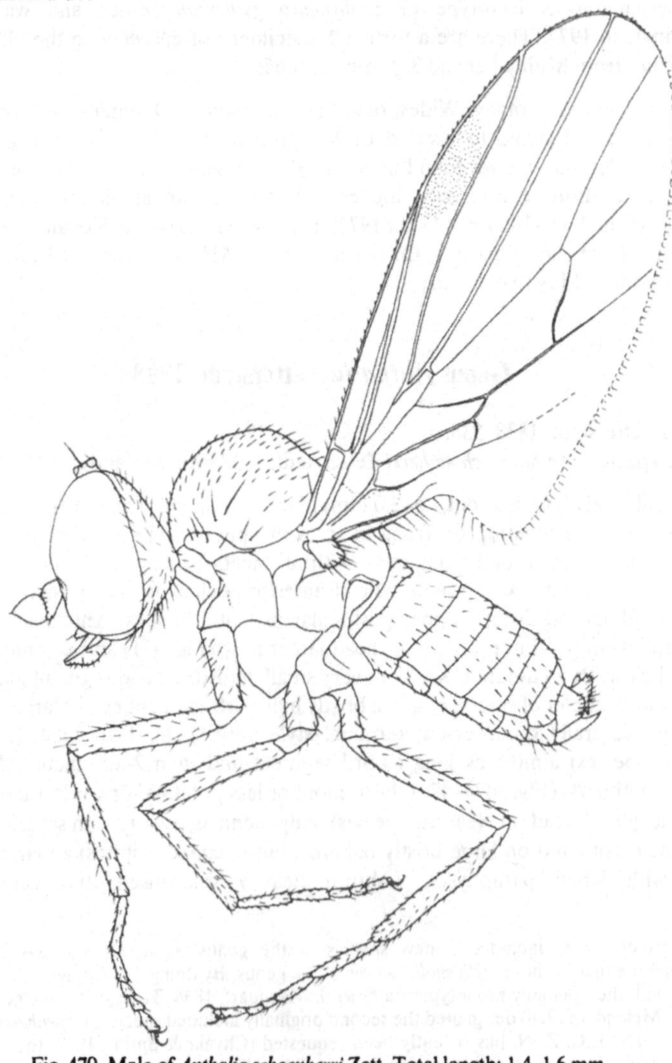

Fig. 470. Male of *Anthalia schoenherri* Zett. Total length: 1.4–1.6 mm.

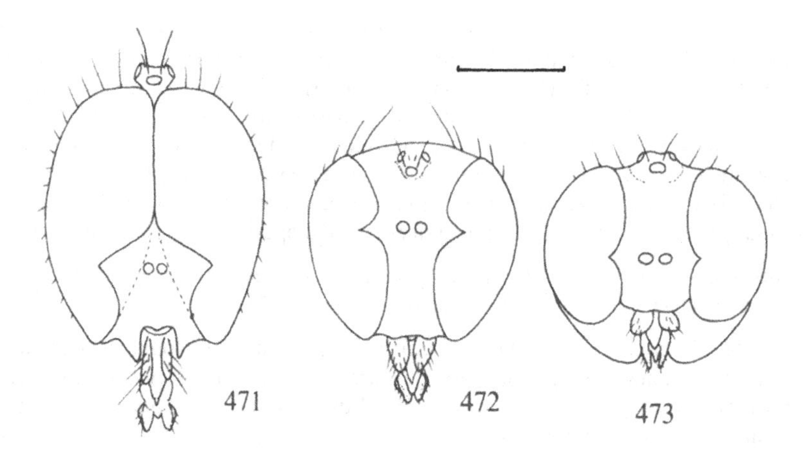

Figs. 471–473. Heads in anterior view. – 471: *Anthalia schoenherri* Zett. ♂; 472: same species ♀; 473: *Allanthalia pallida* (Zett.) ♂. Scale: 0.2 mm.

Wings (Fig. 374) clear or slightly dusky, large, with a distinct costal bristle, stigma long ovate but not reaching tip of vein R_{2+3}. Costa running to tip of M_1, Sc closely approximated to R_1 but fading away before reaching costa, veins R_{4+5} and M_1 parallel. Discal cell large, emitting 3 veins to wing-margin, vein M_2 incomplete in European species, sometimes complete in Nearctic species. Both basal cells equally long and narrow (2nd basal cell not broadened as in *Euthyneura*), anal cell not very much shorter than 2nd basal cell and equally deep. The vein closing anal cell very slightly recurrent, anal vein faint and disappearing before reaching wing-margin.

Legs rather short and slender, devoid of distinct bristles, clothed with minute hairs only. Fore tibiae and hind femora rarely thickened, hind femora never with ventral spines towards tip as in *Oedalea*. Tibial gland on fore tibiae very indistinct.

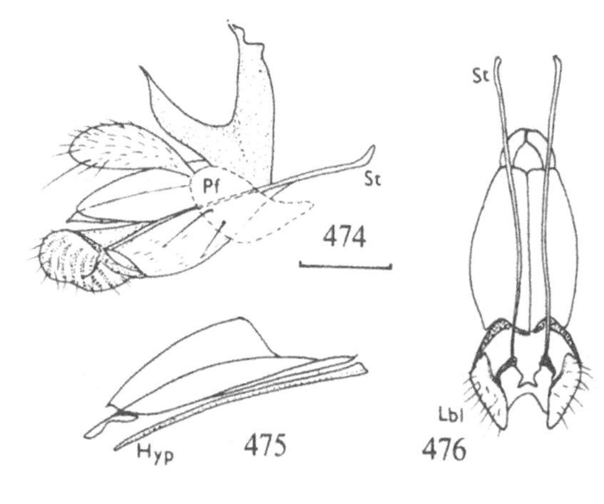

Figs. 474–476. Mouthparts of *Anthalia schoenherri* Zett. ♂. – 474: lateral view; 475: labrum and hypopharynx; 476: labium and maxillae in dorsal view. Scale: 0.1 mm. For abbreviations see p. 13.

Abdomen rather cylindrical and somewhat laterally compresed in ♂, clothed with rather short fine hairs, no distinct bristles; 8th segment small and mostly concealed within the 7th. Genitalia (Figs. 480–487) rather small, asymmetrical and rotated towards right, lamellar processes connected by narrow folds with hypandrial bridge. Aedeagus with a long coaxial apodeme, bifid apically and loosely attached to the inner wall of the asymmetrical hypandrium, which has a conspicuous hook-like process at middle. Abdomen in ♀ not laterally compressed, broad on 1st to 5th segments, apical segments gradually narrowed, becoming slender and ovipositor-like. Cerci small, ovate.

Distribution. An entirely Holarctic genus; 11 species are known from North America, of which one, *A. schoenherri*, also occurs in Scandinavia.

Biology. The adults are nectar (or pollen) feeders, and are often found in large numbers on blossoming bushes and trees. Swarming activity in *A. schoenherri*, with aggregations of many individuals, was observed by Frey (1956) in Finland. Immature stages and development unknown.

35. ***Anthalia schoenherri*** Zetterstedt, 1838
Figs. 374, 470–472, 474–477, 480–487.

Anthalia Schoenherri Zetterstedt, 1838: 539.

A small black species with a short projecting proboscis, short antennae with 3rd segment almost circular and bearing a minute terminal style, wings with vein M_2 abbreviated.

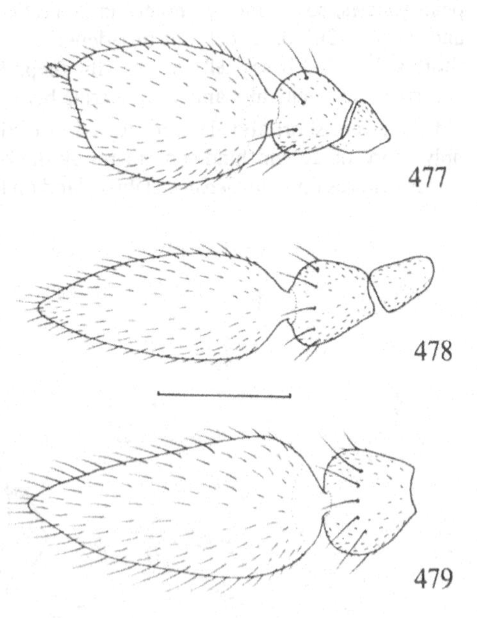

Figs. 477–479. Antennae. – 477: *Anthalia schoenherri* Zett. ♂; 478: *Allanthalia pallida* (Zett.) ♂; 479: same species ♀. Scale: 0.1 mm.

♂. Eyes meeting on frons (Fig. 471) with upper facets considerably enlarged, at least twice as large as those below; face rather short, almost as deep as 3rd ant.s., dull black and parallel-sided. A pair of long black ocellar bristles, upper postocular bristles shorter. Antennae (Fig. 477) with 3rd segment broadly ovate, apically with a very short 1-articulated style. Proboscis (Fig. 474) short, as long as 3rd ant. s., palpi black, the dark preapical bristle apparently longer than palpus. Thorax subshining black, pleura more densely grey dusted, all hairs and bristles black; acr broadly irregularly 6-serial, narrowly separated from multiserial dc, all short but rather coarse; 1 long humeral, 1 posthumeral, 2 notopleurals, 1 postalar, 1 pair of long prescutellar dc, and 2 pairs of scutellars, inner pair longer.

Wings (Fig. 374) almost clear with dark brown veins, stigma faintly brownish; vein M_2 abbreviated on apical half. Squamae and halteres blackish. Legs uniformly dark brown to blackish brown, rather short and simple except for the spindle-shaped fore tibiae. Legs covered with only minute dark hairs, hind femora with longer hairs dorsally and ventrally, and still with 2 longer curved av bristly hairs before tip.

Abdomen subshining black, dorsum very thinly silvery grey pollinose, pubescence short and dark, hind marginal hairs at sides somewhat longer. Genitalia (Figs. 480–487) small and distinctly upturned.

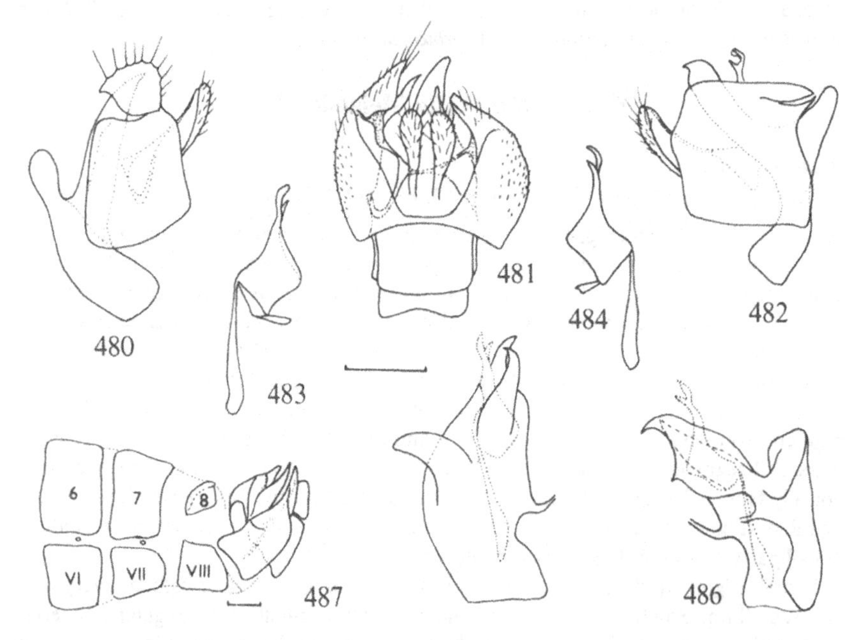

Figs. 480–487. Male genitalia of *Anthalia schoenherri* Zett. – 480: right lateral view; 481: dorsal view; 482: left lateral view; 483, 484: aedeagus in two views, 485: hypandrium in laterodorsal view, aedeagus dotted; 486: the same in lateroventral view; 487: postabdomen. Scale: 0.1 mm. For abbreviations see p. 13.

Length: body 1.4–1.6 mm, wing 1.7–2.0 mm.

♀. Very much like male but eyes separated by a very broad frons (Fig. 472) that widens above, all facets equally small, squamae and halteres yellowish, all coxae and femora with a tendency to be lighter brown, and apical abdominal segments ovipositor-like.

Length: body 1.5–2.0 mm, wing 1.9–2.0 mm.

Lectotype designation. There are 4 ♀ of *A. schoenherri* from Gransele in the Ins. Lapp. Coll. at Lund, all in very poor condition. In the Dipt. Scand. Coll. there are 2 ♂ and 4 ♀ under the same name, labelled "Lapp." and bearing the same small black label as the species in the Ins. Lapp. Coll. All specimens are conspecific and are undoubtedly syntypes. One male from the Dipt. Scand. Coll., labelled "*A. Schönherri* ♂ Lapp.", with abdomen partly eaten by dermestids but well-preserved genitalia, was labelled in 1977 and is herewith designated as lectotype of *A. schoenherri* Zett.

Distribution and biology. Northern species; in Sweden only from Nb. (Storsund), but rather common in Finland from the south (Ab, N) north to Ks; also Russian Karelia and north to the Kola Peninsula (Kantalaks). – Apart from Fennoscandia and the NW of European USSR it has also been recorded from North America, but I have not studied any Nearctic specimens. – June and July; according to Tuomikoski (1952) very common on flowers of mountain-ash (*Sorbus aucuparia*).

Genus *Allanthalia* Melander, 1928

Allanthalia Melander, 1928: 61.

Type-species: *Anthalia pallida* Zetterstedt, 1838 (orig. des.).

A small yellow species (Fig. 488), about 1.5 mm long, with simple legs, dichoptic eyes in both sexes, and antennae without style. Eyes separated by a broad frons in both sexes that widens above, all facets equally small. Face (Fig. 473) as deep as frons above antennae, but in dried specimens eyes often meeting below, antennal excisions inconspicuous. Facial part prominent, distinct broad cheeks below eyes. Anterior pair of ocellar bristles rather small and fine, as long as small postocular bristly hairs, posterior pair minute. Antennae (Fig. 478) inserted slightly above middle of head in profile, basal segment very small, 2nd segment globular and armed with a circlet of small preapical hairs, 3rd segment long ovate, about twice as long as deep near base, slightly pointed apically (somewhat conical); terminal style absent. Mouthparts very small, practically invisible, hidden in the mouth-cavity. General structure of mouthparts as in *Anthalia* but all parts shorter and palpi short ovate.

Thorax slightly arched on dorsum, with distinct humeri, postalar calli and very convex scutellum. Small thoracic hairs fine and rather small, acr irregularly 4-serial, separated from at least 3-serial dc, the hairs of inner row becoming longer posteriorly and ending in 2 long prescutellar pairs. Large bristles in full number: 1 humeral, 1 posthumeral, 2 notopleurals, 1 postalar and 2 pairs of scutellars. Prosternum small, *Hybos*-like.

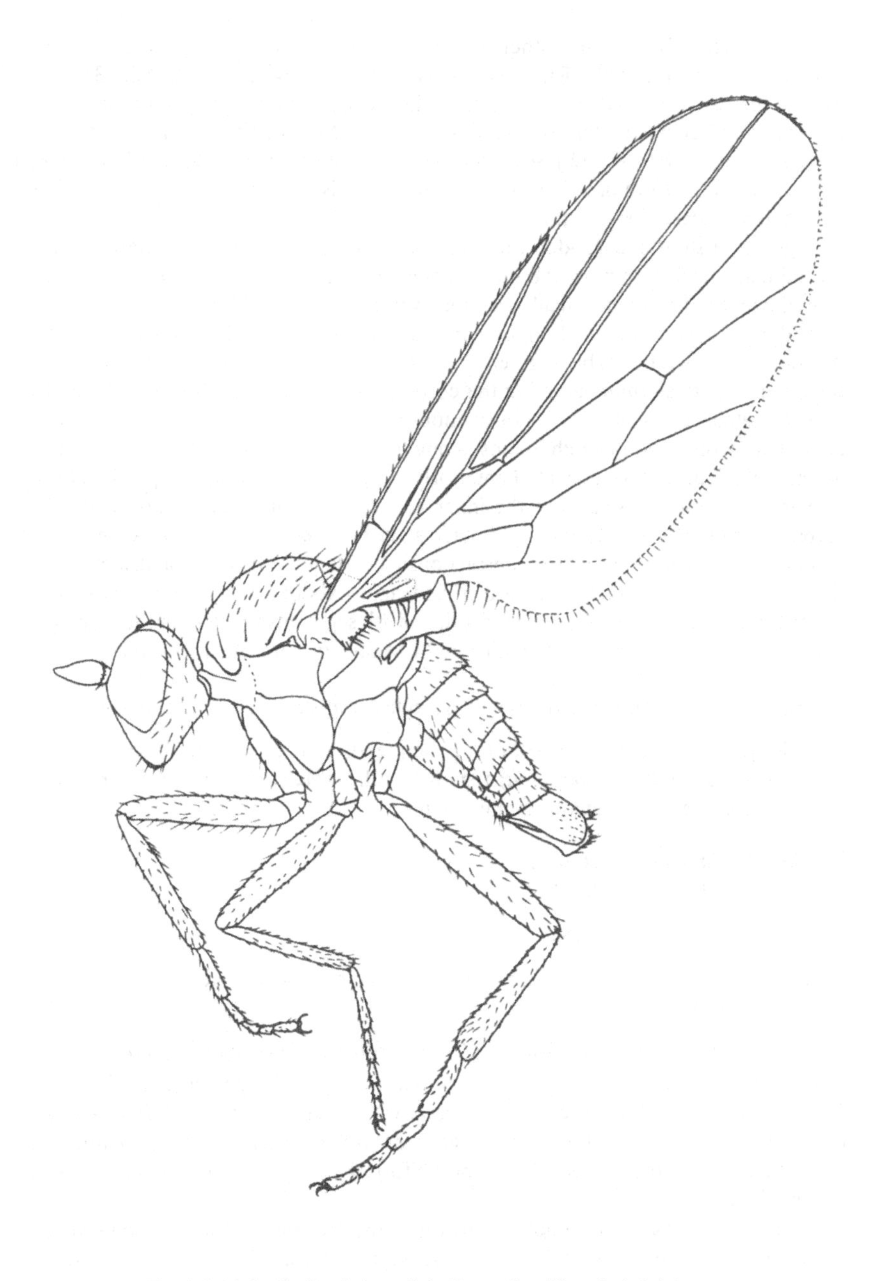

Fig. 488. Male of *Allanthalia pallida* (Zett.). Total length: 1.4–1.6 mm.

Wings (Fig. 375) long and rather narrow, a long costal bristle, stigma as in *Anthalia* but very pale and indistinct. Radial sector as long as in previous genus, about 3 times as long as the distance between its origin and humeral crossvein. Venation as in *Anthalia* but discal cell decidedly longer and narrower, vein M_2 practically reaching the wing-margin, at most abbreviated just before tip, and anal vein very indistinct beyond anal cell, rather in the form of a fine fold. Axillary lobe distinct, although not as prominent as in *Anthalia,* axillary excision very obtuse.

Legs rather short and slender, fore tibiae slightly spindle-shaped with only a trace of the tubular gland near base. All pubescence very short and fine, no distinct bristles. Apical two tarsal segments on all pairs dorsoventrally flattened.

Abdomen cylindrical, finely pubescent, hind marginal hairs longer and bristle-like. Abdomen in ♂ with 6 visible segments, 7th and 8th segments mostly concealed within the large 6th, 8th sternum large but the corresponding tergum in the form of a narrow ring. Genitalia (Figs. 489–494) conspicuously large, elongated, about 1/4 length of abdomen, twisted towards right almost along the longitudinal axis. Periandrial lamellae asymmetrical, narrowly connected anteriorly, very firmly joined with hypandrial bridge by narrow sclerites arising from the inner wall of terminal processes (as in *Anthalia*). Hypandrium very elongate and asymmetrical, its broadest terminal process very firmly connected with aedeagus, which conspicuously overlaps the tip of hypandrium; the latter very cornet-like, closing the base of aedeagus. Abdomen in ♀ gradually tapering towards tip, basal five segments broad, 6th and 7th segments smaller but the ovipositor-like tip of abdomen formed by the 8th and 9th segments; cerci elongate-oval.

Distribution. A monotypic genus with a single (supposedly) Holarctic species.

Biology. The adults are nectar or (?) pollen feeders, and are often found on flowers of mountain ash *(Sorbus aucuparia)* together with *Anthalia schoenherri.* Swarming activity not observed, and immatures unknown.

36. ***Allanthalia pallida*** (Zetterstedt, 1838)
Figs. 375, 473, 478, 479, 488–494.

Anthalia pallida Zetterstedt, 1838: 539.

Very small yellowish species with eyes dichoptic in both sexes and short brownish antennae without style.

♂. Frons and occiput yellowish to yellowish brown, ocellar tubercle often darkened; all bristles on head brownish. Base of antennae yellow, 3rd segment (Fig. 478) mostly dark brown, apical style absent. Mouthparts very small, brownish. Thorax yellow to yellowish brown, mesonotum often darkened or with a darker median stripe down the rows of acr; entire thorax very thinly and uniformly silvery grey dusted, thoracic hairs and bristles pale.

Wings (Fig. 375) clear with light brownish veins, squamae and halteres pale yellow. Legs uniformly pale yellow, apical two tarsal segments scarcely darker, all pubescence pale.

Abdomen yellowish brown, dorsum often extensively darkened and dulled by silver-grey dusting, covered with minute pale hairs, sides and hind margins of terga with longer hairs. Genitalia (Figs. 489–494) yellow with tip of left lamella blackish.

Length: body 1.4–1.6 mm, wing 2.0–2.1 mm.

♀. Very much like male, but abdomen tapering apically and covered with short hairs everywhere, thorax and abdomen sometimes extensively darkened, almost blackish-brown (wholly yellow in central European populations).

Length: body 1.6–1.8 mm, wing 2.1–2.4 mm.

Lectotype designation. In the Ins. Lapp. Coll. at Lund there is only a label under *A. pallida* "Ins. Lapp. ♂", but no specimens. In the Dipt. Scand. Coll. there is one ♂ in good condition, labelled "*A. pallida* ♂ Lycksele", which is herewith designated as lectotype of *Allanthalia pallida* (Zett.) and was labelled accordingly in 1977.

Distribution and biology. A species with the same type of distribution in Fennoscandia as *Anthalia schoenherri*, but everywhere rather rare; from N. Sweden (Ly. Lpm.) and throughout Finland from Ab and N north to Ks (although no records from C. Finland available), also in Russian Karelia and the Kola Peninsula (Kantalaks). – Mountains of C. Europe (Krkonoše Mts., Czechoslovakia) and North America. – All specimens with dates recorded are from July.

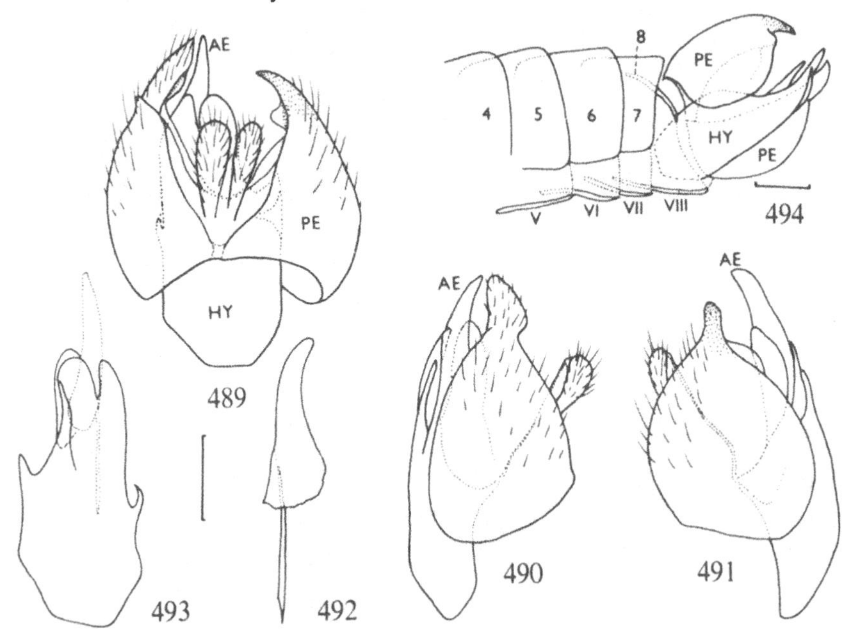

Figs. 489–494. Male genitalia of *Allanthalia pallida* (Zett.). – 489: dorsal view; 490: right lateral view; 491: left lateral view; 492: aedeagus; 493: hypandrium, aedeagus dotted; 494: postabdomen. Scale: 0.1 mm. For abbreviations see p. 13.

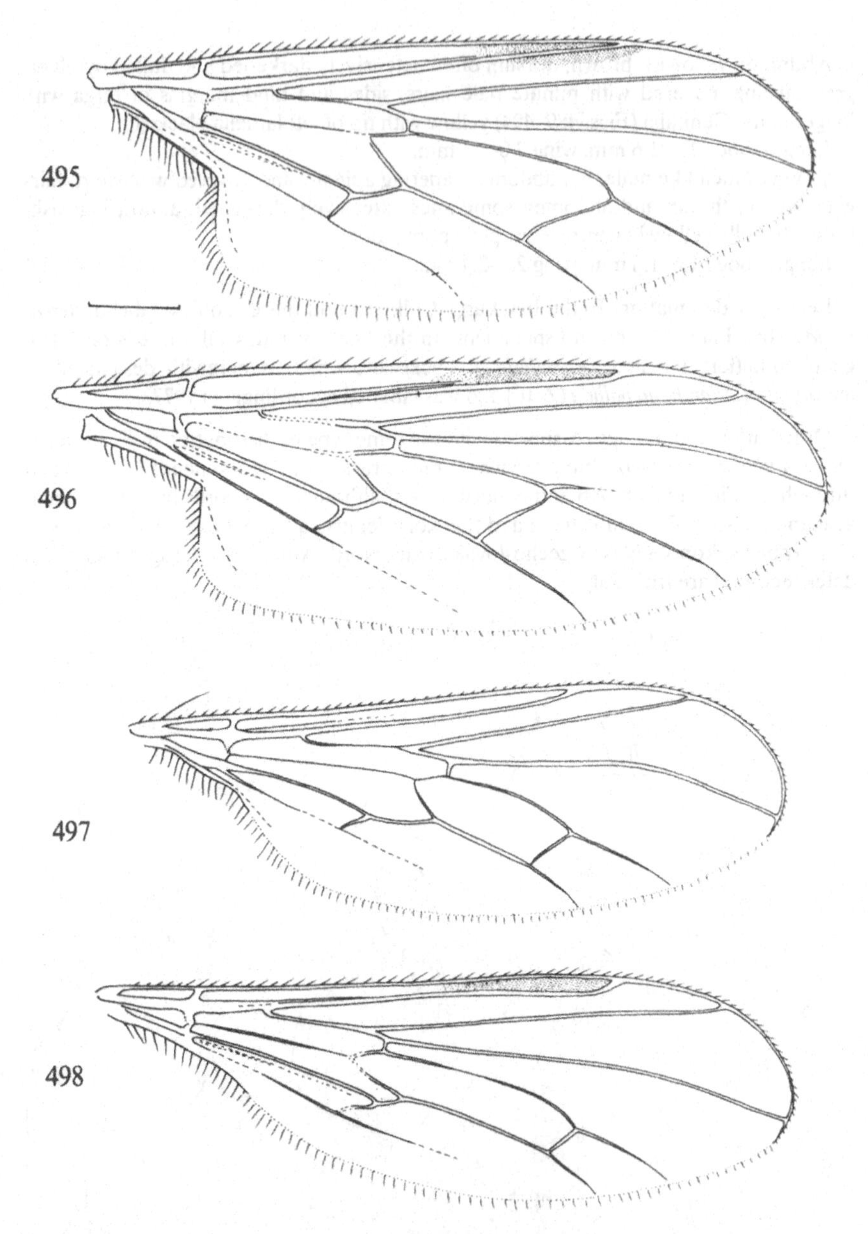

Figs. 495–498. Wings of Ocydromiini. – 495: *Ocydromia glabricula* (Fall.); 496: *Leptopeza flavipes* (Meig.); 497: *Leptodromiella crassiseta* (Tuomik.); 498: *Oropezella sphenoptera* (Loew). Scale: 0.5 mm.

Tribe Ocydromiini

Head holoptic or very narrowly dichoptic, face (if present) linear; arrangement of eyes generally identical in both sexes. Antennae with 3rd segment small, style arista-like, bare and very long, terminal and with a small basal article; in *Ocydromia* arista supra-apical and 1-articulated. Proboscis short and pointing downwards. Mesonotal hairs small and inconspicuous, acr biserial, dc uniserial (multiserial in *Leptopeza*). Discal cell emitting 2 veins to wing-margin, or with a short stump of the third upper vein (M_1). Hind legs generally elongated and raptorial, tibiae bristled except in *Ocydromia*. Male genitalia rotated through 90° towards right but not upturned, asymmetrical, with large appendages on periandrial lamellae; hypandrium rather small and almost symmetrical. Aedeagus rather short with curious long terminal appendages, no postgonites. Female abdomen ovipositor-like only in *Leptopeza*. Adults predaceous; larvae in dung (*Ocydromia* larviparous), or decaying vegetable matter or rotten wood.

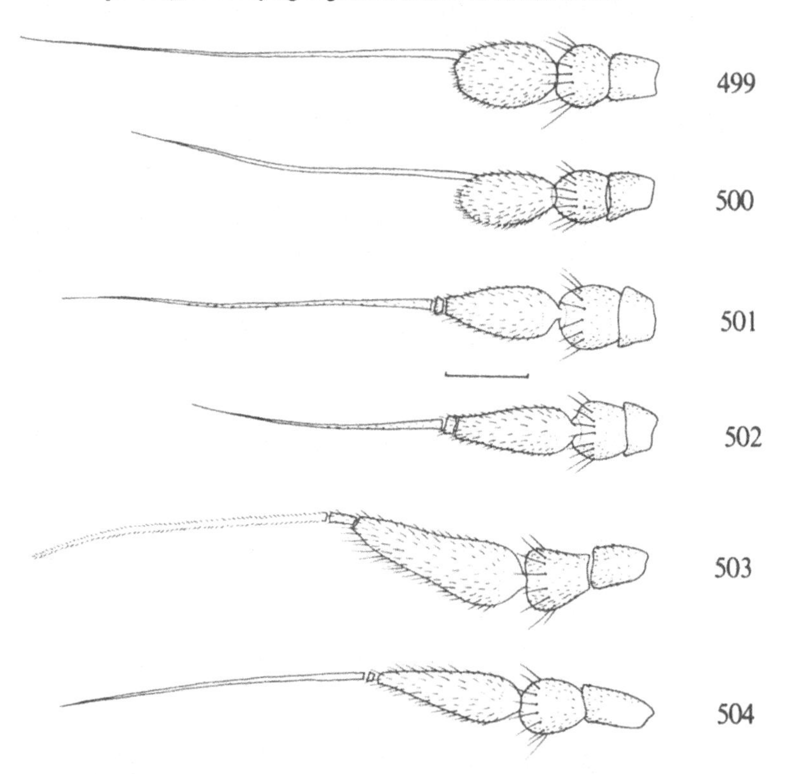

Figs. 499–504. Antennae of Ocydromiini. – 499: *Ocydromia glabricula* (Fall.) ♂; 500: *Ocydromia melanopleura* Loew ♂; 501: *Leptopeza flavipes* (Meig.) ♂; 502: *Leptopeza borealis* Zett. ♂; 503: *Leptodromiella crassiseta* (Tuomik.) ♂; 504: *Oropezella sphenoptera* (Loew) ♂. Scale: 0.1 mm.

Genus *Ocydromia* Meigen, 1820

Ocydromia Meigen, 1820: 351.
Type-species: *Empis glabricula* Fallén, 1816 (des. Westwood, 1840).
Euscinesia v. Gistl, 1848: x (unjustified replacement name for *Ocydromia*).
Type-species: isogenotypic with *Ocydromia*.

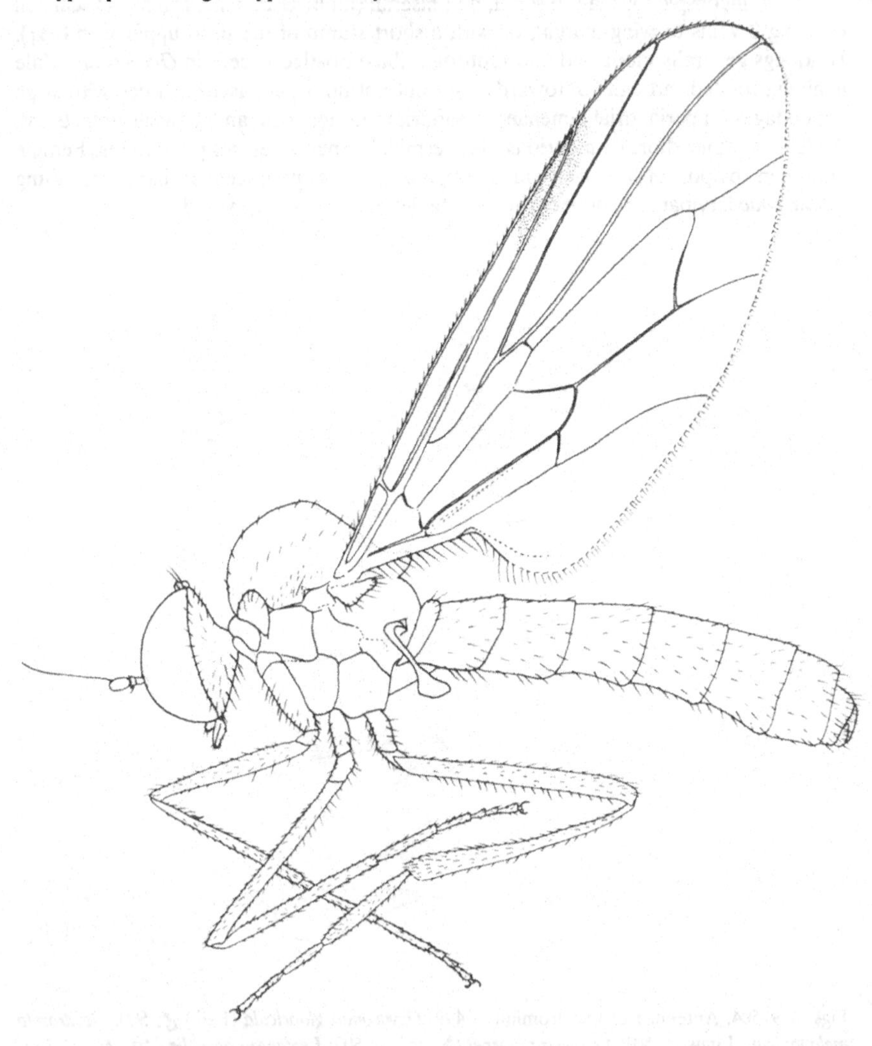

Fig. 505. Male of *Ocydromia glabricula* (Fall.). Total length: 3.2–4.5 mm.

Medium-sized (body length about 3–4.5 mm) long-legged species (Fig. 505) with body polished black to yellow. Very variable in colour, and almost devoid of distinct bristles. Head (Fig. 511) nearly globular, eyes meeting on frons in ♂, very narrowly separated in ♀, leaving frons very linear. Upper facets indistinctly enlarged in ♂, uniformly small in ♀. Face linear in both sexes, scarcely broader than frons in ♀. Occiput convex both above and below neck, covered with fine hair-like bristles, anterior pair of ocellar bristles long but weak, posterior pair minute. Antennae (Figs. 499, 500) inserted at about middle of head in profile; basal segment small, scarcely longer than deep, 2nd segment globular with a circlet of small preapical hairs; 3rd segment ovate, about as long as basal two segments combined, with a very long, bare, supra-apical, 1-articulated, arista-like style. Mouthparts (Figs. 506–510) short and pointing downwards, palpi flattened and rather long ovate, apically with several (usually three) preapical bristly hairs, attached to bristled (2 bristles) palpifers. Maxillae reduced to slender stipites, laciniae absent. Labrum and hypopharynx sclerotised, labium soft with large membraneous labellae with pseudotracheae.

Thorax almost globular, polished and nearly bare, very arched above with small but well-developed humeri and postalar calli. Small thoracic hairs indistinct, acr irregularly 2-serial, dc uniserial and scarcely longer posteriorly, large bristles fine: 3 notopleurals, 1 postalar and 1–5 pairs of scutellars, but only apical pair long. Scutellum, notopleural depression and metapleura covered with a fine dense pile-like pubescence.

Wings (Fig. 495) large, iridescent, usually much darker in ♂, entirely covered with microtrichia. Costa ending at tip of vein R_{4+5}, vein Sc closely approximated to R_1 and fading away before reaching costa; a distinct elongate stigma at tip of vein R_1. Radial sector about as long as the distance between its origin and humeral crossvein. A large discal cell emitting 2 veins to wing-margin, vein M_1 absent, rarely in the form of a short

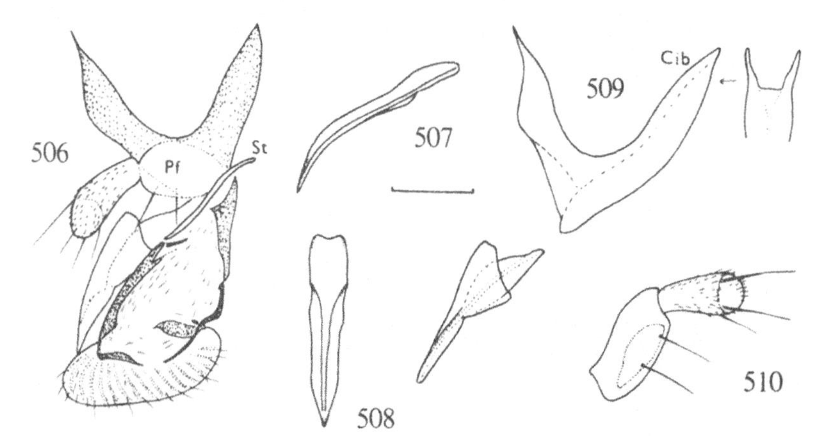

Figs. 506–510. Mouthparts of *Ocydromia glabricula* (Fall.) ♂. – 506: lateral view; 507: hypopharynx in lateral view; 508: the same in ventral view; 509: basal sclerites with labrum; 510: palpus. Scale: 0.1 mm. For abbreviations see p. 13.

stump. Basal cells equally long but 2nd basal cell broader, anal cell shorter than basal cells, the vein closing it slightly recurrent or nearly at right angles to anal vein, which is distinct throughout except for extreme tip. Axillary lobe well-developed but axillary excision obtuse. Costal bristle absent, no alula.

Legs long and slender, covered with only fine hairs, devoid of distinct bristles. Hind tibiae towards tip and hind metatarsus sometimes slightly thickened. The sense organ (Figs. 512, 513) on fore tibiae near base very distinct and its opening armed with a specifically characteristic tuft of hairs.

Abdomen long and slender, distinctly compressed laterally and covered with only weak hair-like bristles. 8th segment in ♂ well visible, tergum not very much smaller than preceding terga but sternum on the contrary broader than the very narrow preceding sterna. Genitalia (Figs. 514–520) twisted round through almost 90° to the right along longitudinal axis, periandrial lamellae asymmetrical with simple large terminal process, which is firmly connected at base by narrow sclerotised folds to hypandrial bridge. Hypandrium almost symmetrical, small and distinctly convex, with a broad hypandrial bridge; cerci rather large. Aedeagus with a short leaf-shaped coaxial apodeme and a curious bifid terminal appendage; base of aedeagus firmly attached to the hypandrial bridge. Female abdomen (Figs. 76, 77) blunt-tipped, not at all ovipositor-like (thus resembling *Leptodromiella* and *Oropezella*), cerci short ovate.

Distribution. Only a few *Ocydromia* species are known with certainty, several others attributed to *Ocydromia* undoubtedly belong to other ocydromiine genera. Two species occur in the Palaearctic including Scandinavia, one of which *(O. glabricula)* is apparently Holarctic; two species are known from Africa (Smith, 1969) and perhaps a further two from the Oriental region.

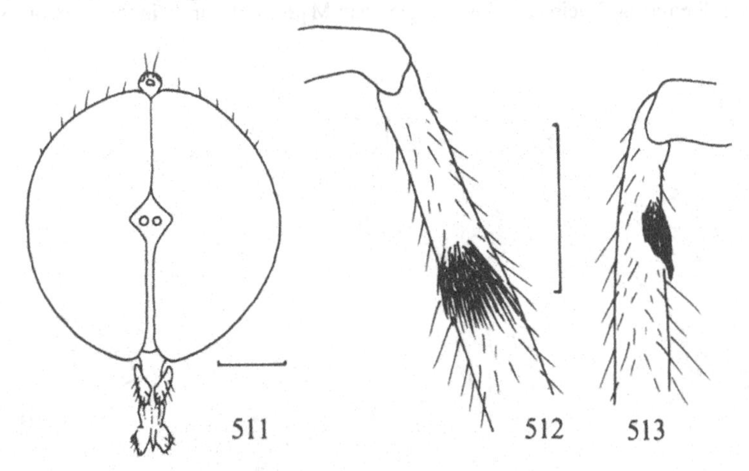

Fig. 511. Head in frontal view of *Ocydromia glabricula* (Fall.) ♂.
Figs. 512–513: Base of fore tibia in *Ocydromia;* anteroventral view, with the sense organ. – 512: *glabricula* (Fall.) ♂; 513: *melanopleura* Loew ♂. Scale: 0.2 mm.

Biology. The adults occur for a long period from early spring to late autumn, on vegetation in shady places or slowly hovering above the ground. Collin (1961) observed hovering males of *glabricula* before evening at 5 p. m. in England, and Tuomikoski (1939, 1952) recorded male swarms during the evening hours. Immatures of *glabricula* have repeatedly been described; the larvae live in dung and decaying vegatable matter and females are viviparous, scattering larvae when flying over dung (Grunin, 1953; Hobby & Smith, 1962). The second Palaearctic species, *O. melanopleura,* is very probably also viviparous, as I have seen small dead larvae attached to the tip of the ♀ abdomen in several dried specimens.

Key to species of *Ocydromia*

1 Larger, about 3.5–4 mm long, scutellum with 3–5 pairs of scutellar bristles. Thorax mostly black in ♂, more or less yellow in ♀. 3rd ant.s. rather small (Fig. 499); the sense-organ on fore tibiae anteriorly near base in the form of a blunt-tipped brush of adpressed hairs (Fig. 512) .... 37. *glabricula* (Fallén)

\- Smaller, about 3 mm long, scutellum with only 1 pair of black terminal bristles, other pairs indistinguishable from short black pubescence. Thorax including pleura black in both sexes. 3rd ant.s. larger (Fig. 500); the sense-organ on fore tibiae in the form of an apically pointed and narrower brush of slightly erect hairs (Fig. 513) .... 38. *melanopleura* Loew

37. ***Ocydromia glabricula*** (Fallén, 1816)
Figs. 52, 77, 495, 499, 505, 506–512, 514–522

Empis glabricula Fallén, 1816: 33.
Ocydromia rufipes Meigen, 1820: 353.
Ocydromia scutellata Meigen, 1820: 354.
Ocydromia fuscipennis Macquart, 1823: 147.
Ocydromia dorsalis Meigen, 1830: 334.
Ocydromia coxalis v. Roser, 1840: 53.
Ocydromia peregrinata Walker, 1849: 488.

Generally larger, about 3.5–4 mm, with thorax polished black to almost wholly yellow (in pale ♀), scutellum with at least 3 pairs of distinct bristles.

♂. Antennae (Fig. 499) black, 3rd segment rather small, almost as long as basal segments. Palpi blackish, proboscis yellowish brown. Thorax wholly polished black, with all intermediates through to specimens with thoracic pleura, humeri, postalar calli and scutellum yellowish, but mesonotum (in comparison with ♀) always black; 6–10 scutellar bristles brownish to almost black, terminal pair the longest.

Wings (Fig. 495) more or less brownish clouded, sometimes almost blackish, always very iridescent. Stigma on very darkened wings slightly visible. At least coxae, trochanters and base of femora yellowish, otherwise legs black; in pale specimens the black coloration restricted to tips of tibiae and tarsi. The sense-organ (Fig. 512) on fore tibiae anteriorly near base armed with a blunt-tipped brush of dark adpressed hairs.

Abdomen polished black to brownish black, covered with scattered fine dark bristly hairs, hind marginal bristles longer. Genitalia (Figs. 514–520) small, polished black, aedeagus with unequally long terminal appendages, the longer one finely serrate.

Length: body 3.2–4.5 mm, wing 2.8–4.2 mm.

♀. Generally much paler than male and eyes narrowly separated by a linear polished black frons. Legs usually light brown even at tips of tibiae and tarsi, thorax entirely yellowish, sometimes with a black median line on mesonotum, or mostly black on

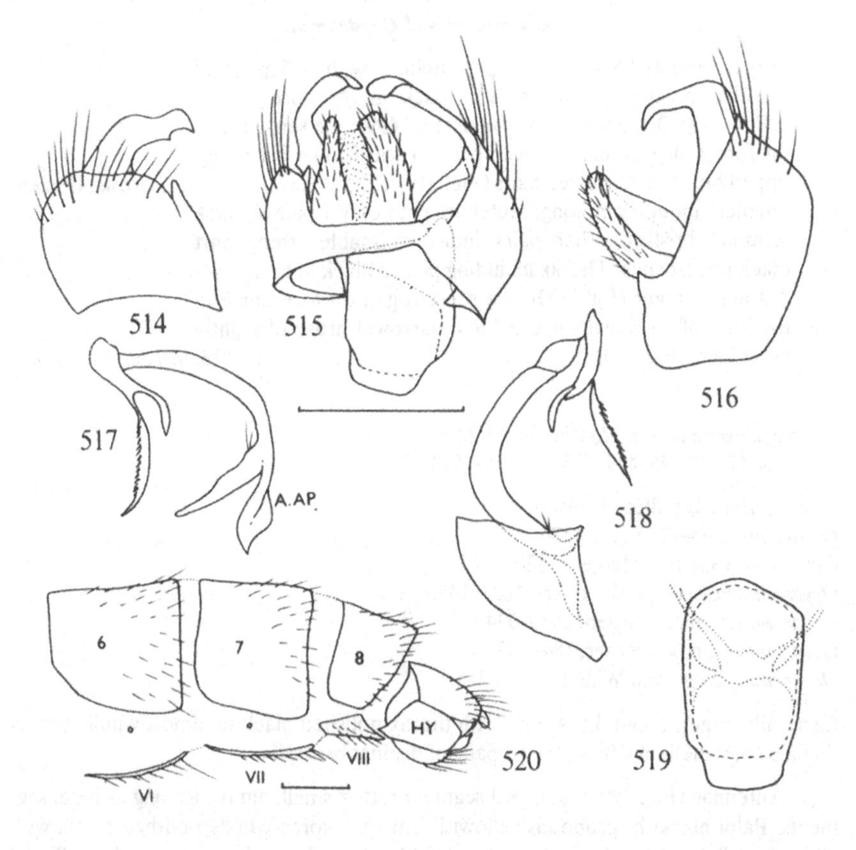

Figs. 514–520. Male genitalia of *Ocydromia glabricula* (Fall.). – 514: right periandrial lamella; 515: hypopygium in dorsal view, aedeagus omitted; 516: left periandrial lamella with cercus; 517: aedeagus; 518: hypandrium with aedeagus, lateral view; 519: hypandrium in ventral view, hypandrial bridge dotted; 520: postabdomen (macerated). Scale: 0.2 mm. For abbreviations see p. 13.

dorsum; some specimens (particularly in the north) with thorax entirely black even on pleura (as in *melanopleura*). Halteres yellowish brown and scutellar bristles paler. Wings almost clear, rarely clouded. Abdomen yellowish on venter, sides of terga rather yellowish brown, leaving dorsum polished black. Cerci pale or blackish (in dark specimens).

Length: body 3.2–4.2 mm, wing 3.8–4.6 mm.

Variation. A very variable species in colour, particularly in ♀, as can be seen from the above list of synonyms; these are often used as names for individual colour varieties. The extremely black females are more common in the northern areas of Norway, Sweden, Finland and in Scotland, and they resemble *O. melanopleura* very closely.

Distribution and biology. Very common throughout Scandinavia including the extreme north, also the Kola Peninsula. – Widespread and common in Europe, south to Spain and the Caucasus, east to Iran and Central Asia (Kirgizia, Tien shan), also North America. – From the end of April (Denmark, NEZ) until the second half of October (Sweden, Hall.).

38. ***Ocydromia melanopleura*** Loew, 1840
Figs. 76, 500, 513, 521–525.

Ocydromia scutellata Meigen var.d (♀) Zetterstedt, 1842: 238.
Ocydromia melanopleura Loew, 1840: 19; Tuomikoski, 1937: 18.

Generally smaller, about 3 mm, thorax entirely polished black in both sexes. Scutellum with only 1 pair of terminal bristles.

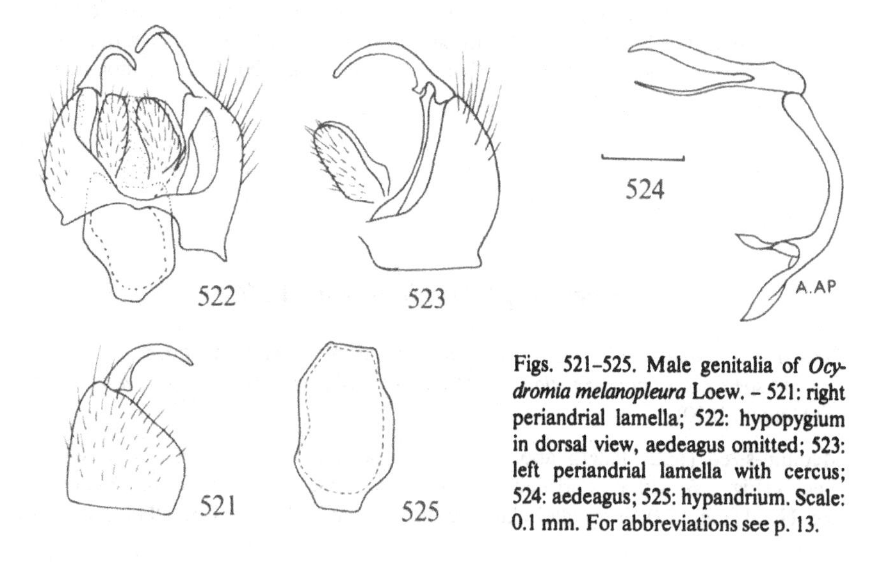

Figs. 521–525. Male genitalia of *Ocydromia melanopleura* Loew. – 521: right periandrial lamella; 522: hypopygium in dorsal view, aedeagus omitted; 523: left periandrial lamella with cercus; 524: aedeagus; 525: hypandrium. Scale: 0.1 mm. For abbreviations see p. 13.

♂. Antennae (Fig. 500) larger than in *glabricula*, 3rd segment largely ovate, distinctly longer than the length of 3rd segment of fore tarsi (almost equally long in *glabricula*). Palpi and proboscis blackish, small. Thorax entirely polished black, pleura rarely translucent blackish brown. Only a single pair of black terminal scutellar bristles, other pairs scarcely distinguishable from the dark pilosity.

Wings brownish black clouded, rather narrower than in *glabricula*, axillary lobe less distinct and rather rounded (more prominent and practically rectangular in *glabricula*). Squamae and halteres very darkened, knobs of halteres almost black. Coxae, trochanters and base of femora yellowish, rest of legs blackish brown to black, tarsi quite black. Hind legs with metatarsi and tibiae towards tip much more slender than in *glabricula*. The sense-organ (Fig. 513) on fore tibiae anteriorly near base covered by a narrower and apically pointed brush of somewhat erect hairs.

Abdomen shining black, venter paler. Genitalia (Figs. 521–525) larger than in *glabricula*, lateral lamellae narrower and with more slender dorsal appendages; the two branches of the bifid appendage at tip of aedeagus equally long but unequally stout, the broader one without serrations.

Length: body 2.5–3.3 mm, wing 2.7–3.3 mm.

♀. Very much like male with constantly polished black thorax, differing in the paler legs (tibiae in particular), the less clouded wings, and the paler side-margins of terga. Apical two abdominal segments and the short blunt cerci black.

Length: body 2.6–3.4 mm, wing 3.0–4.0 mm.

Distribution and biology. Widespread in Scandinavia but much less common than *glabricula*, especially in the north. In Finland north to Kb and Sb, in Sweden to Nb. and Lu. Lpm., in Norway to TRi, approximately 68°N; also in Russian Karelia. – Rather common in Scotland, very rare in England (Collin, 1961), in C. Europe only in mountains (Carpathians, Krkonoše Mts., Alps) and everywhere rather rare, found by Kovalev (in litt.) in central parts of European USSR. – May to September, dates ranging from 15 May (Sweden, Hall.) to 8 September (Denmark, NEZ), in the mountains of C. Europe until the beginning of October.

Note. For a long time *O. melanopleura* was only taken for a dark variety of *glabricula*, and the name was particularly attributed to extensively black coloured females. The two species were correctly differentiated by Tuomikoski (1937).

Genus *Leptopeza* Macquart, 1827

Leptopeza Macquart, 1827: 143 (as *Lemtopeza*, misprint; emend. Macquart, 1834: 320).

Type-species: *Lemtopeza flavipes* Macquart, 1827 (mon.) = *Leptopeza flavipes* (Meigen, 1820).

Medium-sized (body length about 3–6 mm), polished black or yellow (usually ♀) species (Fig. 526) with legs bristled at least on mid tibiae, and abdomen in ♀ ovipositor-like. Head (Fig. 527) almost globular, eyes holoptic in both sexes with all facets equally small and eyes also very approximated below antennae, face almost linear. Occiput convex, usually with long bristles on vertex and postocular margin, anterior pair of ocellar bristles long, posterior pair minute. Antennae (Fig. 501) inserted at middle of head in profile; basal segment indistinctly separated from the larger 2nd globular segment, which has a circlet of small preapical hairs; 3rd segment conical, longer than deep and very pointed, with a very long, bare, 2-articulated terminal arista; the small basal article finely bristled, slightly stouter than rest of arista, looking more like a con-

Fig. 526. Male of *Leptopeza flavipes* (Meig.). Total length: 3.8–5.4 mm.

tinuation of 3rd ant.s. Mouthparts (Figs. 528, 529) short and pointing downwards, generally resembling those of *Ocydromia;* however, palpi more slender and covered with numerous long fine bristly hairs, palpifer with 3 fine bristles (in the type-species) and hypopharynx more downcurved at tip, rather like the anterior end of maxillary stipites.

Thorax only slightly arched above, mesonotum covered with distinct (often long) numerous hairs and bristles; acr multiserial, usually in 6 rows, dc uniserial, prescutellar pairs often large. Usually several pairs of scutellar bristles of which at least one pair is very long and strong, as well as 1 or 2 notopleurals and 1 postalar.

Wings (Fig. 496) more or less clouded but not iridescent as in *Ocydromia,* and entirely covered with microtrichia. Costa ending at tip of vein R_{4+5}, Sc closely approximated to R_1 but abbreviated at tip. Stigma elongate and, compared with *Ocydromia,* only reaching tip of vein R_1, filling the costal cell and not spreading out beyond vein R_1; radial sector as in *Ocydromia.* Only 2 veins from end of discal cell but a stump of the third (M_1) often present. Discal cell rather short and apically broadened, about twice as long as deep before tip. Basal cells equally long and almost equally deep, anal cell shorter, the vein closing it very slightly recurrent and usually fading away before reaching the very indistinct and apically abbreviated anal vein. Axillary lobe well-developed but axillary excision obtuse, more distinctly so in females. A distinct long costal bristle.

Legs long and slender, hind tibiae towards tip and two basal segments of hind tarsi more or less dilated. Mid tibiae with several long bristles in ad and pd rows, smaller

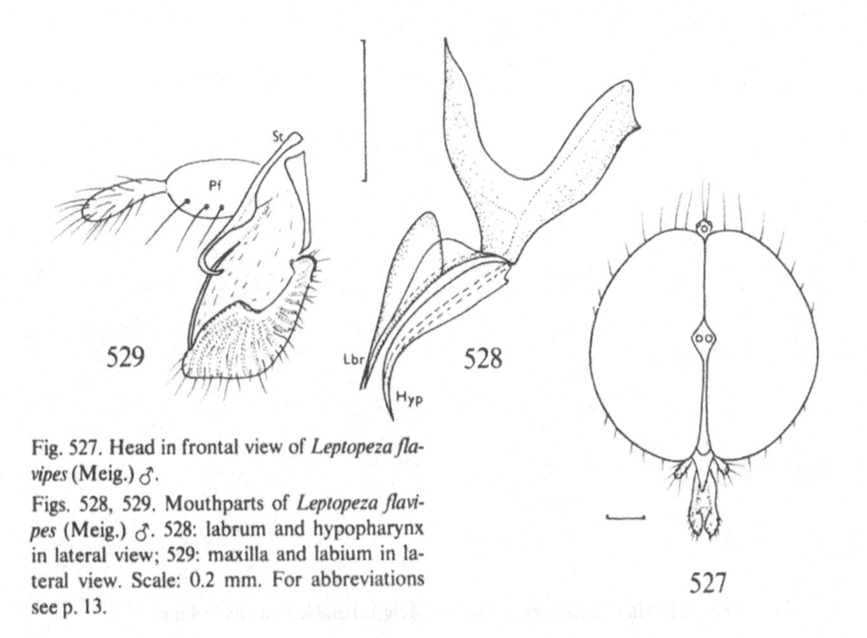

Fig. 527. Head in frontal view of *Leptopeza flavipes* (Meig.) ♂.

Figs. 528, 529. Mouthparts of *Leptopeza flavipes* (Meig.) ♂. 528: labrum and hypopharynx in lateral view; 529: maxilla and labium in lateral view. Scale: 0.2 mm. For abbreviations see p. 13.

bristles also on hind tibiae dorsally. Fore tibiae with a sense-organ anteriorly at base.

Abdomen long and evenly cylindrical, often densely long pubescent, hind marginal hairs bristle-like. 8th segment in ♂ (Fig. 535) much smaller than preceding segments and partly concealed within the 7th, but 8th sternum conspicuously broader than the others. Genitalia (Figs. 530–534) rather large but not broader than abdomen; rotated towards right through nearly 90° and lying along longitudinal axis (not upturned). Periandrial lamellae asymmetrical, large and densely bristled, both with curious dorsal processes; cerci fine, slender and finely bristled. Periandrium firmly connected with hypandrial bridge by narrow sclerotised folds, hypandrium rather small and not distinctly asymmetrical, bristled like periandrial lamellae. Aedeagus firmly attached by its base and a short coaxial apodeme to hypandrium, apically with a curious, simple ribbon-like long appendage. Abdomen in ♀ (Fig. 74) with apical two segments very prolonged and ovipositor-like, cerci slender.

Distribution. The genus includes only a few species; two are known from Europe including Scandinavia (both are supposedly Holarctic), one from Japan (see the note under *flavipes*), and Collin (1941) recorded "? *L. tibialis* Zett." from Ussuri. A further three species are Nearctic. The three "*Leptopeza*" species described from the Oriental region are very probably not congeneric, and the same is probably true of the few species recorded from Australia, New Zealand and the South American Cape Horn.

Biology. The adults occur in shaded humid places, mainly in spring and early summer. They are obviously predators as they have never been collected on flowers. Males of *L. flavipes* that were frequently hovering in small swarms of 5 to 10 individuals just above low vegetation *(Rubus)* were observed by the author at about noon in E. Slovakia, Czechoslovakia, in lowland deciduous forests along the river Latorica in May 1974. Kato (1971) recorded small male swarms under branches of trees in *L. flaviantennalis*. The immatures are unknown, but in view of the long ovipositor-like abdomen females are almost certainly oviparous (compared with *Ocydromia*), as is also shown in Fig. 75. Adults of *flavipes* were reared from a rotten log (Chandler, 1972).

Key to species of *Leptopeza*

1 Larger, 4–5 mm in length. Palpi yellow, thorax black in ♂, yellowish in ♀; acr 8- to 10-serial, short and pale. ♂: halteres pale, abdominal pubescence whitish .... 39. *flavipes* (Meigen)

\- Smaller, about 3 mm in length. Palpi blackish, thorax black in both sexes; acr 6-serial, black, very long in ♂, short in ♀. ♂: halteres dark, abdominal pubescence blackish .... 40. *borealis* Zetterstedt

39. ***Leptopeza flavipes*** (Meigen, 1820)
Figs. 74, 75, 496, 501, 526, 527–529, 530–535.

Ocydromia flavipes Meigen, 1820: 353.
Ocydromia ruficollis Meigen, 1820: 353.

Lemtopeza flavipes Macquart, 1827: 143.
Leptopeza tibialis Zetterstedt, 1842: 242.
Leptopeza flavimana Zetterstedt, 1842: 244 – **syn.n.**
Leptopeza nigripes Zetterstedt, 1842: 244 – **syn.n.**
Leptopeza ruficollis var. *unicolor* Strobl, 1900: 202 – **syn.n.**

Shining black (♂) or yellow (♀) species with rather short multiserial pale acr, palpi yellowish. Legs usually mostly yellowish in both sexes. Larger, about 4–5 mm.

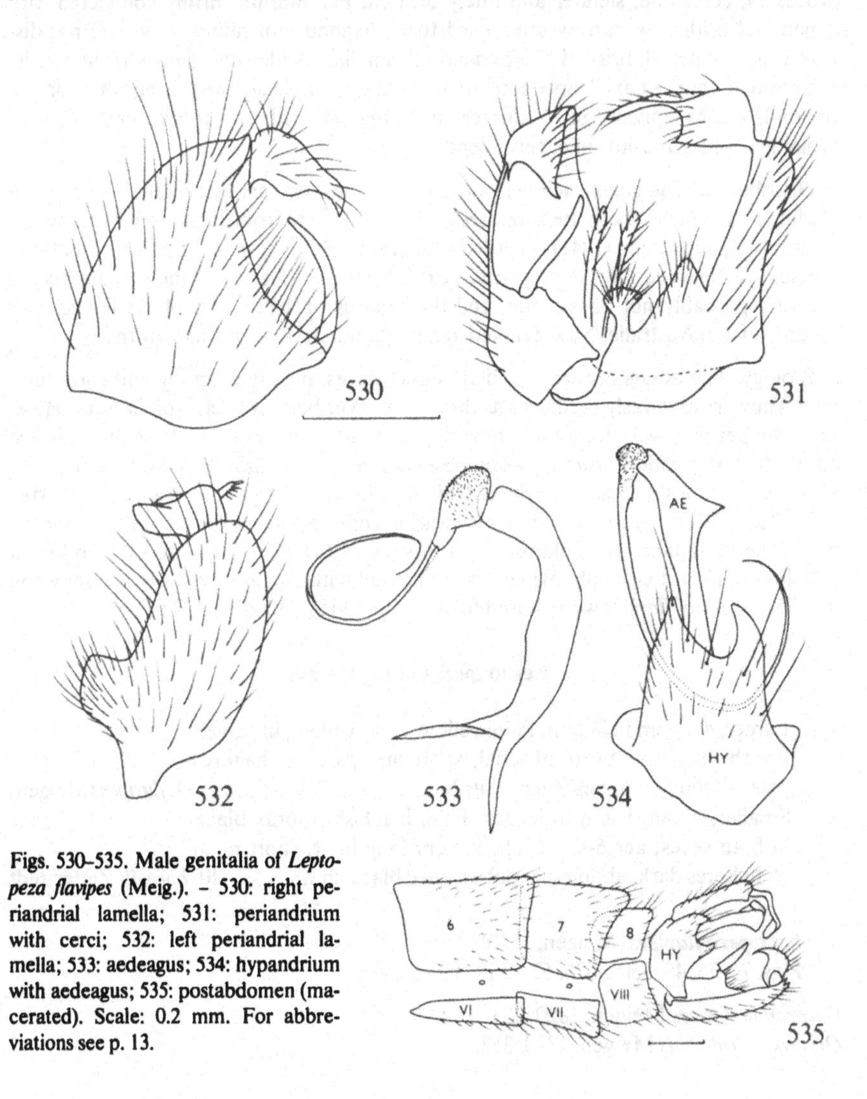

Figs. 530–535. Male genitalia of *Leptopeza flavipes* (Meig.). – 530: right periandrial lamella; 531: periandrium with cerci; 532: left periandrial lamella; 533: aedeagus; 534: hypandrium with aedeagus; 535: postabdomen (macerated). Scale: 0.2 mm. For abbreviations see p. 13.

♂. Occiput greyish dusted with a row of long black postocular bristles. Antennae (Fig. 501) black, 3rd segment at least twice as long as deep, arista more than 1,5 times as long as antenna, terminal. Palpi pale yellow, labellae of proboscis yellowish brown. Thorax black, all thoracic hairs and bristles pale; acr 8- to 10-serial, dc uniserial, all small, shorter than 3rd ant.s.; 8–12 scutellars, 2 or 3 pairs long and darker.

Wings (Fig. 496) faintly brownish grey infuscated, discal cell often with the stump of a third vein, or at least with a fold indicating upper vein M_1. Squamae and halteres pale yellow, latter often light brownish, a fine pale costal bristle. Legs mostly pale yellow, but posterior coxae, apex of hind femora, most of hind tibiae and practically all tarsi brown to blackish, rarely legs extensively darkened (*nigripes* Zett.). Legs covered with pale or dark (on dark areas) hairs, mid tibia with at least 3 conspicuous black ad bristles, similar paler pd bristles, and some shorter black bristles also on hind tibiae dorsally.

Abdomen polished black, covered with conspicuously long pale hairs, genitalia as in Figs. 530–535.

Length: body 3.8–5.4 mm, wing 3.9–4.8 mm.

♀. Postocular ciliation shorter and paler. Thorax mostly yellow, mesonotum often brownish in contrast to yellow scutellum, metanotum black or, in southern areas (C. Europe), often quite yellow (var. *unicolor*); mesonotum sometimes with a dark median stripe or extensively darkened, or whole of thorax mostly black (*tibialis* Zett. of extreme north and ? Asia). All thoracic bristles (including scutellars) paler, acr and dc much shorter. Legs generally paler, posterior coxae and sometimes also hind femora quite pale. Wings clearer, halteres pale. Abdomen shorter pubescent, dorsum almost polished black but sides of terga and venter yellowish. Apical three abdominal segments extensively blackish, 8th segment ovipositor-like, yellowish at base.

Length: body 3.6–4.6 mm, wing 4.0–4.9 mm.

Synonymies. The species is very variable in colour, particularly the legs in ♂ and thorax in ♀. *L. tibialis* Zett. was synonymised with *flavipes* by Frey (1956) in "Lindner", and the syntypic specimens in Zetterstedt's Coll. at Lund are extremely black females, with thorax quite black leaving only humeri, legs, venter of abdomen and "ovipositor" yellowish. *L. flavimana* Zett. and *L. nigripes* Zett. var.b are males of *flavipes* with extensively darkened legs; the single ♂ of *nigripes* var.a from Norway (Suul) has legs practically black, except for yellow apical third of fore femora and basal half of fore tibiae. The holotype ♀ of var. *unicolor*, described by Strobl (1901: 32) from "Kärnten", is not present in Strobl's Coll. at Admont, according to Morge (letter comm., 12.iv.1979), but it is undoubtedly a common form of *flavipes* in temperate Europe.

L. flaviantennalis Kato, 1971b, described from Japan, may well be conspecific with *L. flavipes* as well, as the figured ♂ genitalia are almost identical with those of *flavipes*. The paler antennae, the larger size (body length up to 6.7 mm) and the more numerous (10–16) scutellar bristles may well be within the range of variability of *flavipes*.

Distribution and biology. Widespread and common in Scandinavia including the extreme north, also the Kola Peninsula. – Throughout Europe including the south, very common in temperate regions, also North America and (?) Asia. – Second half of May to August, rarely to September (8.ix. – Zetterstedt, 1842), mainly in June and July. In C. Europe as early as the beginning of May.

Note. The name *ruficollis*, under which Meigen (1820) described the female of *flavipes* (males), is still used for this species by American authors and in recent catalogues of North and South America (Melander, 1965 and Smith, 1967 respectively). I am following Collin (1961) and retain for this species the name *flavipes* which is in general use in Europe, as Collin first revised Meigen's types and the grounds for his decision seem to be quite sound.

40. ***Leptopeza borealis*** Zetterstedt, 1842
Figs. 502, 536–540.

Leptopeza borealis Zetterstedt, 1842: 243.

Smaller, about 3 mm, polished black species in both sexes, palpi blackish. Dark multiserial acr extremely long in ♂. Legs blackish in ♂, pale in ♀.

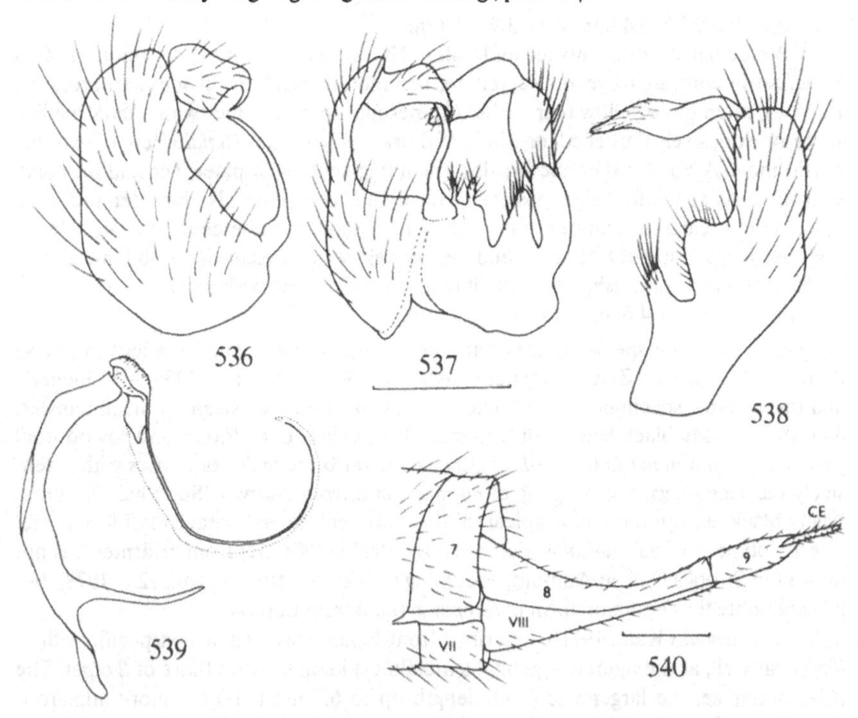

Figs. 536–539. Male genitalia of *Leptopeza borealis* Zett. – 536: right periandrial lamella; 537: periandrium with cerci; 538: left periandrial lamella; 539: aedeagus.
Fig. 540. Female "ovipositor" of *Leptopeza borealis* Zett. Scale: 0.2 mm. For abbreviations see p. 13.

♂. A long dark postocular ciliation, antennae (Fig. 502) black, terminal arista distinctly shorter than in *flavipes*, about as long as antenna, scarcely longer. Palpi blackish with black hairs, proboscis black. Thorax wholly black, polished on mesonotum and with the same greyish dusting as in *flavipes*. All thoracic hairs and bristles black; acr and dc upstanding and very long, much longer than 3rd ant.s., acr 6-serial, dc uniserial, anteriorly with additional long hairs.

Wings brownish clouded, more extensively so along costal margin, squamae and halteres blackish; a long black costal bristle. Legs polished black to blackish brown, only tips of anterior four femora and base of tibiae translucent yellowish, all hairs and bristles dark. Large bristles confined to 2 black ad and pd pairs on mid tibia and less distinct dorsal bristles on hind tibia.

Abdomen polished black and covered with long blackish pubescence. Genitalia (Figs. 536–539) differing from that of *flavipes* particularly in the more pointed process of right lamella, and left lamella with two distinct lobes anteriorly.

Length: body 2.5–3.8 mm, wing 2.8–3.5 mm.

♀. Resembling male very closely but all hairs and bristles on head and thorax shorter and finer, squamae and halteres pale. Legs extensively yellow on coxae and anterior four femora and tibiae; tarsi, anterior tibiae at tip, most of hind tibiae and apical half of hind femora brown to brownish black. Abdomen quite black or translucent brownish at sides and on venter, pubescence rather long but much paler than in ♂, 8th segment (Fig. 540) in the form of a long blackish "ovipositor".

Length: body 2.8–3.2 mm, wing 2.8–3.3 mm.

L. borealis is hardly a variable species in colour when compared with *flavipes*. Collin (1961) described the ♂ legs as mostly tawny yellow, but the single ♂ sent to him by Oldenberg from Germany was probably an immature specimen.

Lectotype designation. There are 4 ♀ from the original syntypic series in the Dipt. Scand. Coll. at Lund; a female labelled "Mullfj. 27 Jul." is herewith designated as lectotype of *Leptopeza borealis* Zett. and was labelled accordingly in 1977. The single male of *borealis* in the same Collection originates from Coll. Boheman, taken on 20.v. 1842 in Ostrogothia, and is not a syntype.

Distribution and biology. A species with a northern distribution, rather common throughout Finland, also in N. Norway and Sweden, south to Ög. Absent in Denmark and S. Sweden. – Very rare in Scotland (1 ♀) and by no means common in mountains of C. Europe, south to Romania (Mt. Rodnei) and N. Italy (Vallombrosa); recorded also from central parts of European USSR (Kovalev, in litt.) and North America (Melander, 1965). – June and July, in the mountains of temperate Europe also in August.

Genus *Leptodromiella* Tuomikoski, 1936

Leptodromiella Tuomikoski, 1936: 187.

Type-species: *Oropezella crassiseta* Tuomikoski, 1932 (orig. des.).

Rather smaller species (Fig. 541), about 2.5 mm long, resembling primitive *Platypalpus*

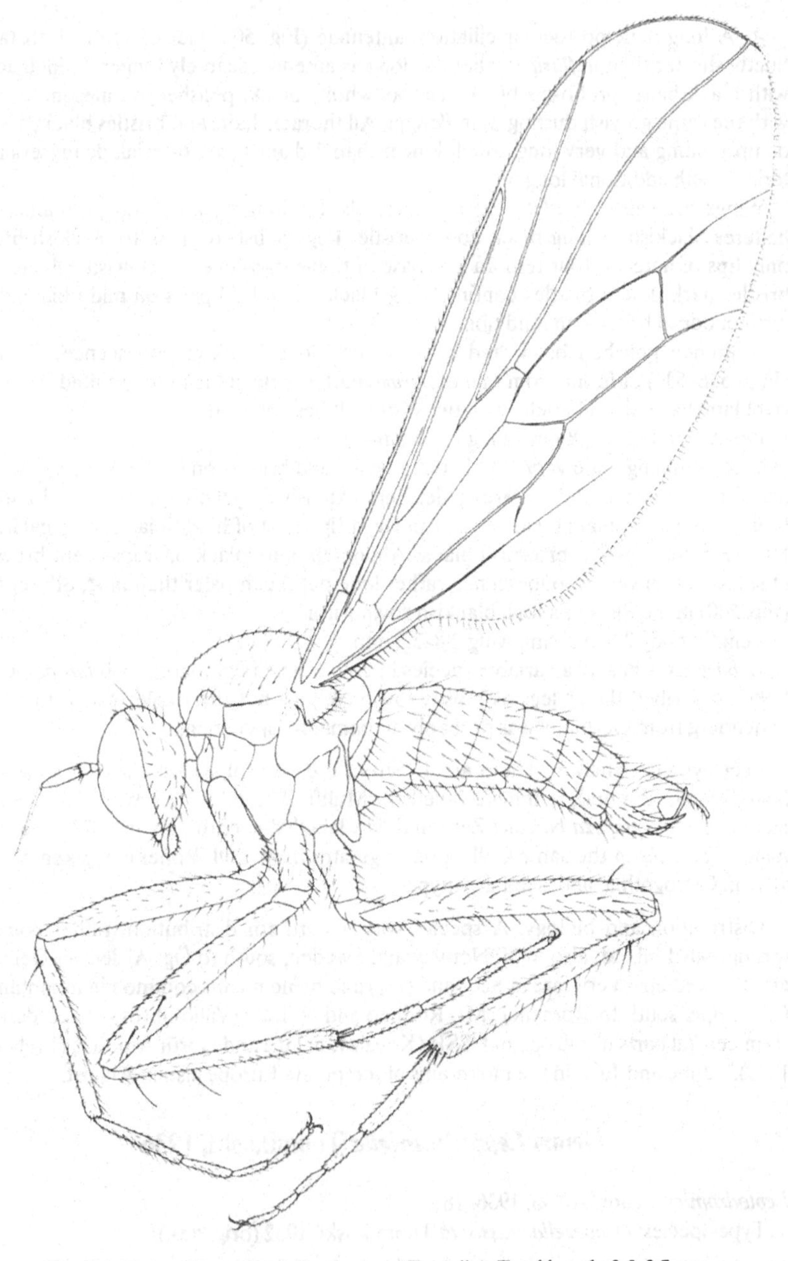

Fig. 541. Male of *Leptodromiella crassiseta* (Tuomik.). Total length: 2.5–2.7 mm.

species in appearance. Eyes dichoptic in both sexes, frons rather narrow, about as deep as anterior ocellus, but widening out above, face linear (Fig. 542). Anterior facets below very slightly enlarged in ♂, all facets equally small in ♀. 2 or 3 pairs of upper postocular bristles prominent, anterior pair of ocellar bristles much smaller, posterior pair minute. Antennae (Fig. 503) inserted about middle or scarcely above middle of head in profile, very conspicuous because of the somewhat thickened whitish arista, in this respect resembling species of the *Platypalpus albiseta*-group. Basal antennal segment slightly longer than deep, broadly connected with the 2nd globular segment, which bears a circlet of small preapical hairs. 3rd segment evenly conical, about 2.5 times as long as deep and apically pointed; terminal arista 2-articulated, basal article dark like the tip of 3rd segment, otherwise arista whitish, as deep as the small basal article but bare, much longer than 3rd segment. Mouthparts as in *Leptopeza* in general structure, short and pointing downwards, but palpi with only a single preapical bristle.

Thorax slightly arched on dorsum and, compared with *Leptopeza*, rather bare; biserial acr and uniserial dc small and fine, no distinct humeral bristle (only several small fine hairs) but 1 or 2 notopleurals, 1 postalar, 1 pair of prescutellar dc and the inner pair of 6 scutellars strong and very distinct. Prothoracic episterna with a long downcurved bristle, prosternum small, *Hybos*-like.

Wings (Fig. 497) with axillary lobe less developed than in *Leptopeza*, axillary excision very obtuse, and stigma in the form of a faint clouding at tip of costal cell. Sc inconspicuous, fading away not far beyond the origin of radial sector, costa ending at tip of R_{4+5} which is more decidedly bowed than in *Leptopeza*. Discal cell emitting two veins to wing-margin, M_1 absent, about 3.5 times as long as deep, longer than in *Leptopeza* but much shorter than in *Oropezella*. Basal cells equally long but 2nd basal cell broadened apically, anal cell much shorter; the vein closing it very recurrent and incomplete, anal vein rather distinct but disappearing before reaching the wing-margin. A distinct long costal bristle.

Legs apparently not as long and slender as in other ocydromiine genera, hind tibiae slightly swollen before tip but metatarsi slender. All femora with rows of anterior and posterior bristly hairs, posterior four tibiae with several long ad and pd bristles and distinct spine-like preapical bristles.

Abdomen cylindrical, rather densely pubescent, posterior hind marginal bristles not differentiated. Genitalia (Figs. 543–546) asymmetrical and rotated towards right, not broader than abdomen. Periandrial lamellae rather small, narrowly connected anteriorly, dorsal processes simple and slender, cerci small. Aedeagus, which has a long ribbon-like simple appendage apically as in *Leptopeza*, firmly attached on its whole length to the inner wall of the conspicuously large, convex and apically pointed cornet-like hypandrium. 8th segment in ♀ narrowed but not so ovipositor-like as in *Leptopeza*, cerci long and slender.

Classification. Collin (1961: 275) doubted the generic status of *Leptodromiella* (he mentioned *crassiseta* under *Oropezella*), but the generic diagnosis given above emphasises its validity. *Leptodromiella* is an intermediate generic taxon between *Leptopeza* and

Oropezella, and quite recently Kovalev (1979b) has independently reached the same conclusion as to its generic status.

Distribution. A monotypic genus, with a single rare western Palaearctic species originally described from Scandinavia.

Biology. The development and biology of the single known species are still unknown but they are probably not very different from the related *Leptopeza* and *Oropezella* species. The somewhat enlarged anterior facets in males may well indicate both aerial swarming habits and predatory activity.

41. ***Leptodromiella crassiseta*** (Tuomikoski, 1932)
Figs. 497, 503, 541–546.

Oropezella crassiseta Tuomikoski, 1932: 49.

Black species with yellow legs, thorax silvery-grey dusted with two broad polished black stripes on mesonotum. Antennae black with a long whitish terminal arista.

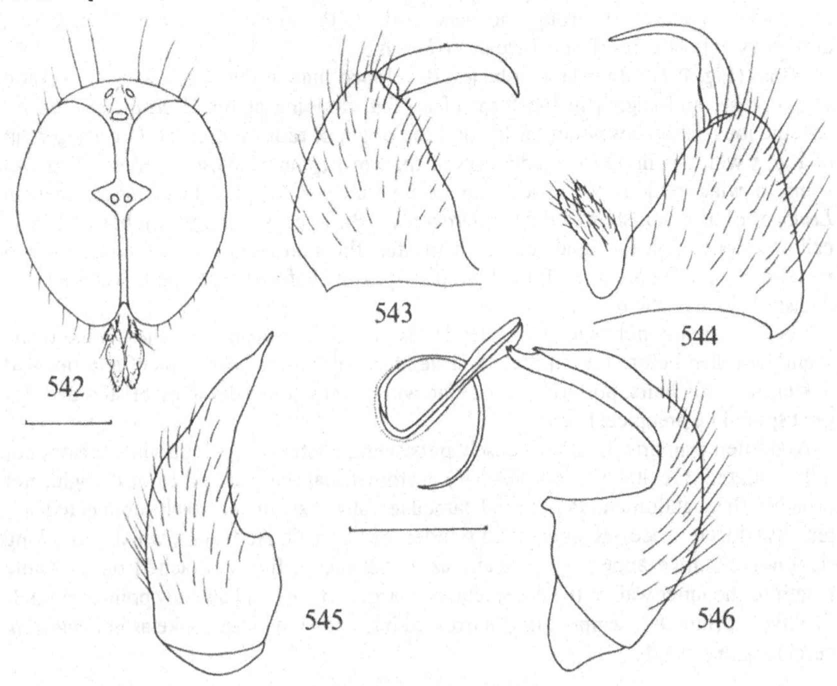

Fig. 542. Head in frontal view of *Leptodromiella crassiseta* (Tuomik.) ♂.

Figs. 543–546. Male genitalia of *Leptodromiella crassiseta* (Tuomik.). – 543: right periandrial lamella; 544: left periandrial lamella with cerci; 545: hypandrium in ventral view; 546: hypandrium with aedeagus in lateral view. Scale: 0.2 mm.

♂. Head black, rather densely greyish dusted, frons short and narrow, face very linear. A pair of small dark anterior ocellar bristles, 3 pairs of long pale postverticals, otherwise occiput clothed with long pale hairs. Antennae (Fig. 503) black, 3rd segment long and pointed, terminal arista whitish, slightly longer than antenna. Palpi pale yellowish with a pale terminal hair, proboscis light brownish. Thorax black, uniformly rather densely silvery-grey dusted, mesonotum with 2 broad polished black longitudinal stripes between acr and dc. All thoracic hairs and bristles pale.

Wings (Fig. 497) almost clear, a distinct pale costal bristle. Squamae and halteres pale yellow. Legs pale yellow including coxae, posterior femora and tibiae more brownish, tarsi darkened. All pubescence and large bristles on legs pale except for dark tarsal pubescence. In addition to the long bristly hairs anteriorly and posteriorly on all femora, there is a distinct dorsal bristle on hind femur before tip, and posterior four tibiae in addition to a comb of spur-like apical bristles with 2 irregular pairs of ad and pd bristles.

Abdomen subshining brownish black and rather densely covered with fairly long pale hairs. Genitalia (Figs. 543–546) concolorous with abdomen but the narrow and apically pointed dorsal processes of periandrial lamellae polished black, much darker in contrast to paler and dusted lamellae. Hypandrium clothed with rather long pale hair-like bristles, like the lamellae.

Length: body 2.5–2.7 mm, wing 3.3–3.4 mm.

♀. Resembling male very closely but eye-facets equally small and the pale abdominal pubescence shorter. Abdomen more shining with 8th segment much narrower than preceding segments, cerci slender.

Length: body 2.5–2.6 mm, wing 3.2–3.4 mm.

Holotype identification. The holotype ♀, labelled "Padasjoki, 16.6.1929, Tuomikoski, No. 4809", is in perfect condition in the Finnish Collection of the Zool. Museum, Helsinki.

Distribution and biology. Rare species, known only from Finland (Ta, Padasjoki and N, Helsinki) and Sweden (Hall., Enslöv). – Besides the Fennoscandian specimens only a further 2 ♀ are known, taken in a moist fir-wood in the central part (Moscow region) of European USSR (Kovalev, 1979b). – June.

Genus *Oropezella* Collin, 1926

Oropezella Collin, 1926: 214.

Type-species: *Leptopeza sphenoptera* Loew, 1873 (orig. des.).

Leptometopiella Melander, 1928: 70.

Type-species: *Leptopeza sphenoptera* Loew, 1873 (orig. des.).

Medium-sized polished and finely bristled species (Fig. 547) with long slender wings without axillary lobe. Head higher than deep, with a pair of long postvertical bristles, a pair of shorter anterior ocellar bristles (posterior pair very minute) and only short

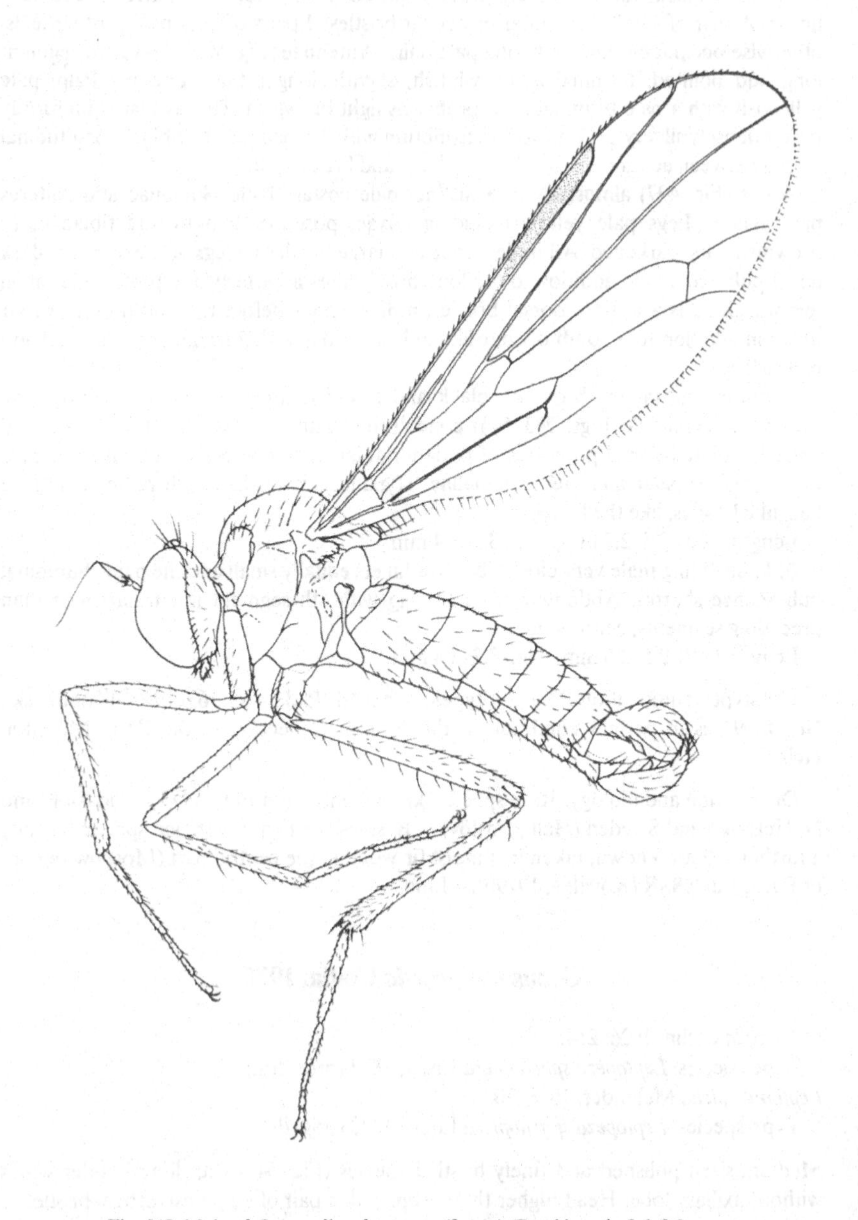

Fig. 547. Male of *Oropezella sphenoptera* (Loew). Total length: 2.6–3.3 mm.

postocular ciliation. Eyes narrowly separated by a short frons in both sexes, meeting for a long distance below antennae, with anterior facets below antennae distinctly enlarged in both sexes. Occiput slightly convex above neck, concave below. Antennae (Figs. 548, 549) inserted much above middle of head in profile, basal segment (Fig. 504) cylindrical, longer than deep, 2nd segment globular and with a circlet of small preapical bristly hairs. 3rd segment rather long and slender, about 3 times as long as deep near base, apically very pointed; arista 2-articulated, bare, long and slender, distinctly longer than antenna, basal article small and bare. Mouthparts (Figs. 550–552) small and pointing downwards, palpus as in *Leptodromiella* with a single long preapical bristly hair, palpifer armed with 3 fine bristles. Labium with large fleshy labellae armed with fine sensory hairs and pseudotracheae.

Thorax almost globular, although only slightly arched on dorsum. Mesonotum practically bare and polished, small hairs fine and inconspicuous, acr narrowly biserial, dc uniserial, prescutellar pairs not differentiated. Large bristles confined to one long notopleural, a smaller postalar and a long pair of apical scutellars, outer two pairs very small.

Wings (Fig. 498) almost clear or slightly clouded, more or less iridescent and entirely covered with microtrichia. Axillary lobe not developed, axillary excision consequently absent and wings conspicuously narrowed on basal third. Vein Sc closely approximated to and almost confluent with R_1, costa ending at tip of R_{4+5}. Stigma rather distinct, occupying the whole apical third of the narrow costal cell right up to tip of vein R_1. Two veins from end of discal cell, which is very long and narrow, about 5 times as long

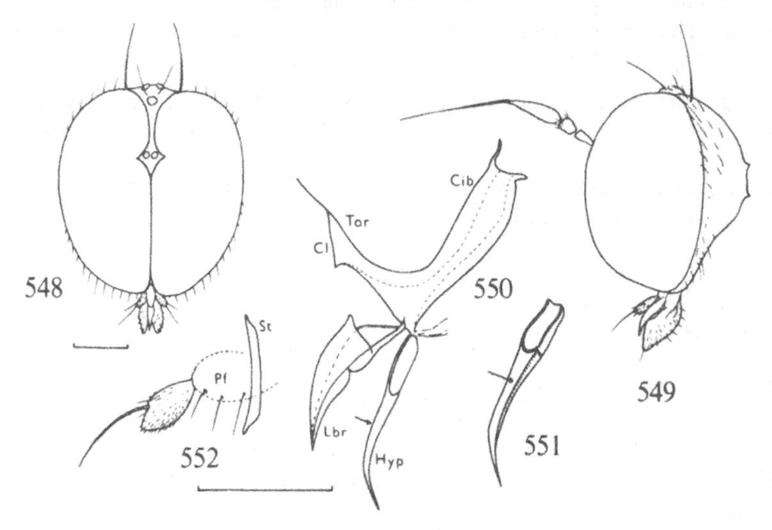

Figs. 548, 549. Head of *Oropezella sphenoptera* (Loew) ♂. – 548: frontal view; 549: lateral view.
Figs. 550–552. Mouthparts of *Oropezella sphenoptera* (Loew) ♂. – 550: labrum and hypopharynx in lateral view; 551: hypopharynx in anterolateral view; 552: maxilla with palpus, lateral view. Scale: 0.2 mm. For abbreviations see p. 13.

as deep; second basal cell slightly longer than first basal cell, anal cell as long as first basal cell and all three cells about equally deep. The vein closing anal cell slightly recurrent and S-shaped, anal vein fine and incomplete, or very distinct. Costal bristle absent.

Legs long and slender, hind tibiae scarcely dilated towards tip, posterior four tibiae with several distinct bristles dorsally and/or ventrally, preapical bristles well-developed. Posterior metatarsi not dilated but finely bristled beneath.

Abdomen long and rather cylindrical, covered with fine hairs, hind marginal bristles differentiated at least on basal segments. 8th tergum in male narrowed but 8th sternum unmodified. Genitalia (Figs. 553–555) large and asymmetrical, rotated towards right along longitudinal axis through about 90°. Periandrial lamellae very convex, narrowly connected anteriorly and with large flat terminal processes; cerci small and slender. Hypandrium rather long and cornet-like, resembling that of *Leptodromiella*. Aedeagus with a coaxial apodeme and a bifid appendage at tip very much resembling that of the *Ocydromia* species, its smaller lower branch directed backwards. Abdomen in ♀ (Fig. 556) with 8th segment rather narrowed, but the ovipositor-like structure of *Leptopeza* not developed; cerci slender.

Distribution. Only two species are known from Europe and the Palaearctic (the Austrian *O. rugosiventris* still from the holotype ♀), and none from North America. About a dozen species are recorded from the southern hemisphere (South America, New Zealand), but the generic assignment of some others from Australia and Tasmania needs clarification, particularly as regards the related genera *Hoplopeza* Bezzi, *Scelolabes* Phil. or *Pseudoscelolabes* Coll.; the same applies to the Australasian "*Leptopeza*" species. One species occurs in Scandinavia.

Biology. The adults are obviously predaceous, as is indicated by the arrangement of eyes with enlarged anterior facets in both sexes. Flower visiting was newer observed. The adults occur for a long period from late spring to early autumn in shady places, resting on low herbage or slowly hovering above the ground; regular swarms were not observed. The immatures are unknown.

42. ***Oropezella sphenoptera*** (Loew, 1873)
Figs. 498, 504, 547–556.

Leptopeza sphenoptera Loew, 1873: 215.
Leptopeza lonchoptera Loew; Pokorny, 1887: 394 (lapsus).

Mesonotum polished black, prothorax largely silvery dusted. Antennae inserted above middle of head, wing very narrow at base, axillary lobe not developed and discal cell very long.

♂. Frons (Fig. 548) short and polished black, as deep as anterior ocellus. Occiput greyish dusted, large bristles black, otherwise occiput clothed with whitish hairs. Antennae black, 3rd segment narrowly pointed, a bare terminal arista longer than antenna

(Fig. 504). Palpi very small, blackish, apical bristly hairs dark, labellae yellowish brown. Thorax black, polished on mesonotum, pleura very thinly greyish dusted but prothorax remarkable by its dense silver dusting. Humeri, postalar calli and sides of scutellum often yellowish. A few large thoracic bristles and small mesonotal hairs black. Prothoracic episterna with 2 whitish hair-like bristles.

Wings (Fig. 498) faintly brownish grey clouded, anal vein very indistinct and fading

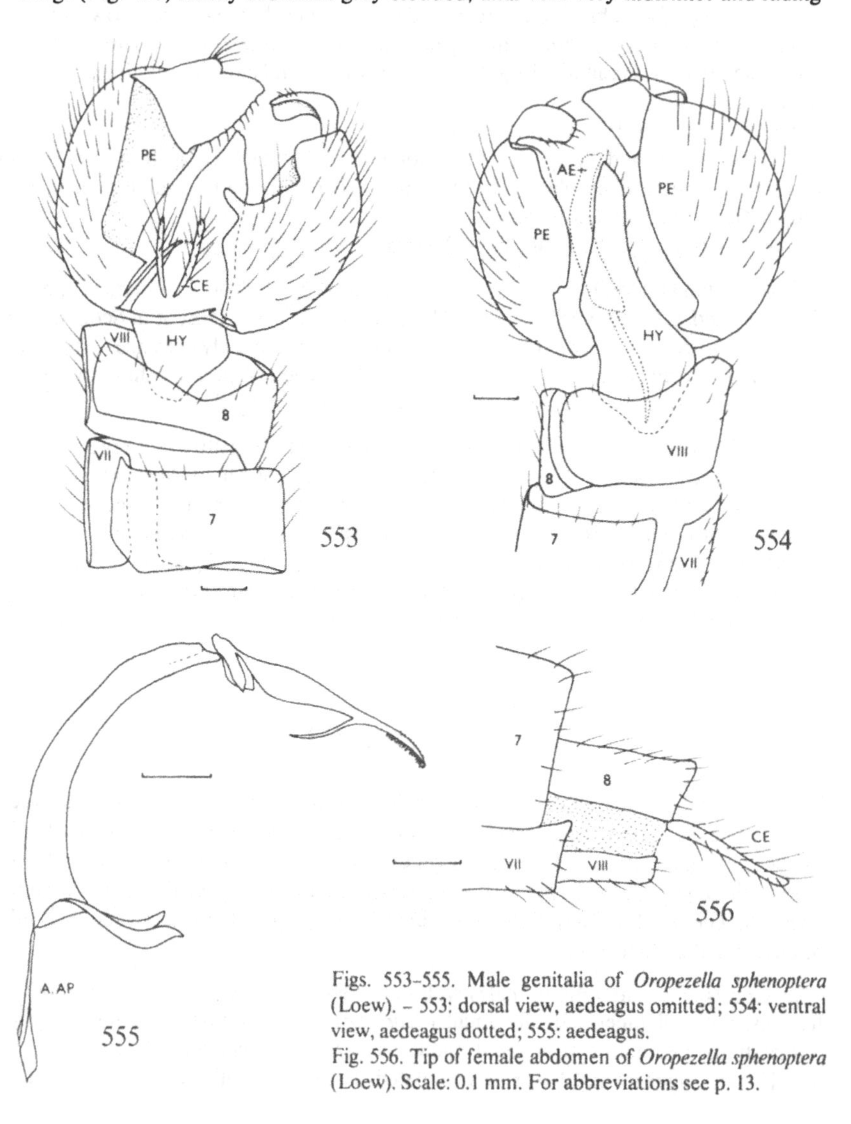

Figs. 553–555. Male genitalia of *Oropezella sphenoptera* (Loew). – 553: dorsal view, aedeagus omitted; 554: ventral view, aedeagus dotted; 555: aedeagus.
Fig. 556. Tip of female abdomen of *Oropezella sphenoptera* (Loew). Scale: 0.1 mm. For abbreviations see p. 13.

away far before reaching the wing-margin. Squamae and halteres pale yellow. Legs pale yellow (coxae in particular), tarsi and sometimes posterior femora and tibiae darkened at tip. Short pubescence dark, mid tibiae with 2 long ad and pd bristles, and a single pv bristle in apical third; hind tibiae (besides the preapical bristles) with a pair of smaller but strong black dorsal bristles near base. Posterior four femora with a row of dark av bristly hairs, those on hind femur very long and bristle-like.

Abdomen polished black on dorsum, sides and venter brownish to yellowish brown, basal terga often pale. Pubescence pale and inconspicuous, hind marginal bristles longer and darker. Genitalia (Figs. 553–555) very large and globose, much broader than abdomen.

Length: body 2.6–3.3 mm, wing 3.6–4.0 mm.

♀. Resembling male very closely, abdomen wholly black or uniformly brownish, leaving 8th tergum deep black; 8th segment smaller than the preceding segments but unmodified, cerci long and slender.

Length: body 2.3–3.2 mm, wing 3.5–3.8 mm.

Distribution and biology. A southern species known only from Denmark (EJ, NEZ – 5 ♂, 9 ♀) and S. Sweden (Sk., Hall. – 14 ♂, 11 ♀). – England (absent in Scotland) and temperate Europe, everywhere rather uncommon, south to Italy (Firenze), Yugoslavia (Dalmatia), Corfu (Kérkira), the Caucasus and N. Africa (Algeria). – June to September, in C. Europe as early as May, in the Mediterranean in April.

Note. The central European *Leptopeza rugosiventris* Strobl, 1910 (♀) is a second European *Oropezella* species. The holotype female (in Coll. Strobl at Admont) is still the only known specimen, and differs from *sphenoptera* by the larger size (body 4.6 mm, wing 4.7 mm), the wrinkled punctured and densely whitish grey pubescent abdomen, by the more brownish clouded wings with 2nd basal cell longer, by the longer and more cylindrical 3rd ant.s., by the more distinct anal vein and much shorter anal cell, by the almost bare thorax without a silver patch anteriorly, and by the differently bristled tibiae.

Family Atelestidae

This new family is a well-founded monophyletic group of three genera with only a few recent species, which were previously either placed in the cyclorrhaphous family Platypezidae (*Atelestus*), or included as genera of "uncertain systematic position" close to or in the Hybotinae (*Meghyperus, Acarteroptera*), particularly because of the enlarged anal cell. The family, and its presumed common origin with the Cyclorrhapha, is fully discussed in the General part.

Diagnosis. Eyes holoptic in males with upper facets enlarged, broadly dichoptic in females, with all facets equally small (*Acarteroptera* ♀ unknown). Antennae (Figs. 557–560) inserted below or at about middle of head in profile, 3rd segment short, terminal style pubescent, 2- or 3-articulated (with 2 small basal articles as in

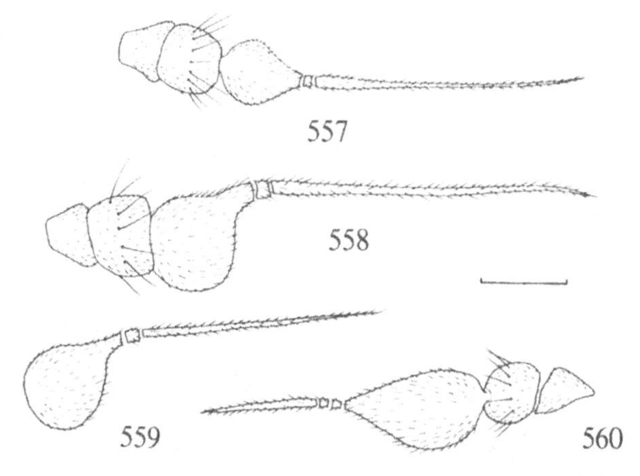

Figs. 557–560. Antennae of Atelestidae. – 557: *Atelestus pulicarius* (Fall.) ♂; 558: *Meghyperus sudeticus* Loew ♂; 559: *Meghyperus sudeticus* Loew ♀; 560: *Meghyperus* sp. ♂ (USA, Calif.). Scale: 0.1 mm.

cyclorrhaphous Diptera), long in *Atelestus*, very short in *Acarteroptera*. Proboscis projecting obliquely forwards or almost horizontal, of varying lengths, with fully developed maxillae; palpi cylindrical, labium soft, large labellae with pseudotracheae. Thorax slightly arched above, a small isolated *Hybos*-like prosternum, acr (at most 4-serial) separated from dc. Wings with well-developed axillary lobe and alula (latter more prominent in males of *Atelestus* and *Meghyperus*), wholly covered with small microtrichia. Costa ending at M_1, Sc incomplete, radial sector short and 2-branched (R_{4+5} unforked), M_{1+2} unforked or, if forked (*Meghyperus*), the veins M_1 and M_2 branched beyond discal cell; discal cell absent in *Atelestus*. Anal cell large, at least as long as basal cells or longer. Ovate stigma extended beyond R_1 but not reaching R_{2+3}. Legs without ordinary bristles, at least hind tibiae dilated (in males hind tibiae and metatarsi often very compressed, Platypezid-like), fore tibiae without tubular gland. Abdomen in male with 8th tergum small and ring-like, 8th sternum large, female abdomen ovipositor-like. Male genitalia quite symmetrical and not rotated, periandrium developed, aedeagus with a coaxial apodeme and postgonites, freely surrounded near base by a firm hypandrial ring, which bears two long proximal apodemes.

Classification. The family as defined here includes the genera *Atelestus* Walk. and *Meghyperus* Loew, and very probably also the Chilean *Acarteroptera* Coll. (not studied by me), for which the morphological data given here has been taken from the published descriptions. *Atelestus* was generally placed in the Platypezidae and was not included in the great monographs of the "Empididae" until Collin (1961), who included it in his broad concept of the Hybotinae. *Meghyperus* was generally placed in the Hybotinae, or treated as an aberrant "empidid" genus of the hybotine-complex. *Acarteroptera* was originally considered by Collin (1933) to be related to *Meghyperus*. Tuomikoski (1966)

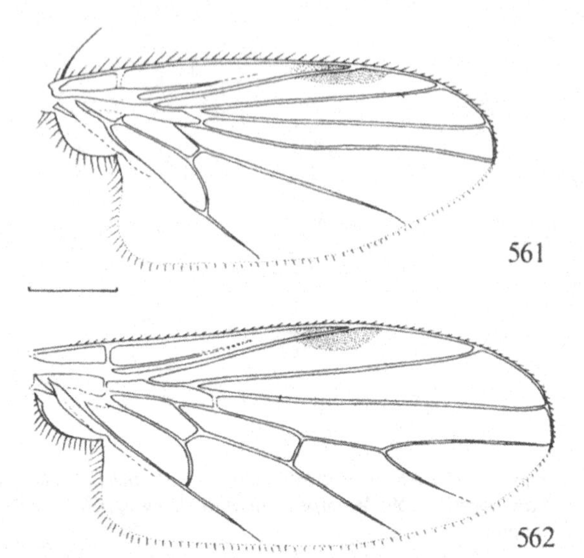

Figs. 561, 562. Wings of Atelestidae. – 561: *Atelestus pulicarius* (Fall.) ♂; 562: *Meghyperus sudeticus* Loew ♀. Scale: 0.5 mm.

was actually the first who pointed out that these three genera have a discrete systematic position within the former "Empididae", and he thought that they might have to be removed from the family. Hennig (1970) erected the subfamily Atelestinae for this group of genera within his subfamily-group Ocydromioinea of the "Empididae" and this classification has been followed until now.

Distribution. The genus *Atelestus* includes two European species, *Meghyperus* is Holarctic in distribution with one Palaearctic and two (possibly three) Nearctic species, and *Acarteroptera* is represented by two Neotropical (Chilean) species.

Key to genera of Atelestidae

1 Discal cell absent, vein M_{1+2} unforked (Fig. 561). Small mesonotal hairs (acr and dc) long and bristle-like .... *Atelestus* Walker (p. 230)
– Discal cell present, vein M_{1+2} apically forked into M_1 and M_2 (Fig. 562). Small mesonotal hairs very fine and inconspicuous .... *Meghyperus* Loew (p. 237)

Genus *Atelestus* Walker, 1837

Atelestus Walker, 1837: 229.

Type-species: *Atelestus sylvicola* Walker, 1837 (mon.) = *Atelestus pulicarius* (Fallén, 1816).

Platycnema Zetterstedt, 1838: 534.
Type-species: *Empis pulicarius* Fallén, 1816 (mon.).

Small, body length about 2–2.5 mm, blackish grey and conspicuously black bristled species (Fig. 563) with a very distinctive wing-venation. Head larger in male (Fig. 564), with eyes holoptic and anterior facets slightly enlarged, smaller in female (Fig. 565), with eyes broadly dichoptic and all facets equally small. Head hemispherical in profile, anteriorly rounded, upper parts of occiput almost flat; antennae inserted below middle of head in profile, more distinctly so in male. Anterior pair of ocellar bristles very long and strong, posterior pair much smaller, scarcely as long as distinct postocular ciliation. Face very short and rather broad in both sexes, males with antennal excisions triangular. Ocellar tubercle small, although distinct in both sexes. Frons in ♀ very broad, as deep at middle as eyes, widening above, with 2 pairs of frontoorbital bristles and another pair at middle pointing forwards, and 1 pair of equally long postverticals. Antennae (Fig. 557) with short basal segments, 2nd segment globular and armed with a circlet of bristles, 3rd segment pointed but almost spherical; terminal style very long

Fig. 563. Male of *Atelestus pulicarius* (Fall.). Total length: 2.2–2.7 mm.

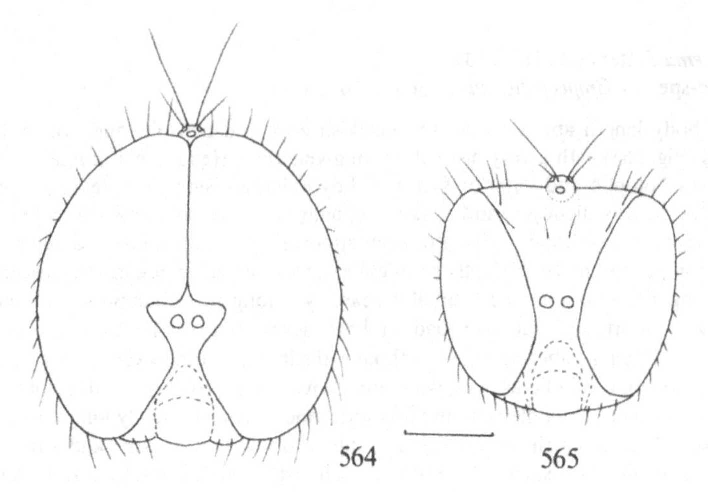

Figs. 564, 565. Head in frontal view of *Atelestus pulicarius* (Fall.). – 564: male; 565: female. Scale: 0.2 mm.

(much longer than rest of antenna) and microscopically pubescent, 2-articulated, basal article very small. Mouthparts (Fig. 563) pointing forwards but very small, practically hidden in the mouth cavity, sometimes only the large setulose labellae partly visible. Clypeus (Fig. 566) joined by a fine short membrane with antennae, labrum (Fig. 570) short with one-pointed apex, hypopharynx (Fig. 571) equally long, cornet-like, maxillary palps (Fig. 569) long cylindrical and clubbed apically, loosely connected with fully developed maxillae; no palpifers. Labium with conspicuously large labellae with distinct pseudotracheae and small sensory setae.

Thorax rather higher than long, mesonotum convex, humeri (placed unusually low), postalar calli and scutellum prominent. Prosternum small, *Hybos*-like. Mesonotal bristles very strong, 1 humeral, 1–2 posthumerals, 3 notopleurals, 1 postalar and 2 pairs of scutellars, inner pair the strongest; acr and dc biserial and not very much smaller than other large bristles, latter ending in 2 or 3 pairs of strong prescutellar bristles. Thoracic pleura bare, a small bristle on prothoracic episterna.

Wings (Fig. 561) clear or almost clear, entirely covered with microtrichia, axillary lobe very prominent and a distinct alula. A large ovate stigma extending beyond tip of vein R_1. Costa reaching vein M_{1+2}, although less distinct beyond R_{4+5}, short bristled on basal half and with a strong costal bristle. Sc closely approximated to R_1 but disappearing at half the distance from its base to costa, radial sector short, veins R_{4+5} and M_{1+2} unforked. Discal cell absent, basal cells rather slender, anal cell distinctly longer than basal cells, the vein closing it arched, joining anal vein at right angles, and anal vein almost complete. Crossvein m-m absent, crossvein r-m at about basal third of wing.

Legs distinctly bristled, anterior two pairs and hind femora rather slender, hind tibiae laterally compressed and apically swollen, much broader at tip than corresponding femora, also hind metatarsus slightly swollen and cylindrical. All metatarsi lengthened, almost as long as rest of corresponding tarsi. Fore tibiae without the tubular gland of the Hybotidae.

Male abdomen (Fig. 572) rather short-cylindrical, covered with distinct bristles forming a median row of shorter bristles and a hind marginal row of more conspicuous bristles. Anterior 7 abdominal segments unmodified, 8th tergum in the form of a very narrow ring joined on each side beneath with a posteriorly projecting lobe, which bears a distinct bristle at tip. 8th sternum formed by two lobes, each armed with two bristles at tip. Genitalia (Fig. 574) symmetrical and not rotated, lying along longitudinal axis. Periandrial lamellae narrowly connected anteriorly above (due to the unrotated position), broadly separated below; terminal processes scarcely distinguishable, but there are distinct lobes (? gonostyli) on the inner wall posteriorly (Fig. 577). Hypandrium small, triangular in shape and not completely sclerotised, firmly connected with the sclerites arising from inner periandrial wall, and forming a well-developed hypandrial bridge round the aedeagus. Cerci slender. Aedeagus conspicuously long and slender with an equally deep and long coaxial apodeme, apically with 2 slender, firmly sclerotised and microscopically pubescent whitish postgonites, covering laterally the tip of aedeagus. Base of hypandrium produced into paired slender apodemes lying along the aedeagal apodeme. Female abdomen (Fig. 573) rather broad and somewhat

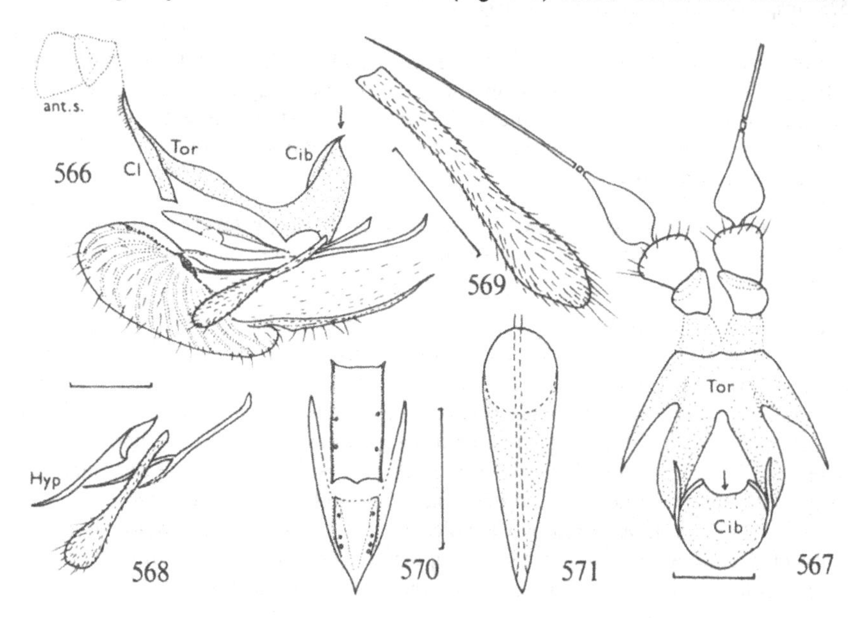

Figs. 566–571. Mouthparts of *Atelestus pulicarius* (Fall.) ♂. – 566: lateral view; 567: dorsal view, labium and palpi omitted; 568: maxilla with palpus and hypopharynx; 569: maxillary palpus; 570: labrum in dorsal view; 571: hypopharynx in dorsal view. Scale: 0.1 mm. For abbreviations see p. 13.

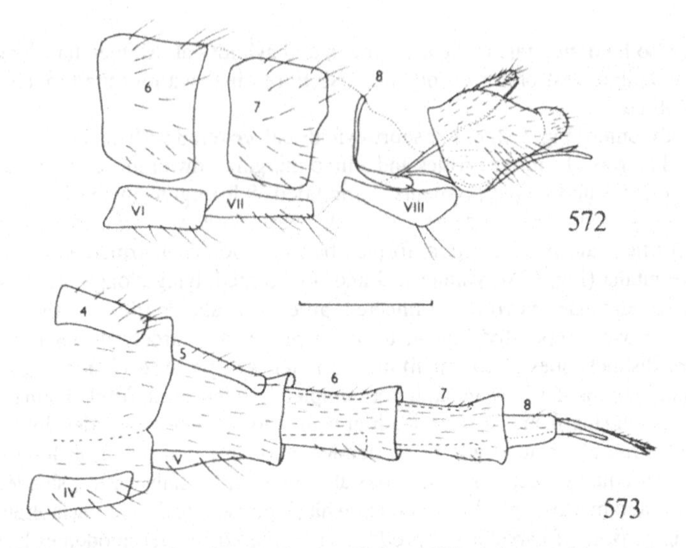

Figs. 572, 573. Abdomens of *Atelestus pulicarius* (Fall.). – 572: male postabdomen (macerated); 573: "ovipositor"-like tip of female abdomen. Scale: 0.3 mm.

dorsoventrally compressed on basal four segments, 5th to 8th segments very narrowed, telescopic, almost tubular and ovipositor-like. Terga of apical four segments incompletely sclerotised, reduced to only darker lines of narrow lateral margins. Whole abdomen with fewer and shorter bristles than in male, with rather broad (and paler) intersegmental membrane.

Distribution. The genus is known only from Europe: two species are recognised from the temperate and cooler zone but, for the time being, only one has been found in Scandinavia.

Biology. The immatures are unknown, and no precise data on the biology of the adults are available. The adults are most often swept from low herbage or from the foliage of trees in deciduous forests and, in view of the very short forwardly directed proboscis, they are very probably predators of other insects, although there is no evidence of such a habit from direct field observations; flower visiting has not been observed. The remarkably holoptic eyes of the males clearly indicate swarming activity, as has been verified by field observations in both *A. dissonans* (as *pulicarius*) by Verrall (1901) and *A. pulicarius* by Chandler (in litt.). Aerial mating was not observed, although according to Chandler females appeared to enter the male swarm. The adults are rather local in occurrence but often common in suitable biotopes.

Key to species of *Atelestus*

1 Generally larger, 2.2–2.7 mm. ♂: tibiae distinctly bristled,

mid tibia with at least one strong ad bristle at middle longer than other bristles. ♀: frons and mesonotum dull greyish dusted .... 43. *pulicarius* (Fallén)
- Generally smaller, 1.7–2.3 mm. ♂: tibiae with shorter bristles, mid tibia with all ad bristles almost equally long. ♀: frons and mesonotum rather extensively shining black .... *dissonans* Collin

43. ***Atelestus pulicarius*** (Fallén, 1816)
Figs. 53, 557, 561, 563–579.

Empis pulicaria Fallén, 1816: 33.
Atelestus sylvicola Walker, 1837: 229.
Platycnema tibiella Zetterstedt, 1842: 333 – **syn.n.**

Somewhat larger, with completely long black bristles on thorax and legs in male, frons and mesonotum in female thinly grey dusted.

♂. Eyes (Fig. 564) holoptic, all occipital hairs and bristles black. Thorax almost shining black when viewed from above, distinctly greyish dusted in anterior view. All bristles black, long and strong, acr biserial, slightly shorter than dc, which are biserial anteriorly, posteriorly becoming longer in the inner row only and ending in 2 or 3 very strong prescutellar pairs.

Wings (Fig. 561) faintly brownish clouded with a dark ovate stigma at tip of vein R_1, veins blackish brown, costal bristle prominent. Halteres brownish to brownish black. Legs varying from dark brown to light brownish, in dark-legged specimens at least "knees" and base of tarsi paler. All hairs and bristles on legs black, anterior four femora with distinct ad and pv bristly hairs, hind femora with conspicuous ad and av bristles, anterior tibiae with distinct ad bristles, at least one on mid tibia at about middle remarkably long. Hind tibiae laterally compressed and dilated towards tip, hind metatarsus slightly swollen, cylindrical, as long as the following tarsal segments combined.

Abdomen almost subshining black, all bristles black, hind marginal bristles conspicuously long. Genitalia as in Figs. 574–579.

Length: body 2.2–2.7 mm, wing 2.4–2.9 mm.

♀. All bristles on head, thorax and legs much shorter than in male, abdomen covered with mostly pale hairs, hind marginal bristles inconspicuous. Legs mostly yellowish, leaving tarsi (except metatarsi) and often also hind tibia blackish. The very broad frons (Fig. 565) and mesonotum rather dull, thinly greyish dusted, halteres rather blackish. Abdomen generally paler (due to weak sclerotisation) and rather dorsoventrally flattened, ovipositor-like on apical half.

Length: body 2.3–2.6 mm, wing 2.0–2.8 mm.

Distribution and biology. Rather a rare southern species (8 ♂ and 7 ♀ examined) known from Denmark (LFM, Bogø, 1.vii.1917, W. Lundbeck), S. Sweden (Sk., Gtl.), according to Zetterstedt (1842: 333) also Sm., and SW. Finland (Al, Ab, N). – Great

Britain, NW and central European USSR, C. Europe, south to France and Yugoslavia, locally common. - July (dates from Finland not available), outside Scandinavia from the end of May to mid August.

Synonymy. There were no doubts as to the synonymy of Walker's *sylvicola* with *pulicarius*, but Zetterstedt's *tibiella* remained unclear as Collin (1961: 233) did not find the type at Lund. In the Dipt. Scand. Coll. at Lund there are 3 ♂ and 1 ♀ of *A. pulicarius* under the name *Platycnema pulicaria* from Esperöd (Sk.) and 1 ♀ labelled "*P. tibiella* ♀ Gottl.", which is undoubtedly the holotype ♀ of *Platycnema tibiella* Zetterstedt, 1842; the specimen was labelled accordingly in 1977, and it is undoubtedly conspecific with *pulicarius;* it is a mouldy specimen with hind tibiae distinctly compressed and with thinly greyish dusted frons and mesonotum, and therefore not identical with *A. dissonans.* In the Göteborgsamlingen at Lund there is another ♀ of *pulicarius* under "*Microsania stigmaticalis* Zett.", labelled "♂ Finl. Bonsd.", a female sent to Zetterstedt by Bonsdorf.

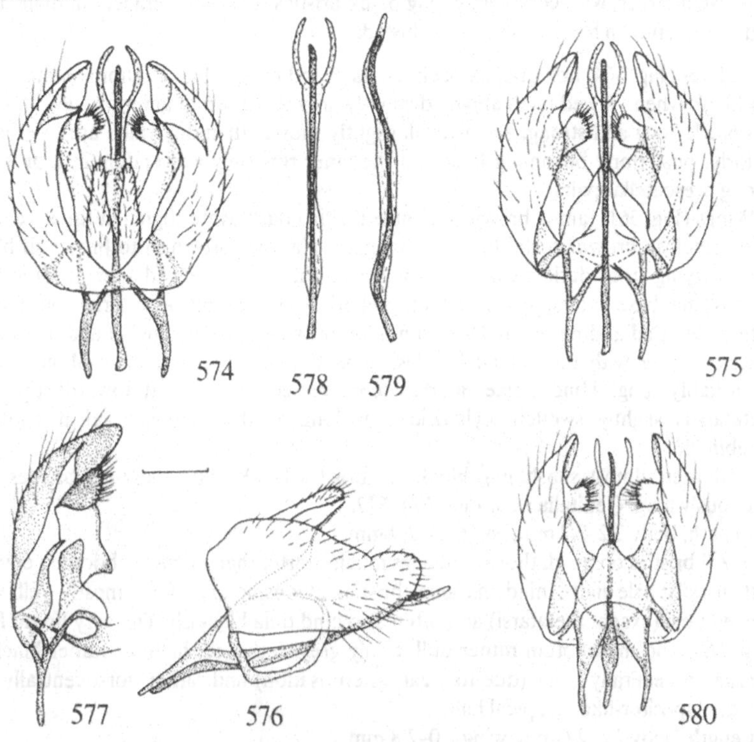

Figs. 574–579. Male genitalia of *Atelestus pulicarius* (Fall.). – 574: dorsal view; 575: ventral view, cerci omitted; 576: lateral view; 577: inner lateroventral view to left periandrial lamella; 578, 579: aedeagus in dorsal and lateral views. Scale: 0.1 mm.
Fig. 580. Male genitalia of *Atelestus dissonans* Coll. in dorsal view. Scale: 0.1 mm.

Atelestus dissonans Collin, 1961
Fig. 580.

Atelestus dissonans Collin, 1961: 233.

Generally a smaller species, differing from *pulicarius*, as follows:

♂. All bristles on head and thorax weaker and smaller, thoracic bristles including acr, dc and the strong marginal and scutellar bristles more equally long, not so differentiated as in *pulicarius*. Wings apparently more greyish and thus paler, costal bristle not very prominent. Legs with shorter hairs and bristles everywhere, ad bristles on mid tibia almost equally long. Genitalia: Fig. 580.

Length: body 1.7–1.8 mm, wing 2.1–2.3 mm.

♀. Resembling ♀ of *pulicarius* very closely but legs apparently darker, femora almost blackish brown. Generally blacker species even on abdomen (abdominal pubescence dark to blackish), and frons and mesonotum almost polished black; mesonotum only thinly greyish dusted in front, even in anterior view.

Length: body 2.2–2.3 mm, wing 2.1–2.2 mm.

Distribution and biology. England and C. Europe (Czechoslovakia); everywhere a more local and rather rare species, its occurrence in Denmark and extreme south of Sweden is possible. Apparently an overlooked species with a wide distribution like *pulicarius*. – June to August.

Genus *Meghyperus* Loew, 1850

Meghyperus Loew, 1850: 303.
Type-species: *Meghyperus sudeticus* Loew, 1850 (mon.).

Small, body length about 2.5–3 mm, polished black species (Fig. 581) practically devoid of distinct bristles, and with a very distinctive wing-venation: discal cell present, vein M_{1+2} apically forked.

Eyes holoptic in ♂ with upper facets considerably enlarged (up to 2.5 times larger above), broadly dichoptic in ♀ (Fig. 582), with all facets equally small. Frons in ♀ polished and very broad, widening above, 1 to 3 pairs of upper frontoorbital bristles very minute. A strong pair of anterior ocellar bristles, posterior pair minute. Face rather broad, clypeus large and conspicuously polished in the type-species. Antennae inserted at about middle of head in profile (or slightly below middle in males), paraantennal excisions broadly triangular in ♂. Basal antennal segments short, 1st segment often concealed, 2nd segment with a circlet of small preapical bristly hairs. 3rd segment short conical or slightly longer than deep, terminal style long in the type-species (Fig. 558), almost twice as long as antenna, 2-articulated, with a small basal article on the hook-like tip of 3rd segment which looks more like a continuation of the style; in Nearctic species style much shorter (Fig. 560), only slightly longer than 3rd segment, more decidedly terminal and 3-articulated, with 2 small basal articles; style microscopically pubescent except for the extreme bristle-like tip. Proboscis pointing forwards or obliquely forwards, very short and almost hidden in the mouth-cavity in the type-species (Fig. 583), very long and projecting in Nearctic species (Fig. 590). Mouthparts (Figs. 584–591) in general of similar structure to *Atelestus*, labrum one-pointed apically, maxillae stylet-like with well-developed lacinia and rather long cylindrical palpi, labium with soft labellae with distinct pseudotracheae.

Thorax only slightly arched above, humeri, postalar calli and scutellum well-developed. All thoracic hairs and bristles rather small and inconspicuous; acr 2- to 4-serial, dc uniserial, becoming longer behind and ending in one strong prescutellar pair; 1–2 humerals, 1 posthumeral, 2–3 notopleurals, 1 postalar and 4–6 scutellars. Prosternum small, *Hybos*-like, prothoracic episterna with 2 upcurved hairs and several small bristles on pronotum (prothoracic collar). Thoracic pleura bare.

Wings (Fig. 562) more or less faintly infuscated, densely covered with microtrichia, stigma short ovate round the tip of vein R_1, not reaching R_{2+3}. Axillary lobe well-developed but axillary excision obtuse, alula distinct (larger in ♂). Costal bristle absent or very small and fine, costal ciliation minute. Costa ending at tip of vein M_1, Sc closely

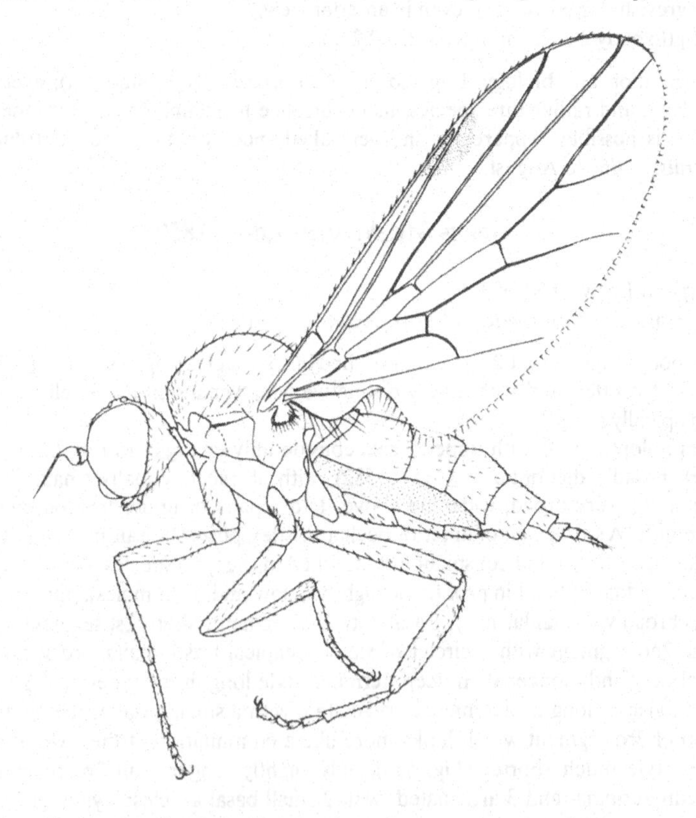

Fig. 581. Female of *Meghyperus sudeticus* Loew. Total length: 2.3–3.1 mm.

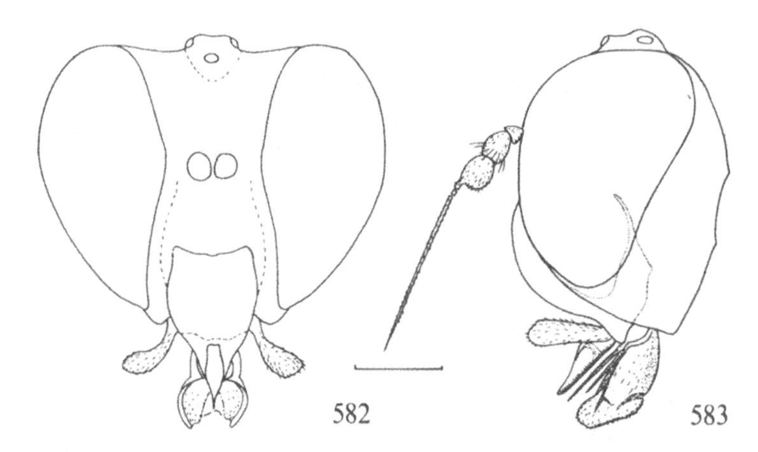

Figs. 582, 583. Head of *Meghyperus sudeticus* Loew ♀. - 582: frontal view; 583: lateral view. Scale: 0.2 mm.

approximated to vein R_1 but disappearing before reaching costa. Radial sector short, as long as the distance between its origin and the humeral crossvein. Vein R_{4+5} unforked, vein M_{1+2} forked halfway along its apical section, the lower branch (M_2) abbreviated at tip. Discal cell present (crossvein m-m developed), basal cells equally long and 2nd basal cell scarcely broader. Anal cell (Fig. 54) longer than basal cells or, in Nearctic species, subequal; the vein closing it arched as in *Hybos* species, reaching anal vein at right angles, anal vein complete, even if fine at tip.

Legs without ordinary bristles, clothed with short pubescence or at most with long fine hairs, especially on femora. Anterior two pairs slender, hind femora slightly stouter but hind tibiae and metatarsi very compressed and dilated in male, in females only hind tibiae more or less thickened, hind metatarsi cylindrical. All metatarsi lengthened but generally shorter than rest of tarsi. Fore tibiae without a tubular gland anteriorly at base.

Abdomen without ordinary bristles, hind marginal bristles absent, but venter in ♂ with fine outstanding bristle-like hairs as in *Microphorus*. Male abdomen (Fig. 598) rather long and cylindrical, often laterally compressed, seven basal segments unmodified. 8th segment very much as in *Atelestus*, tergum in the form of a narrow ring produced below distally, sternum large and heavily sclerotised. Genitalia (Figs. 592–597) with a very close resemblance in general to those of *Atelestus*, but gonostyli absent and aedeagus very distinctive, more closely resembling *Syneches;* postgonites present as in *Atelestus* but pointing downwards. Hypandrium convex, menbraneous and very indistinct apically. Female abdomen nearly bare, four basal segments broad and rather dorsoventrally compressed, apical segments narrow and ovipositor-like; cerci long and slender.

Distribution. A Holarctic genus with a single Palaearctic species, known also from

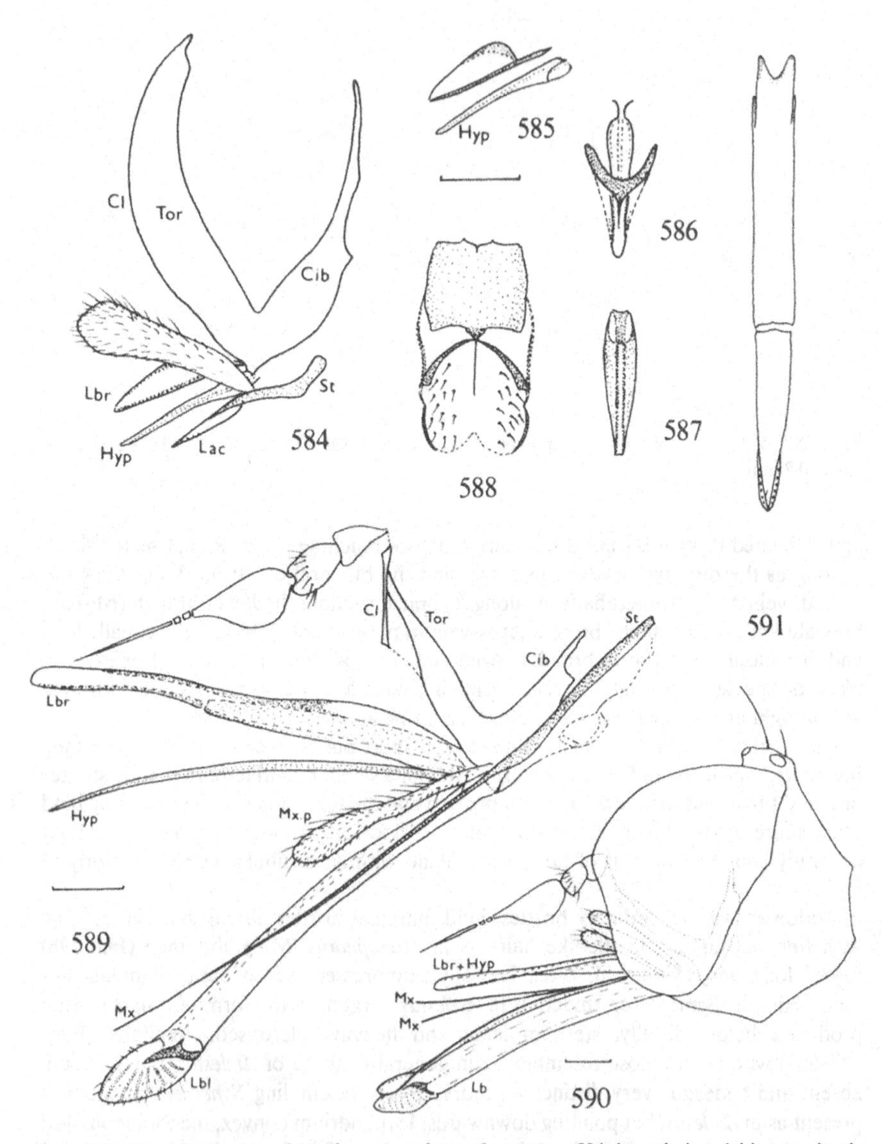

Figs. 584–588. Mouthparts of *Meghyperus sudeticus* Loew ♀. – 584: lateral view, labium omitted; 585: labrum and hypopharynx in lateral view; 586: labrum in dorsal view; 587: hypopharynx in dorsal view; 588: labium in ventral view.

Figs. 589–591. Mouthparts of *Meghyperus* sp. ♀ (USA, Calif.). – 589: lateral view; 590: the same in natural position in head; 591: labrum in dorsal view. Scale: 0.1 mm. For abbreviations see p. 13.

Scandinavia, and two (probably three) Nearctic species, *M. nitidus* Mel. and *M. occidens* Coq. from Ohio and California respectively.

Biology. No information on the immature stages and development is available, nor are any observations on the feeding habits of the adults, although they are supposed to be predaceous. In Europe adults are rarely swept from vegetation or accidentally found in diverse situations, as for instance on windows. On the other hand, their swarming activity is well known; Wilder (letter comm., Dec. 1978) frequently observed swarms of *Meghyperus* sp. (obviously an undescribed species, and not *occidens*) in the springtime in California, USA, and Frey (1956) wrote of *M. sudeticus* that he "beobachtete diese Art über eine Lehmbank in dem kleinen Bache Viisjoki auf der Karelischen Landenge am 9.VII.1933 in grosser Anzahl schwärmend". There is a unique series of 47 ♀ in the Zoological Museum, Helsinki, taken by Frey at Metsäpirtti (Vib), which was undoubtedly collected from this swarm. Apart from this series, I have only seen single specimens of *sudeticus* in various collections, mostly females.

Taxonomic note. There seem to be some important morphological differences between the Palaearctic type-species (*sudeticus*) and the Nearctic species. Of the latter, I had at my disposal only one unidentified species from California, obviously not *M. occidens*. The proboscis is very long in the Nearctic species but conspicuously short in *sudeticus;* however, there are no fundamental structural differences, and the slight distinctions may well be the result of lengthening of the mouthparts in the Nearctic species. (The proboscis of the single available Nearctic species is equally long in both sexes, although Melander (1928: 47) gives as one of the important features of sexual dimorphism in the Nearctic species the fact that the proboscis is almost 3 times as long in females as in males.) An important differential feature seems to be the 3-articulated antennal style in the Nearctic species and the 2-articulated style in *sudeticus* (usually at least a generic character within the Brachycera); however, the hook-like projection of 3rd ant.s. in *sudeticus* (Fig. 558) may well indicate the way in which the additional second basal article has become differentiated.

44. ***Meghyperus sudeticus*** Loew, 1850
Figs. 54, 558, 559, 562, 581–588, 592–594.

Meghyperus sudeticus Loew, 1850: 303.

Polished black species with inconspicuous thoracic hairs and bristles, wings with discal cell and vein M_{1+2} apically forked.

♂. Eyes holoptic with upper facets considerably enlarged, face deep, clypeus conspicuously large, convex and polished black, occupying most of the facial part. Antennae (Fig. 558) black, 3rd segment almost spherical and apically produced, style almost twice as long as antennae, finely pubescent. Mesonotum polished black, anteriorly narrowly brown dusted. All thoracic hairs and bristles black, acr and dc fine and inconspicuous, former narrowly biserial, dc uniserial, ending in 1 longer prescutellar pair. 4–6 scutellars stronger than 2 notopleurals and a postalar.

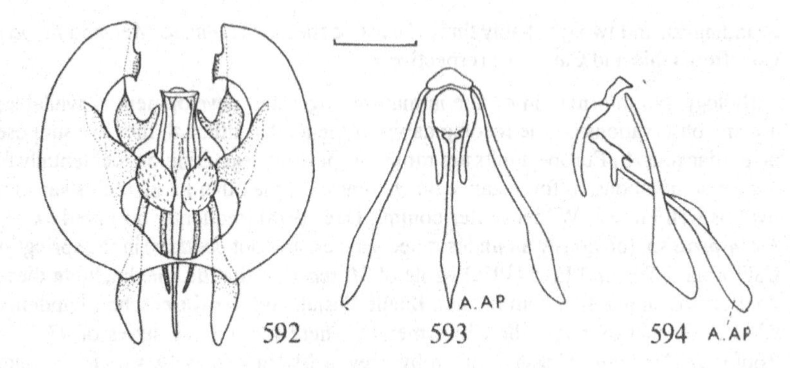

Figs. 592–594. Male genitalia of *Meghyperus sudeticus* Loew. – 592: dorsal view; 593: aedeagus in dorsal view; 594: the same in lateral view. For abbreviations see p. 13.

Wings faintly brownish on costal half, greyish and paler below, stigma dark brown, short ovate; costal bristle absent. Anal cell longer than 2nd basal cell, vein M_{1+2} forked beyond discal cell, and alula well-developed. Squamae and halteres black. Legs almost polished black with "knees" yellowish, devoid of distinct hairs or bristles, only hind femora with somewhat long hairs above and below. Hind tibiae evenly dilated, as deep as hind femur, with a concave excision dorsally before tip.

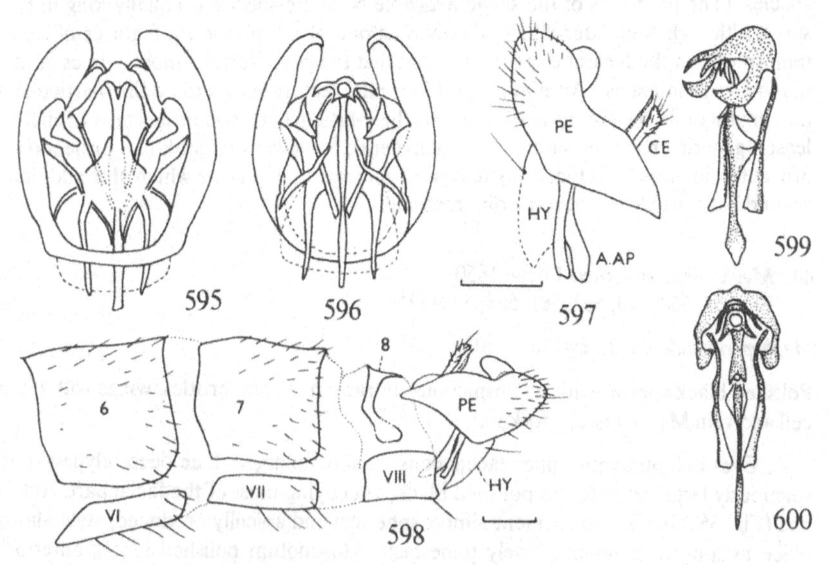

Figs. 595–600. Male genitalia of *Meghyperus* sp. (USA, Calif.). – 595: dorsal view, cerci omitted; 596: ventral view; 597: lateral view; 598: postabdomen (macerated); 599: aedeagus, laterodorsal view, 600: same, dorsal view. Scale: 0.1 mm. For abbreviations see p. 13.

Abdomen cylindrical and devoid of distinct bristles, subshining black, genitalia as in Figs. 592–594.

Length: body 2.8 mm, wing 2.7 mm.

♀. Head (Fig. 582) polished black on frons and face, eyes broadly separated by a dorsally widening frons which has 3 pairs of minute upper frontoorbital bristles. Abdomen almost bare, subshining black on anterior five segments, the very narrowed ovipositor-like apical segments and cerci dull grey.

Length: body 2.3–3.1 mm, wing 2.6–3.0 mm.

Distribution and biology. Very local, known only from S. Finland (Sa, Joutseno, 18.vi.1860, 1 ♀, Thunberg) and the south of Russian Karelia (Vib, Metsäpirtti, 9.vii.-1933, 47 ♀, Frey; 24.vi.1934, 1 ♀, Krogerus). – Euroasian species, described from the Krkonoše Mts. (Riesengebirge), Czechoslovakia; further records are from the Netherlands, Austria, Czechoslovakia, the two Germanies, Hungary and Yugoslavia; east from the Ussuri region (Collin, 1941). Males everywhere very rare. – June and July, in C. Europe from May to August.

Family Microphoridae

The newly-erected family Microphoridae only includes about forty recent species, and is mainly Holarctic in distribution. The Microphoridae are morphologically well-separated and distinct in general appearance from the former "Empididae" and, as discussed in the section on Phylogeny, it is suggested that they form together with the Dolichopodidae a monophyletic subgroup of the Empidoidea.

Diagnosis. Males holoptic with upper facets enlarged (Microphorini), or broadly dichoptic with all facets equally small (Parathalassiini) as in all females. Antennae inserted at or below middle of head in profile (Microphorini), or much above middle (Parathalassiini). Antennae with 3rd segment long conical to almost pear-shaped, terminal style microscopically pubescent, varying in length, 2-articulated in the Microphorini, 1-articulated in the Parathalassiini. Proboscis projecting forwards or almost vertical (in which case always very short), mouthparts either with fully developed maxillae including lacinia (Microphorini), or lacinia absent (Parathalassiini). Palpi elongate, membraneously connected with maxillary stipites, and labium with distinct labellae with pseudotracheae. Head closely set upon thorax, which is higher than long or deep, more or less arched above, usually with well-developed large thoracic bristles; acr 2- to 4-serial, or even absent (some *Microphorella*). Prosternum *Hybos*-like and, even if enlarged (*Microphorella*), always well-separated from episterna by membraneous areas. Wings very distinctive because of the basal origin of radial sector and the very small basal cells, anal cell equally small and conspicuously rounded at tip (except *Parathalassius*); anal vein very incomplete. Costa running around the wing, at least as an ambient vein on posterior wing-margin, Sc complete (ending in costa) or nearly complete; radial sector 2-branched (R_{4+5} not forked) and, although very short,

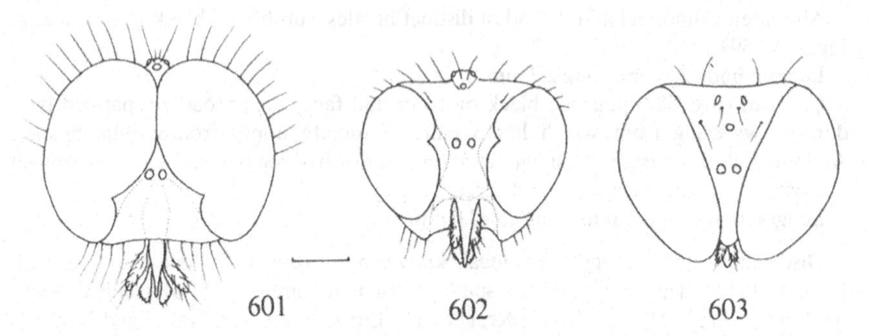

Figs. 601–603. Heads in frontal view of Microphoridae. – 601: *Microphorus holosericeus* (Meig.) ♂; 602: same species ♀; 603: *Microphorella praecox* (Loew) ♂. Scale: 0.2 mm.

originating well proximally, opposite humeral crossvein. Discal cell emitting 3 veins to wing-margin, M_1 and M_2 arising separately from discal cell. Alula not developed but alular fringes long, axillary lobe prominent in Microphorini, quite absent in *Microphorella*. Wings wholly covered with microtrichia, costal stigma at tip of vein R_1 (absent in Parathalassiini). Legs rather short and covered with mostly weak bristles, usually simple (always in females), hind tibiae and metatarsi often dilated in males (*Microphorus*) or hind tarsi curiously shaped (*Microphorella*). Male genitalia asymmetrical but obviously not rotated, their inverted right-hand position alongside abdomen resulting from the compression and rotation of the two pregenital abdominal segments. Hypopygium of very complicated structure, with a remarkably enlarged hypandrial part and curious aedeagal armature, the capsule-like hypopygium conspicuously large in the Parathalassiini. Female abdomen either telescopic with small terminal cerci (most Microphorini), or conspicuously blunt-ended, with 9th tergum apically cleft and spinose, heavily sclerotised and polished like the two or three preceding segments.

Classification. The family Microphoridae includes two tribes: the Microphorini, with two recent genera *Microphorus* Macq. and *Schistostoma* Beck., and the Parathalassiini, also with two recent genera, *Parathalassius* Mik and *Microphorella* Beck. The genera were originally placed in the subfamily Empidinae of the former "Empididae", and then more recently as a separate "Microphorus-group" of genera, because of the many distinctive characters in the wings and genitalia. Lundbeck (1910) placed the genus *Microphorus* in the subfamily Ocydromiinae and Collin (1961) assigned it to his unnatural subfamily Hybotinae, although he noted that the peculiar structure of the male genitalia and the manner in which the terminalia are borne in *Microphorus* and allied genera, might well be enough to justify their inclusion in a separate subfamily. This view was supported by Tuomikoski (1966) who suggested that the many distinctive morphological characters in the mouthparts and male terminalia "definitely place the *Microphorus*-group outside the whole ocydromioid group of subfamilies". Colless

(1963), when studying the genus *Microphorella* in Australia, first showed the close relationship between the *Microphorus*-group of genera and the family Dolichopodidae, stating that this group may well represent "an empidid stock from which the Dolichopodidae have been derived", and this view has been recently supported by Chvála (1981a). Hennig (1970, 1971), followed by Chvála (1976, 1981a), separated the *Microphorus*-group of genera into a subfamily Microphorinae within the former "Empididae"; however, it was Collin (1960) who was actually the first to mention (without comment) a separate subfamily Microphorinae when describing a new *Microphorus* species from Palestine. The Afrotropical genus *Edenophorus*, originally placed by Smith (1969: 308) in the *Microphorus*-group, is not accepted here as belonging to the Microphoridae (see above under the Ocydromiinae).

Distribution. The family with its four recent genera is mainly Holarctic in distribution. The genus *Microphorus* includes 15 Nearctic and 8 Palaearctic species (if the Palestine *M. proboscideus* Coll. is identical with *M. rostellatus* Loew), *Schistostoma* includes 1 Nearctic and at least 4 Palaearctic (mostly Mediterranean) species, *Parathalassius* 3 Nearctic, 1 Palaearctic and 1 Afrotropical species, and *Microphorella* 5 Nearctic, 3 Palaearctic and 1 Australian species. Three species of *Microphorus* and one of *Microphorella* have been found in Scandinavia.

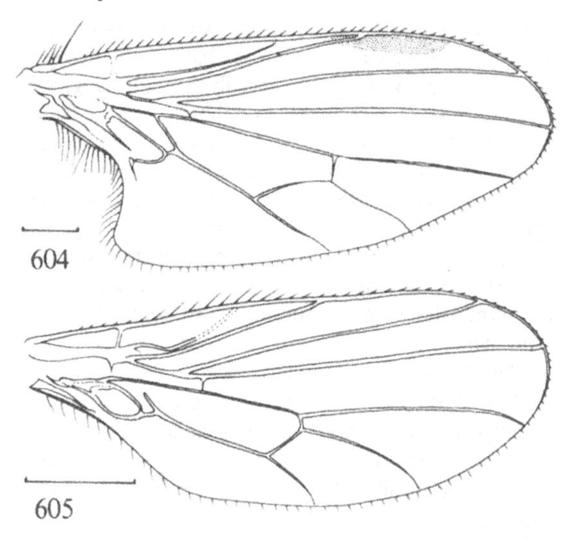

Figs. 604, 605. Wings of Microphoridae. – 604: *Microphorus holosericeus* (Meig.) ♂; 605: *Microphorella praecox* (Loew) ♂. Scale: 0.3 mm.

Key to genera of *Microphoridae*

1 Eyes bare, holoptic in ♂ (Fig. 601), broadly dichoptic in ♀ (Fig. 602). Antennae inserted at or below middle of head in profile, mouthparts long and directed forwards or obliquely forwards. Wings (Fig. 604) with a distinct

stigma, axillary lobe well-developed. Mesonotum with distinct acrostichals and 6–8 scutellars .... *Microphorus* Macquart (p. 246)

– Eyes microscopically pubescent, broadly separated on frons in both sexes (Fig. 603). Antennae inserted well above middle of head in profile, mouthparts short, directed obliquely forwards or nearly vertical. Stigma and axillary lobe on wings (Fig. 605) not developed. Acrostichals minute or absent, 2 scutellars .... *Microphorella* Becker (p. 257)

Genus *Microphorus* Macquart, 1827

Microphorus Macquart, 1827: 139 (as *Microphor*; emend. Macquart, 1834: 345).
Type-species: *Microphor velutinus* Macquart, 1827 (des. Rondani, 1856) = *Microphorus holosericeus* (Meigen, 1804).
Microphora Zetterstedt, 1842: 253 (unjustified emend.), nec Kröber, 1912: 245 (= Therevidae).
Holoclera Schiner; Melander, 1902: 333.

Small, body length about 2–2.5 mm, black or grey species (Fig. 606) with distinct projecting proboscis and well-developed axillary lobe on wing. Eyes bare, males holoptic (Fig. 601) with upper facets considerably enlarged, females broadly dichoptic (Fig. 602) and all facets equally small. Head rather large, closely set upon thorax, occiput flat or slightly concave above, convex below. Three ocelli on a distinct tubercle, anterior pair of ocellar bristles long, posterior pair very minute; females with a small hair on upper fronto-orbitals on each side. Face very broad, conspicuously so in ♂. Antennae inserted at about middle of head in profile, more below in ♂. Basal ant. segment (Fig. 611) very small, 2nd larger and globular, armed with a circlet of preapical bristly hairs. 3rd segment elongate, laterally compressed, very broad at base and evenly tapering to a long point, finely pubescent like a terminal style; style shorter than 3rd segment (rarely longer in Nearctic species) usually half as long as 3rd segment, 2-articulated, with a small basal article. Proboscis rather long, often as long as head is high, pointing forwards or obliquely forwards. Mouthparts (Figs. 607–610) with labrum (4-pointed at tip) and hypopharynx equally long, maxillae with well-developed lacinia almost as long; palpi cylindrical to almost long clubbed, well bristled, membraneously joined to maxillary stipites, not directly connected with maxillae. Labium with rather small but distinct soft labellae with pseudotracheae.

Thorax higher than long, very arched above, acr 2- to 4-serial, dc 1- to 3-serial, often with numerous bristly hairs at sides; large bristles in full number and often very strong (humeral, intrahumeral, posthumeral, notopleural, supra-alar, postalar), scutellum with 4–8 marginal bristles. Prosternum small, *Hybos*-like, prothoracic episternum and pronotum with a small hair on each side, metapleura bare.

Wings (Fig. 604) clear or faintly uniformly clouded, entirely covered with microtrichia; costal stigma (if visible) long, occupying basal half of costal section between R_1 and R_{2+3}. Axillary lobe large but axillary excision very obtuse, wings remarkably broad at base, no alula. One or two costal bristles and distinct costal ciliation. Costa running around wing, although often finer on posterior wing-margin beyond vein M_1. Sc closely approximated to R_1, finer or abbreviated just before

Fig. 606. Male of *Microphorus holosericeus* (Meig.). Total length: 2.0–2.6 mm.

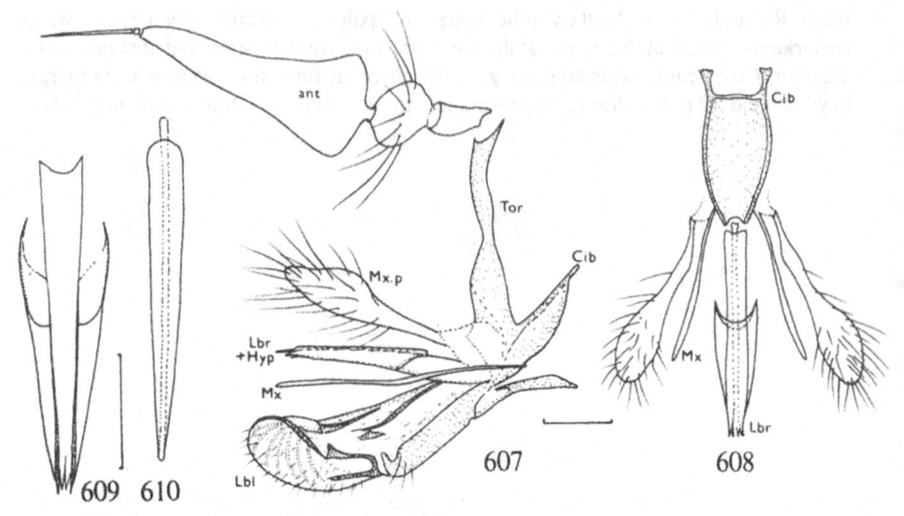

Figs. 607–610. Mouthparts of *Microphorus holosericeus* (Meig.) ♂. – 607: lateral view; 608: dorsal view, tormae and labium omitted; 609: labrum in ventral view; 610: hypopharynx in ventral view. Scale: 0.1 mm. For abbreviations see p. 13.

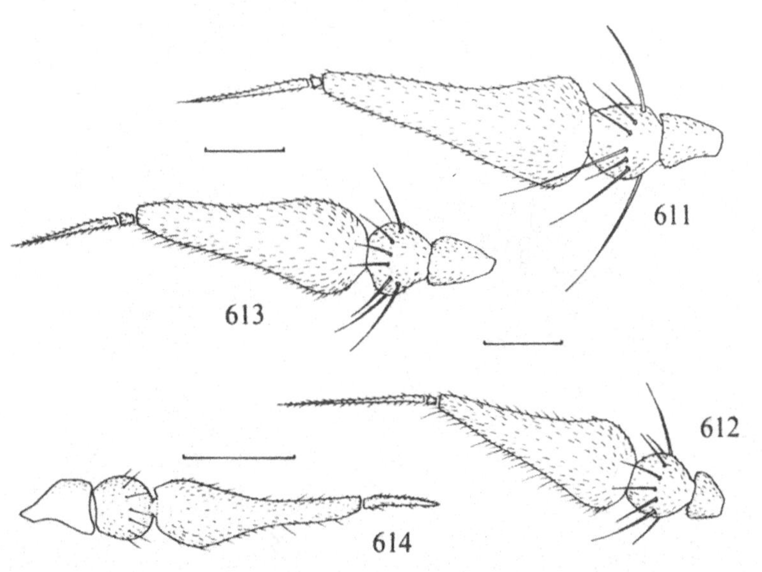

Figs. 611–614. Antennae of Microphoridae. – 611: *Microphorus holosericeus* (Meig.) ♂; 612: *Microphorus anomalus* (Meig.) ♂; 613: *Microphorus crassipes* Macq. ♂; 614: *Microphorella praecox* (Loew) ♀. Scale: 0.1 mm.

reaching costa, less often complete. Veins R_{2+3} and R_{4+5} slightly diverging and both ending nearly at wing-tip. Radial sector rather short and very characteristic because of its basal origin, arising exactly opposite humeral crossvein. Discal cell large, emitting 3 veins to wing-margin. Basal and anal cells very short and small (Fig. 55), the vein closing 2nd basal cell (crossvein tb) complete, anal cell rounded at tip; anal vein (A_1) very reduced in recent species, in the form of a short vestige of a vein bent downwards at its junction with the recurrent branch of vein Cu_{1b} (closing anal cell).

Legs (Figs. 615–617) rather short, hind tibiae and metatarsi sometimes very compressed and dilated in ♂, tibiae and femora usually with distinct rows of bristly hairs above and below, ordinary bristles absent; pubescence much shorter in ♀. Fore tibiae without a tubular gland and tarsi rather short.

Abdomen cylindrical, considerably laterally compressed in ♂, more dorsoventrally flattened in ♀, usually clothed with erect long bristles in male even on venter. 7th segment in ♂ (Fig. 72) somewhat compressed, slightly bent downwards and turned towards right, 8th segment turned towards right through nearly 90° so that 8th sternum

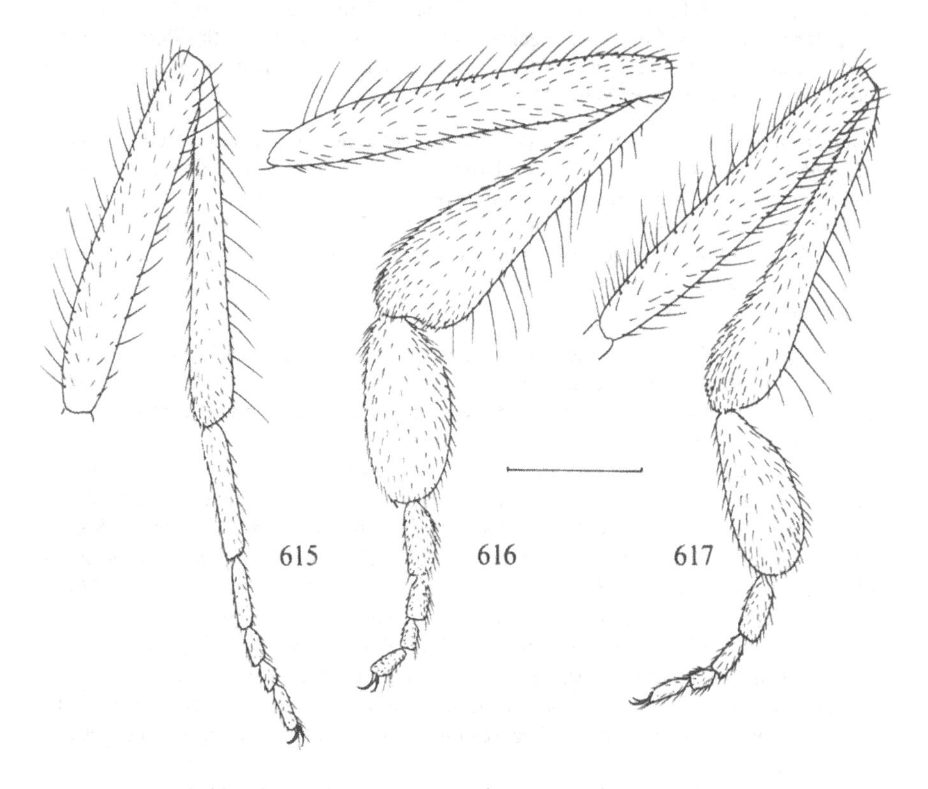

Figs. 615–617. Hind legs in anterior view of *Microphorus*. – 615: *holosericeus* (Meig.) ♂; 616: *anomalus* (Meig.) ♂; 617: *crassipes* Macq. ♂. Scale: 0.3 mm.

lies on the left side of abdomen and hypopygium on the right side, with ventral sclerite (hypandrium) placed dorsally. Hypopygium itself not rotated, the inverted and lateral position caused by rotation and deflexion of the last two pregenital abdominal segments. Genitalia (Figs. 618–623) of very peculiar and complicated structure, and the homology of the individual sclerites rather unclear; a detailed study, with illustrations of the hypopygium of the type-species (as *velutinus*), is given by Hennig (1976). Generally, hypandrium very large, armed with 3 pairs of conspicuous bristles, apically produced into a long asymmetrical projection (Fig. 621) which is attached to the tip of aedeagus. Periandrial lamellae asymmetrical and narrowly connected anteriorly, aedeagus overlapping the lamellae, more or les curved, base of aedeagus enclosed by strikingly peculiar structures (aedeagal armature, sensu Hennig), consisting of a conspicuous curved, flattened and apically pointed sclerite (Fig. 622) (asymmetrical hypandrial appendage, sensu Hennig) and sometimes, as in the type-species, with large lateral cylindrical sclerites (telomeres, sensu Hennig).

Female abdomen rather telescopic at tip, cerci small and slender, usually concealed within the last abdominal segment. I did not find the cleft spinose ♀ 9th tergum (hemitergites) in any European species, a characteristic feature of the Parathalassiini and *Schistostoma*, but Hennig (1976) illustrated such an arrangement in the ♀ of the Nearctic *Microphorus sycophantor* Mel.

Distribution. A Holarctic genus with 15 Nearctic and 8 Palaearctic species. *Microphora angustifrons*, described by Kröber (1912) from New Britain, is a therevid. Three species have been found in Scandinavia.

The separation of *Microphorus* species is rather difficult (particularly in females), and the European species are generally misidentified in collections. For this reason I have given more detailed diagnoses, and the distributions are based exclusively on specimens seen by me.

Biology. The adults are found in various biotopes, preferring lowlands close to water or deciduous forests, and are often swept from meadows. According to Collin (1961), Beling in 1882 described the larva of *M. anomalus* which was found beneath leaves in a beech plantation, but no further data on immatures are available so far I am aware. The adults are predaceous, but are often found also on flowers and are undoubtedly nectar or pollen feeders as well. Females of *Microphorus* have repeatedly been observed to feed on insects caught in spiders' webs (Laurence, 1949; Downes & Smith, 1969), and have also been observed on several occasions (Collin, 1961 and personal observations) to hover in large numbers outside window-panes, which may well be a special habit for hunting other insects. Males often hover in small swarms at the end of overhanging branches of hedges or trees (Collin, 1961 – *M.crassipes*) or in loose swarms of 3 to 10 individuals before sunset in open clearings at a height of up to 3 metres (personal observations – *M.anomalus*); in no case have males ever been observed to carry prey when hovering in aerial aggregations, as is a common habit in most species of the Empidinae.

Key to species of *Microphorus*

1 Eyes meeting on frons – ♂♂ 2
– Eyes widely separated on frons – ♀♀ 4
2(1) Hind tibiae and metatarsi slender, not at all dilated (Fig. 615). Mesonotum dull grey with black stripes on the lines of bristles, acr broadly biserial, dc uniserial, all less numerous and conspicuous .... 45. *holosericeus* (Meigen)
– Hind tibiae and metatarsi remarkably dilated (Figs. 616, 617). Mesonotum not very dull, rather subshining, small mesonotal hairs (acr and dc) smaller and more numerous 3
3(2) Mesonotum rather greyish from some angles, with 3 broad dull black stripes; acr 4–serial, dc 2– to 3–serial, all rather small and numerous. Venter of abdomen with 8–9 pairs of long straggling bristles. Hind legs (Fig. 616) very compressed .... 46. *anomalus* (Meigen)
– Mesonotum rather subshining black and practically without longitudinal stripes; acr irregularly 3– to 4–serial, dc 2–serial (1–serial anteriorly), all rather longer and less numerous. Venter of abdomen with less numerous (6 pairs) long straggling bristles. Hind legs (Fig. 617) somewhat less compressed and more bristled .... 47. *crassipes* Macquart
4(1) Mesonotal bristles conspicuous and few in number, at most 7–8 bristles in one row; acr broadly 2–serial, dc 1–serial. Mesonotum grey dusted, with distinct brown stripes on the lines of the bristles .... 45. *holosericeus* (Meigen)
– Mesonotal bristles rather small, acr 3– to 4–serial, dc at least 2–serial 5
5(4) Acr and dc small and numerous (more than 10 bristles in one row), acr regularly 4–serial and barely separated from 2 to 3–serial dc. Mesonotum brownish grey dusted in anterior view, with 3 indistinct dull black longitudinal stripes (the broad central stripe more distinct in posterior view). Proboscis and palpi rather shorter, palpi with 1 or 2 fine bristly hairs .... 46. *anomalus* (Meigen)
– Acr and dc rather longer and less numerous (about 8 bristles in one row), acr irregularly 3– to 4–serial, dc 2–serial (almost 1–serial anteriorly) and well–separated from acr by bare stripes. Mesonotum rather black, somewhat shining and unstriped. Proboscis and palpi longer, palpi with several black preapical bristly hairs .... 47. *crassipes* Macquart

45. ***Microphorus holosericeus*** (Meigen, 1804)
Figs. 55, 72, 601, 602, 604, 606–611, 615, 618–623.

Empis holosericeus Meigen, 1804: 231.
Microphor velutinus Macquart, 1827: 140.
Microphora fuscipes Zetterstedt, 1852: 4268 (nec Zetterstedt, 1842 = *Trichinomyia fuscipes* Zetterstedt, 1838).
Microphorus vicinus Mik, 1887: 99.
Microphorus dalmatinus Strobl, 1902: 468 – **syn.n.**

Thorax dull grey, lighter grey in ♀, with mesonotal bristles long and less numerous: acr widely 2-serial, dc 1-serial. ♂: hind legs slender.

♂. Eyes meeting on frons, occiput dull blackish grey, densely clothed with short black bristles, postocular bristles very long, almost as long as antennal style. Antennae (Fig. 611) inserted much below middle of head. Proboscis short, palpi with long, densely-set black bristles. Thorax dull blackish grey, mesonotum duller grey with 2 paler narrow stripes between the rows of bristles, subshining or nearly velvety-black in dorsal view. All thoracic bristles black, acr and dc almost as long as upper postocular occipital bristles, arranged in four almost equally distant rows; acr broadly 2-serial, dc 1-serial,

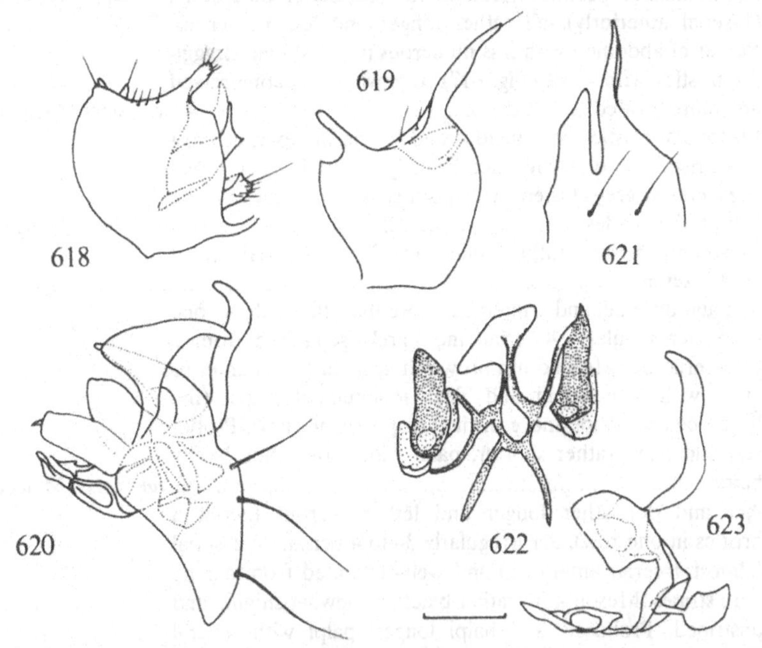

Figs. 618–623. Male genitalia of *Microphorus holosericeus* (Meig.). – 618: right periandrial lamella with cercus; 619: left periandrial lamella; 620: hyandrium with aedeagal complex; 621: tip of hypandrium from above; 622: armature of aedeagus from behind; 623: aedeagus. Scale: 0.1 mm.

becoming longer posteriorly, last pair as long as the strongest scutellars; each row consisting of at most 7–8 bristles.

Wings faintly brownish with a brown ovate stigma. Squamae blackish with dark fringes (not pale as given by Collin, 1961: 321), halteres almost black. Legs blackish, all femora and tibiae slender. Fore femora with pd and pv rows of long black bristles, similar pv and posterior bristles on mid femora, and ad and av bristles on hind femora. Posterior four tibiae with a row of long bristles dorsally.

Abdomen almost velvety-black on dorsum, sides finely greyish dusted and somewhat subshining, hind marginal bristles distinct, black. Sterna with much shorter bristles. Genitalia as in Figs. 618–623.

Length: body 2.0–2.6 mm, wing 2.3–3.0 mm.

♀. Eyes broadly separated on frons, body generally lighter grey than in ♂ and everywhere with shorter hairs and bristles. Mesonotum lighter grey even when viewed from above, with distinct brown stripes on the lines of bristles; acr and dc arranged as in ♂ but shorter. Wings almost clear with a very faint stigma, halteres paler. Abdomen dull greyish brown, somewhat subshining when viewed from above.

Length: body 1.9–2.3 mm, wing 2.2–2.6 mm.

Distribution and biology. Very common in Denmark and southern Fennoscandia, less common towards north, in Sweden to Jmt., in Finland to Ks, in Norway to TRi, approximately 68°N; also Russian Karelia. - Widespread in Europe, south to Romania, Bulgaria, Yugoslavia, France and Spain. - May to July, but mainly in May; in S. Europe as early as the beginning of April, in C. Europe until August.

Synonymy. *Microphorus dalmatinus* Strobl, 1902 was described from a single male, now in the Strobl Coll. at Admont. The holotype ♂ is labelled "Zara ♂ Novak / *Microphorus* n.sp. ? *velut.* v. *dalmatina* ♂" and, although it is more greyish dusted than usual in the male sex, it is undoubtedly identical with *M. holosericeus.*

46. ***Microphorus anomalus*** (Meigen, 1824)
Figs. 612, 616, 624–627.

Platypeza anomala Meigen, 1824: 9.
Microphora tarsella Zetterstedt, 1842: 257.

Mesonotum dark grey with 3 broad dull black stripes, more subshining black in ♀, mesonotal bristles small and very numerous, acr 4–serial, dc 2– to 3–serial. ♂: hind tibiae and metatarsi very compressed and dilated.

♂. Occiput subshining black with scattered short black bristles, those on postocular margin scarcely longer. Antennae (Fig. 612) with a long style not much shorter than 3rd segment. Proboscis short and palpi very small, spically with 1 or 2 long black bristly hairs. Thorax blackish grey, pleura more bluish grey dusted. Mesonotum with 3 broad dull black stripes on the lines of bristles, almost uniformly brownish grey dusted in anterior view, dull black when viewed from above, or with 2 very narrow greyish lines between acr and dc in some lights. All thoracic bristles black; acr very small, 4–serial (irregularly 3-serial in front of prescutellar depression) and very indistinctly separated from 2- to 3-serial dc which are longer posteriorly, ending in 2 long prescutellar pairs; 3 pairs of scutellars, inner and outer pairs half as long as the 2nd pair, which is slightly longer than last pair of prescutellar dc.

Wings practically clear, with blackish veins and distinct ovate brownish stigma. Usually 2 costal bristles, axillary lobe very produced. Squamae blackish with dark fringes, halteres black. Legs blackish to blackish brown, fore femora with a row of distinct pv bristles becoming longer towards tip, mid femora with about 9 equally long pv bristles. Fore tibiae with short hairs, but mid tibiae with 3 dorsal bristles (at base, at middle and before tip) much longer than tibia is deep. Hind femora (Fig. 616) not distinctly deeper than anterior femora, dorsally with a row of long bristles, ad bristles small, and ventrally with 2 rows of fine bristly hairs. Hind tibiae very compressed and dilated towards tip, dorsally with a row of long black bristles. Hind metatarsus evenly compressed, about as deep as tibia at tip and as long as following four tarsal segments; 2nd segment slightly ovate. Anterior metatarsi long, not very much shorter than rest of tarsi.

Abdomen subshining black or almost dull black on dorsum from some angles, hind marginal bristles long, black. Venter more greyish and armed with 8–9 pairs of long black straggling bristles. Genitalia (Figs. 624–627) with rather slender aedeagus and lamellae without conspicuous processes.

Length: body 2.0–2.5 mm, wing 2.2–2.6 mm.

♀. Frons very broad, greyish black, face sligthly broader than in ♂. The dull black mesonotal stripes rather indistinct and mesonotum consequently more subshining black, lighter grey dusted in anterior view. Acr and dc as in ♂ but apparently smaller, almost uniformly covering mesonotum. Legs with shorter hairs and bristles, particularly

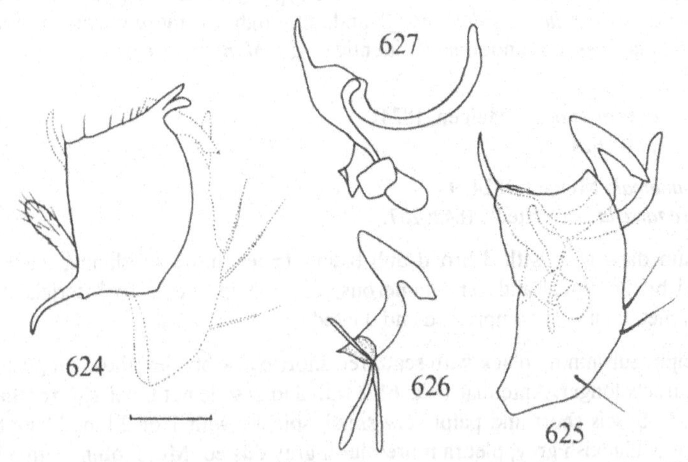

Figs. 624–627. Male genitalia of *Microphorus anomalus* (Meig.). – 624: left periandrial lamella with cercus, hypandrium and aedeagus dotted; 625: hypandrium with aedeagal complex; 626: asymmetrical hypandrial appendage; 627: aedeagus. Scale: 0.1 mm.

on femora, hind femora and tibiae simple although slightly compressed and with somewhat more distinct adpressed hairs dorsally. Mid tibiae with a small black preapical bristle, rarely with another in apical third. Abdomen subshining black, marginal bristles small and venter with only fine small hairs. Halteres yellowish brown, sometimes extensively darkened.

Length: body 1.7–2.3 mm, wing 2.2–2.6 mm.

Distribution and biology. Rather common in Denmark and S. Sweden, north to 57°, not found in Norway and Finland. – Very common in C. Europe, south to Yugoslavia (Dalmatia; Macedonia, Coe, 1962), Romania, Bulgaria, France (Nice), and also recorded from Spain (Strobl, 1909); according to Collin (1961) only a little known species in England! – June to August, on dates ranging from 17 June to 5 August, in C. and S.Europe as early as May.

Note. *Microphora tarsella,* described by Zetterstedt (1842) from Gotland (Nähr, Öja) from the male sex only (females recorded as *M.pusilla* Macq.), is present in the Dipt.Scand.Coll. at Lund under a single *Microphorus* species labelled as "*M.pusilla* Zett.Macq., *crassipes* Meig., *tarsella* Zett.Dipt. Gottl."; 2♀ are *M.crassipes,* 2 ♂ *M.anomalus*, and one ♀ from Stockholm ("ad Holm, Bohem.") is *M. holosericeus.*

47. ***Microphorus crassipes*** Macquart, 1827
Figs. 613, 617, 628–631.

Microphor crassipes Macquart, 1827: 140 (♂).
Microphor pusillus Macquart, 1827: 140 (♀).

More shining black than *anomalus*, mesonotum without dull stripes, acr and dc less numerous and decidedly longer. ♂: hind tibiae and metatarsi dilated as in *anomalus* but the straggling bristles on venter of abdomen less numerous.

♂. Head as in *anomalus* but postocular occipital bristles more differentiated and antennal style shorter, half as long as 3rd segment (Fig. 613). Proboscis rather longer, not very much shorter than head is deep, and palpi almost half as long as labrum, with several bristly hairs dorsally before tip, the hairs longer than other pubescence. Thorax more blackish on mesonotum, without dull longitudinal stripes, dark brownish grey in anterior view, otherwise rather shining black. Pleura blackish grey. All thoracic bristles black; large bristles as in *anomalus* but acr irregularly 4–serial, less numerous and somewhat longer, more distinctly differentiated from dc, which are almost uniserial in front, irregularly 2– to 3-serial at middle and less numerous, about 8 bristles in one row; last 3 prescutellar bristles much longer.

Wings as in *anomalus*; squamae brownish with blackish fringes, halteres almost black. Legs as in *anomalus* but all bristles somewhat longer. Fore and mid legs with similar pv bristles on femora and with 3 distinct bristles on mid tibia, but mid femur with several long av and posterior bristles about as long as femur is deep on basal third, and anterior metatarsi with a long bristle at extreme base beneath (absent in *anomalus*).

Hind legs (Fig. 617) with dorsal bristles on femora and tibiae obviously longer than in *anomalus* and the femoral ad and ventral rows of bristles more erect and consequently more distinct.

Abdomen thinly blackish brown dusted, especially at sides and on venter, dorsum subshining or almost dull velvety-black. Hind marginal bristles long, black, venter with less numerous (6 pairs) long straggling bristles. Genitalia (Figs. 628–631) similar to those of *anomalus* in general but aedeagus apically broadened.

Length: body 1.8–2.3 mm, wing 2.2–2.3 mm.

♀. Differing from *anomalus* ♀ in the almost uniformly rather subshining black mesonotum, the less numerous and somewhat longer acr and dc, which are more distinctly separated by bare lines, dc almost uniserial anteriorly. Labrum and palpi somewhat longer (as in ♂) and posterior four tibiae each with a short black preapical bristle dorsally.

Length: body 1.6–2.2 mm, wing 2.0–2.3 mm.

Note. The separation of *M.crassipes* and *M.anomalus* in the female sex is rather difficult, particularly because of the similarly blackish coloured mesonotum; the less numerous and rather longer acr and dc, as well as the longer proboscis and palpi, are the most decisive differential features.

Distribution and biology. Rare species, only S.Sweden (Gtl., Nähr, 2 ♀, Zetterstedt) and S.Finland (N, Sibbo, 1 ♂ 2 ♂, R.Frey and Helsinge, 1 ♂ 4 ♀, A.Palmén). – Common species in Great Britain (Collin, 1961) but rather uncommon in C.Europe, south to N.Yugoslavia (Serbia) and the French Pyrenees (Vernet), apparently absent further south. – July; in W. and C.Europe from June to the beginning of August.

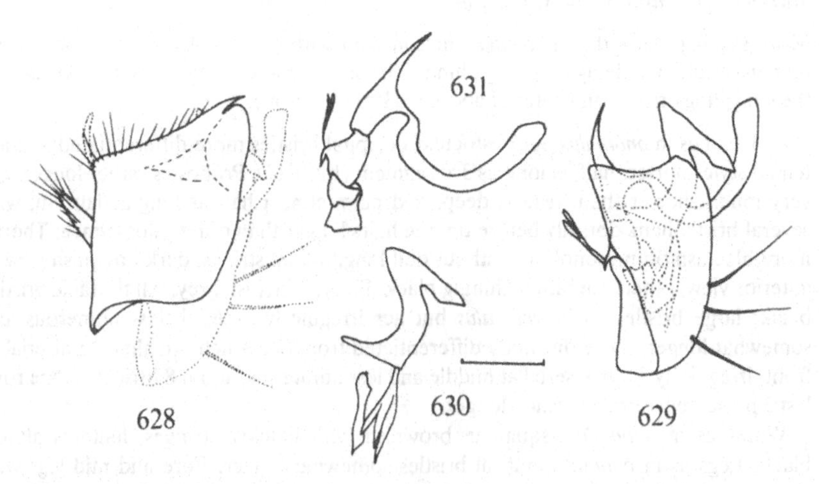

Figs. 628–631. Male genitalia of *Microphorus crassipes* Macq. – 628: left periandrial lamella with cercus, hypandrium and aedeagus dotted, 629: hypandrium with aedeagal complex; 630: asymmetrical hypandrial appendage; 631: aedeagus. Scale: 0.1 mm.

Genus *Microphorella* Becker, 1909

Microphorella Becker, 1909: 28.
Type-species: *Microphorus praecox* Loew, 1864 (orig.des.).

Very small, usually less than 2 mm long, greyish dusted species (Fig. 632) with very small proboscis and wings without axillary lobe. Head (Fig. 603) broadly dichoptic in both sexes, eyes microscopically pubescent with all facets equally small. Occiput concave above neck, convex on longer part below. Frons widening out above, face usually

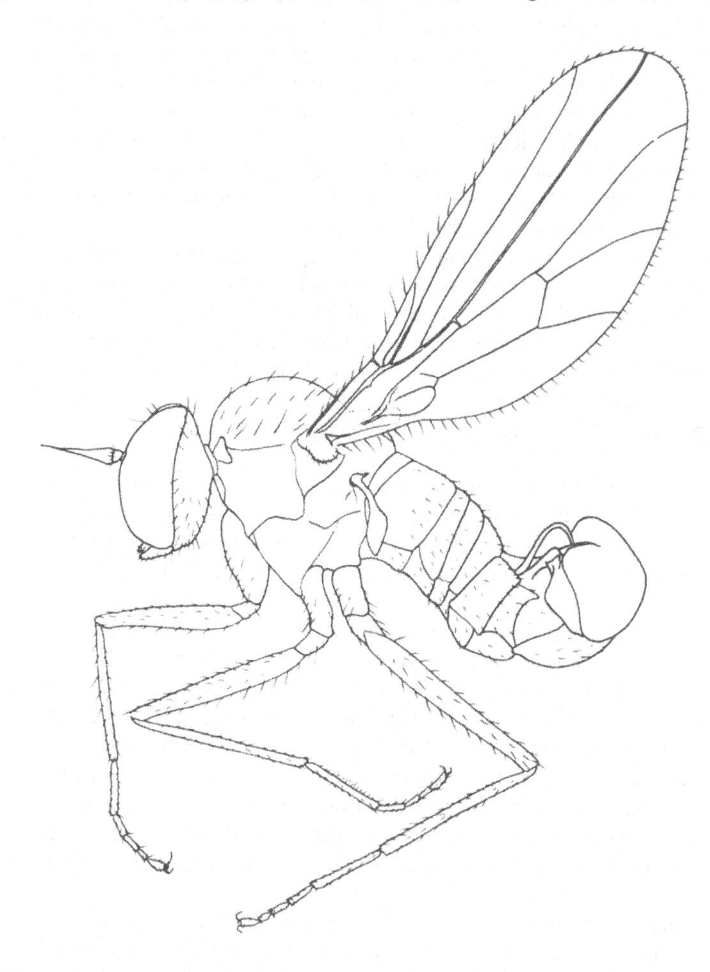

Fig. 632. Male of *Microphorella praecox* (Loew). Total length: 1.2–1.5 mm.

much narrower. A pair of anterior ocellar bristles (directed forwards) about as long as several vertical bristles, posterior pair of ocellars inconspicuous; a pair of distinct upper fronto-orbital bristles directed inwards. No distinct ocellar tubercle, posterior ocelli widely spaced. Antennae inserted above middle of head in profile, basal two segments very small, 2nd segment at most globular and armed with a circlet of small preapical bristles; 3rd segment laterally compressed, evenly conical or almost circular at base, with apical portion very narrowed; style 1-articulated, shorter than 3rd conical segment or much longer and arista-like in antennae with circular 3rd segment, distinctly microscopically pubescent like the 3rd segment. Proboscis very short and small, often entirely concealed within the mouth-cavity, directed obliquely forwards or nearly vertical (but labellae pointing forwards). Mouthparts (Figs. 633–636) consisting of a large clypeus, very short labrum (with anterior edge concave and 2-pointed at tip) and equally short hypopharynx. Maxillary stipites slender and heavily sclerotised posteriorly, laciniae practically absent, anteriorly maxillae confluent with labial paraphyses; palpi club-shaped, usually with only a few fine hairs below, almost bare, membraneously connected with maxillae. Labium short and broad, with distinct soft side labellae obviously without pseudotracheae, anteriorly connected with rod-like slender sclerites (the anterior parts of maxillae).

Thorax distinctly higher than long, but apparently less arched above than in *Microphorus*, with well-developed humeri and postalar calli. Acr either absent (sometimes indicated anteriorly) or 2–serial, but always much smaller than uniserial dc, which are long and bristle-like posteriorly. Humeral bristle absent, at most present as a fine small hair, other bristles confined to 1 posthumeral, 2–3 notopleurals, 1 supra-alar, 1 postalar and 1 pair of long scutellars. Prosternum distinctly larger than in *Microphorus*, and only narrowly separated from episterna and epimera, metapleura bare.

Wings (Fig. 605) clear or finely clouded, very distinctive because of the entire absence of axillary lobe (wing consequently very narrow at base), completely covered with microtrichia. Costal stigma absent, a small costal bristle, but costal ciliation often very distinct. Costa running completely around the wing, Sc broken near base behind humeral crossvein and rather abruptly bent upwards to join costa. R_1 shorter than in *Microphorus*, reaching costa halfway along length of wing, R_{2+3} and R_{4+5} long as in *Microphorus*; radial sector very short, originating opposite humeral crossvein. Discal cell emitting 3 veins to wing-margin, but there are often tendencies to median reduction of vein M_2 and of the posterior crossvein m-m. Basal and anal cells (Fig. 56) very small, as in *Microphorus*, but the vein closing 2nd basal cell (basal crossvein m-cu) incomplete; with a tendency to fusion of the 2nd basal and discal cells as in the Dolichopodidae. The vein closing anal cell very convex as in *Microphorus*.

Legs very much like, and also finely bristled as in, *Microphorus*, but hind tibiae and metatarsi never compressed or dilated; rarely (in Nearctic species) hind tarsi curiously modified.

Abdomen short cylindrical in ♂, more dorsoventrally flattened and telescopic in ♀, clothed with only minute hairs in both sexes, hind marginal or ventral bristles not developed. Male genitalia (Fig. 637) inverted, bent downwards and lying on the right-

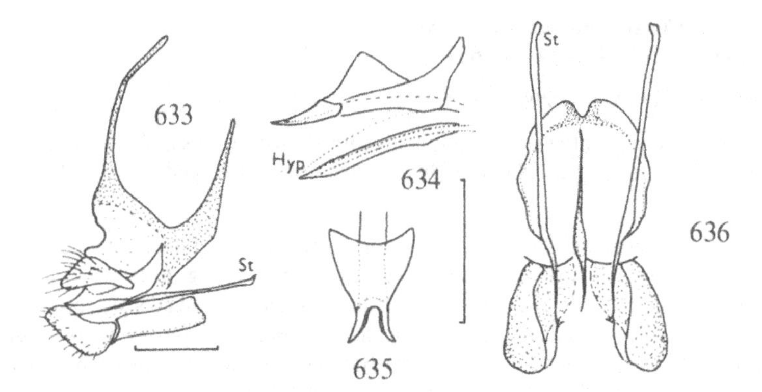

Figs. 633–636. Mouthparts of *Microphorella praecox* (Loew) ♂. – 633: lateral view; 634: labrum and hypopharynx in lateral view; 635: tip of labrum from above; 636: labium with maxillae from above. Scale: 0.1 mm. For abbreviations see p. 13.

hand side along the abdomen as in *Microphorus*, but much larger and reaching far back towards base of abdomen; 5th segment (like the 6th) excised on the right side and consequently narrowed. 8th sternum forming the largest sclerite on the left side of abdomen due to the rotation through almost 90° towards right. Hypopygium itself unrotated but asymmetrical, with conspicuously enlarged hypandrium (in comparison with *Microphorus* not bristled) placed dorsally alongside at least apical half of abdomen. Aedeagus tubular and apically free, sometimes with a brush-like organ at tip. Female abdomen (Fig. 638) unmodified on anterior five segments, 6th segment broadened and partly modified, 7th and following segments mostly concealed within the 6th segment, forming a tubular, apically finely spinose "ovipositor".

Distribution. Mostly a Holarctic genus, with only 9 species: five are North American, one Australian and three species occur in Europe. As well as *M.praecox* (also found in eastern Fennoscandia) and the C.European *M.beckeri* Str., the S.European *Heleodromia curtipes* (Becker, 1910), described as *Sciodromia*, is also a species of *Microphorella*.

Biology. The adults are apparently predaceous. According to Engel (1940), they inhabit dry sandy biotopes in Europe, like *Schistostoma* species, and the specifically adapted spinose female "ovipositor" is an arrangement for laying eggs in sand. Krogerus (1932) was of the same opinion for *M.praecox* in Finland. However, Colless (1963) collected the Australian *M.iota* by sweeping near streams and on moist rocks in a slowly-trickling watercourse, which he suggested to be the natural habit for this species; moreover, Wilder (letter comm., 11 Dec.1978) collected the Nearctic *Microphorella* species in very moist areas, especially on large rocks in streams or seepages; she also recorded rare and small swarms of *Microphorella* sp. about an inch above the rocks. Immatures unknown.

48. ***Microphorella praecox*** (Loew, 1864)
Figs. 56, 603, 605, 614, 632–639.

Microphorus praecox Loew, 1864: 47.

Very small, dull, light grey dusted species with all fine bristles and hairs whitish, acr absent. Antennae inserted above middle of head in profile, and wings without axillary lobe.

♂. Frons conspicuously broad, silvery grey dusted, face much narrower and brownish grey. All bristles on head whitish. Antennae (Fig. 614) black, 3rd segment long and rather slender, about 3 times as long as deep, terminal style somewhat stout, microscopically pubescent, 1/3 as long as 3rd segment. Thorax rather bluish grey dusted, all hairs and bristles whitish: acr absent, dc uniserial and inconspicuous, ending in 1 strong prescutellar pair as long as a pair of scutellar bristles.

Wings (Fig. 605) somewhat brownish infuscated with dark brown veins, no stigma. Discal cell short, apical section of vein M_1 nearly twice as long as discal cell. A small pale costal bristle, squamae and halteres pale. Legs uniformly dark brown, simple, rather long and slender. Pubescence whitish, fore tibiae with a row of long posterior bristly hairs longer than tibia is deep, and all femora ventrally with similar hair-like bristles at most as long as femur is deep. Hind metatarsi long, not much shorter than rest of tarsus.

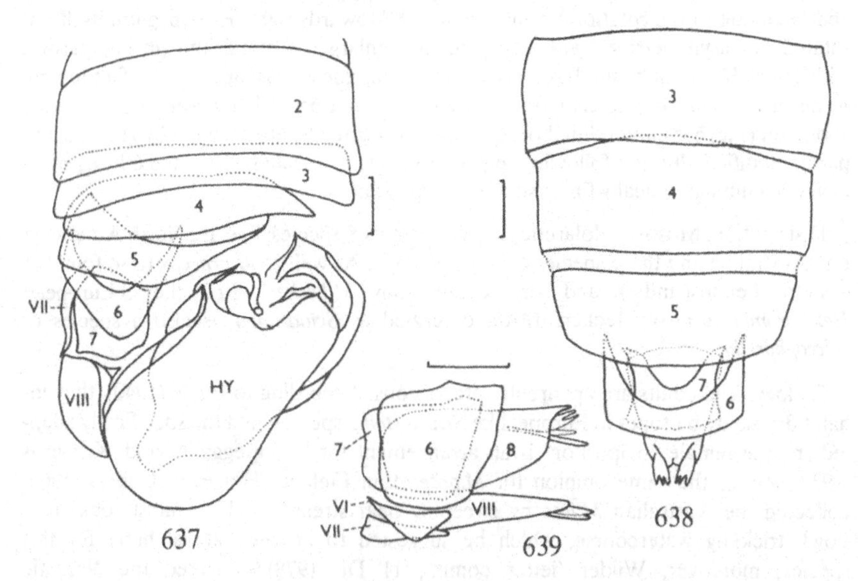

Figs. 637–639. Abdomen and genitalia of *Microphorella praecox* (Loew). – 637: male abdomen in dorsal view; 638: female abdomen in dorsal view; 639: female terminalia in lateral view. Scale: 0.1 mm. For abbreviations see p. 13.

Abdomen also dulled by bluish grey dusting, like thorax, covered with sparse, short pale hairs. Hypopygium (Fig. 637) very large and globose, lying on right-hand side of abdomen and reaching right up to 4th segment.

Length: body 1.2–1.5 mm, wing 1.4–1.6 mm (lectotype body 1.4 mm, wing 1.6 mm).

♀. Frons as in ♂ but face broader, frontal and orbital bristles darkened, legs with shorter hairs. Abdomen more silvery grey dusted and nearly bare on anterior 6 segments, "ovipositor" polished blackish brown, small and slender, concealed in modified 6th segment (Fig. 638).

Length: body 1.3–1.6 mm, wing 1.5–1.6 mm.

Lectotype designation. Described by Loew (1864) from "Schlesien" and Posen. In Loew's collection, now in ZM Berlin, there are 3 ♂ and 3 ♀ (as well as a pin with the remains of a 7th specimen) from Karlowitz, Polish Silesia, with the dates 10.v.1846 (2 ♂ 2 ♀), 1.v. 1841 (♂) and 14.v.1842 (♀); these specimens are all syntypes from Silesia, and the syntypes from Posen are evidently lost. One male in very good condition, labelled by Loew "Karlowitz 10.5.46 / No 10570 / *Microphorus praecox* m." is herewith designated as lectotype of *Microphorella praecox* (Loew) and was labelled accordingly in 1981.

Distribution and biology. Apparently a very rare or overlooked species in Fennoscandia. Krogerus (1932: 112, 182) recorded it from the "Isthmus Karelicus" and from sandy biotopes in Finnish "Bothnia borealis, nördlich von 64° 49'" (obviously ObS), but I did not find any of his specimens in the Helsinki Museum. USSR, Vib: Terijoki, 1 ♂, Hellén; Terijoki, Rajajoki, 3.vi.1926, Krogerus (a pin only). – C.Europe (East Germany, Poland, Czechoslovakia, Hungary). – In sandy biotopes in June, in C.Europe from April to June.

Species	No.	N. Germany	G. Britain	DENMARK											Sk.	Bl.
				SJ	EJ	WJ	NWJ	NEJ	F	LFM	SZ	NWZ	NEZ	B		
Hybos grossipes (L.)	1	●	●	●	●			●	●	●	●		●		●	
H. culiciformis (Fabr.)	2	●	●	●	●	●	●	●	●	●	●	●	●	●	●	
H. femoratus (Müll.)	3	●	●	●	●	●	●	●	●	●	●		●	●	●	
Syndyas nigripes (Zett.)	4	●	●													
Syneches muscarius (Fabr.)			●													
Trichinomyia flavipes (Meig.)	5	●	●	●	●								●		●	
T. fuscipes (Zett.)	6															
Trichina clavipes Meig.	7	●	●			●	●	●				●	●		●	
T. bilobata Coll.	8	●	●	●	●					●			●		●	
T. elongata Hal.	9	●	●	●				●	●				●	●	●	
T. opaca Loew	10		●													
T. pallipes (Zett.)	11	●	●						●							
Bicellaria simplicipes (Zett.)	12	●	●						●	●			●		●	
B. pilosa Lundb.	13	●	●		●	●	●	●					●		●	
B. austriaca Tuomik.	14				●				●				●	●	●	
B. subpilosa Coll.	15		●		●		●			●			●	●	●	
B. bisetosa Tuomik.	16															
B. spuria (Fall.)	17	●	●	●	●			●	●				●	●	●	
B. mera Coll.	18		●							●						
B. sulcata (Zett.)	19	●	●	●	●		●	●	●	●	●		●	●	●	
B. vana Coll.	20		●		●									●	●	
B. intermedia Lundb.	21	●	●		●	●			●				●			
B. nigra (Meig.)	22	●	●	●	●		●		●				●	●	●	
B. nigrita Coll.			●													
Oedalea stigmatella Zett.	23		●	●						●			●		●	
O. freyi sp.n.	24															
O. zetterstedti Coll.	25		●	●	●				●	●			●	●	●	
O. holmgreni Zett.	26		●												●	
O. ringdahli sp.n.	27															
O. tibialis Macq.	28		●												●	
O. flavipes Zett.	29		●	●						●	●		●	●	●	
O. kowarzi Chv.																
O. oriunda Coll.	30		●	●					●							
O. hybotina (Fall.)	31			●						●			●		●	
Euthyneura myrtilli Macq.	32	●	●				●	●		●			●		●	
E. albipennis (Zett.)	33		●													
E. gyllenhali (Zett.)	34	●	●		●				●	●			●		●	
Anthalia schoenherri Zett.	35															

SWEDEN

	Hall.	Sm.	Öl.	Gtl.	G. Sand.	Ög.	Vg.	Boh.	Dlsl.	Nrk.	Sdm.	Upl.	Vstm.	Vrm.	Dlr.	Gstr.	Hls.	Med.	Hrj.	Jmt.	Äng.	Vb.	Nb.	Ås. Lpm.	Ly. Lpm.	P. Lpm.	Lu. Lpm.	T. Lpm.
1	●	●		●		●								●			●		●	●								
2	●			●		●															●							
3	●	●				●		●									●											
4		●	●							●																		
5	●	●										●																
6																			●	●					●			●
7	●	●		●		●																						
8	●	●																		●			●					
9																				●			●					
10																												
11				●						●													●					
12	●			●		●																	●					
13	●																		●	●			●				●	●
14																												
15	●																		●	●	●		●		●		●	●
16																							●				●	●
17	●		●							●					●						●		●		●		●	●
18																												
19			●							●													●					●
20			●																									
21	●		●							●										●								
22	●																			●			●					●
23	●	●		●		●						●		●						●								
24																												
25	●																			●								●
26						●																						
27																									●			
28																												
29																												
30																												
31																				●								
32	●					●														●							●	●
33																				●							●	
34																				●					●			
35																							●					

		NORWAY														
		Ø + AK	HE (s + n)	O (s + n)	B (ø + v)	VE	TE (y + i)	AA (y + i)	VA (y + i)	R (y + i)	HO (y + i)	SF (y + i)	MR (y + i)	ST (y + i)	NT (y + i)	Ns (y + i)
Hybos grossipes (L.)	1			◗				◗		◗	●	◗			◗	
H. culiciformis (Fabr.)	2															
H. femoratus (Müll.)	3										◖					
Syndyas nigripes (Zett.)	4															
Syneches muscarius (Fabr.)																
Trichinomyia flavipes (Meig.)	5															
T. fuscipes (Zett.)	6														◗	◗
Trichina clavipes Meig.	7															
T. bilobata Coll.	8															
T. elongata Hal.	9															
T. opaca Loew	10															
T. pallipes (Zett.)	11															
Bicellaria simplicipes (Zett.)	12															
B. pilosa Lundb.	13				◗					◖	●	◗				
B. austriaca Tuomik.	14								◗		◗					
B. subpilosa Coll.	15			●							●	●				◗
B. bisetosa Tuomik.	16															
B. spuria (Fall.)	17			◖												◗
B. mera Coll.	18															
B. sulcata (Zett.)	19			◖								◖				
B. vana Coll.	20			◗												
B. intermedia Lundb.	21											◗				
B. nigra (Meig.)	22															
B. nigrita Coll.																
Oedalea stigmatella Zett.	23															
O. freyi sp.n.	24															
O. zetterstedti Coll.	25															
O. holmgreni Zett.	26															
O. ringdahli sp.n.	27															
O. tibialis Macq.	28															
O. flavipes Zett.	29															
O. kowarzi Chv.																
O. oriunda Coll.	30															
O. hybotina (Fall.)	31															
Euthyneura myrtilli Macq.	32														◗	
E. albipennis (Zett.)	33														◗	
E. gyllenhali (Zett.)	34															
Anthalia schoenherri Zett.	35															

N O R W A Y

					FINLAND																				USSR		
	Nn (ø + v)	TR (y + i)	F (v + i)	F (n + ø)	Al	Ab	N	Ka	St	Ta	Sa	Oa	Tb	Sb	Kb	Om	Ok	Ob S	Ob N	Ks	LkW	LkE	Le	Li	Ib	Kr	Lr
1					●	●	●	●	●	●	●	●	●	●	●	●	●	●	●	●					●	●	
2					●	●	●	●																		●	
3					●	●	●	●	●	●			●	●	●										●	●	
4					●	●	●		●	●		●	●	●					●							●	
5						●	●			●															●		
6		●			●														●	●	●		●	●		●	●
7					●	●	●	●		●	●	●	●		●	●		●							●	●	
8					●	●	●	●	●	●	●			●	●	●			●	●					●	●	
9				◖.	●	●	●		●	●		●		●						●				●		●	●
10							●																				
11					●	●	●	●	●	●			●			●									●	●	
12							●	●	●									●								●	●
13		◖	◖	●			●	●												●			●	●			●
14																											
15		◗	◖	◖		●	●	●	●	●	●	●	●	●	●	●	●	●	●	●	●	●	●	●	●	●	●
16							●													●				●		●	
17				◖	●	●	●	●	●	●	●	●	●	●	●	●	●	●	●	●	●			●	●	●	●
18																											
19					●	●	●			●	●		●	●	●			●		●			●	●	●	●	●
20																											
21					●	●	●	●	●	●	●	●	●	●						●						●	
22	◖	◖		◖	●	●	●	●	●	●		●		●	●	●				●	●		●	●		●	●
23					●	●	●	●	●	●				●						●					●	●	
24																				●	●		●				●
25				◗		●	●		●												●		●				●
26					●																●						
27																											
28																									●		
29																											
30																											
31					●	●	●		●	●					●				●	●						●	●
32		●			●	●	●	●	●	●			●	●	●		●		●	●	●		●	●	●	●	●
33										●										●	●						●
34						●	●		●	●				●	●		●			●	●					●	●
35						●	●		●	●							●			●						●	●

		DENMARK														
		N. Germany	G. Britain	SJ	EJ	WJ	NWJ	NEJ	F	LFM	SZ	NWZ	NEZ	B	Sk.	Bl.
Allanthalia pallida (Zett.)	36															
Ocydromia glabricula (Fall.)	37	●	●	●	●			●	●	●	●	●	●	●	●	
O. melanopleura Loew	38	●	●						●				●		●	
Leptopeza flavipes (Meig.)	39		●	●	●		●	●	●	●			●	●	●	
L. borealis Zett.	40		●													
Leptodromiella crassiseta (Tuomik.)	41															
Oropezella sphenoptera (Loew)	42	●	●		●								●		●	
Atelestus pulicarius (Fall.)	43	●	●							●					●	
A. dissonans Coll.			●													
Meghyperus sudeticus Loew	44															
Microphorus holosericeus (Meig.)	45	●	●	●				●		●			●	●	●	
M. anomalus (Meig.)	46		●	●				●	●	●	●		●	●	●	
M. crassipes Macq.	47	●	●													
Microphorella praecox (Loew)	48															

		NORWAY														
		Ø + AK	HE (s + n)	O (s + n)	B (ø + v)	VE	TE (y + i)	AA (y + i)	VA (y + i)	R (y + i)	HO (y + i)	SF (y + i)	MR (y + i)	ST (y + i)	NT (y + i)	Ns (y + i)
Allanthalia pallida (Zett.)	36															
Ocydromia glabricula (Fall.)	37										◗	◗			◗	
O. melanopleura Loew	38											◖				
Leptopeza flavipes (Meig.)	39	◗													◗	
L. borealis Zett.	40														◗	
Leptodromiella crassiseta (Tuomik.)	41															
Oropezella sphenoptera (Loew)	42															
Atelestus pulicarius (Fall.)	43															
A. dissonans Coll.																
Meghyperus sudeticus Loew	44															
Microphorus holosericeus (Meig.)	45													◗		
M. anomalus (Meig.)	46															
M. crassipes Macq.	47															
Microphorella praecox (Loew)	48															

SWEDEN

	Hall.	Sm.	Öl.	Gtl.	G. Sand.	Ög.	Vg.	Boh.	Dlsl.	Nrk.	Sdm.	Upl.	Vstm.	Vrm.	Dlr.	Gstr.	Hls.	Med.	Hrj.	Jmt.	Ång.	Vb.	Nb.	Ås. Lpm.	Ly. Lpm.	P. Lpm.	Lu. Lpm.	T. Lpm.
36																									●			
37	●					●				●				●						●			●					●
38	●	●		●																			●				●	
39	●	●		●		●						●							●	●								●
40						●														●	●						●	
41	●																											
42	●																											
43		●		●																								
44																												
45			●					●				●								●								
46			●	●																								
47				●																								
48																												

					FINLAND																				USSR		
	Nn (ø + v)	TR (y + i)	F (v + i)	F (n + ø)	Al	Ab	N	Ka	St	Ta	Sa	Oa	Tb	Sb	Kb	Om	Ok	Ob S	Ob N	Ks	LkW	LkE	Le	Li	Ib	Kr	Lr
36						●	●													●						●	●
37	◗	●	◖	●	●	●	●	●	●	●	●	●	●	●	●	●		●	●	●	●		●	●	●	●	●
38	◖	◗			●	●	●	●	●	●				●	●										●	●	
39	◖			◖	●	●	●	●	●	●	●	●		●	●	●		●	●	●					●	●	●
40					●	●	●			●				●	●		●	●	●	●	●	●	●	●	●	●	●
41							●			●																	
42																											
43					●	●	●																				
44											●														●		
45		◗			●	●	●		●	●	●		●		●					●						●	
46																											
47							●																				
48																		?							●		

Literature

Aczél, M., 1954: Orthopyga and Campylopyga, new Divisions of Diptera. – Ann.ent.Soc.Am., 47: 75–80.

Afanasjeva, O. G., 1980: On the functional morphology of the male genitalia in Diptera. I. *Tipula czizeki* de Jong (Tipulidae) and *Eristalis nemorum* L. (Syrphidae). – Ént. Obozr., 59: 262–268 (in Russian).

Bährmann, R., 1960: Vergleichend-morphologische Untersuchungen der männlichen Kopulationsorgane bei Empididen. – Beitr.Ent., 10: 485–540.

Becker, T., 1909: *Microphorus* Macq. und seine nächsten Verwandten. – Wien.ent.Ztg, 28: 25–28.

Bei-Bienko, G. Ja., 1969: Opredelitel nasekomych evropejskoj casti SSSR (Key to insects of the European part of the USSR), vol. 5, Diptera, Aphaniptera, Part 1, 805 pp., Leningrad (in Russian).

Bezzi, M., 1899: Contribuzioni alla fauna ditterologica Italiana. – Boll.Soc.ent.ital., 30: 121–164.

Bigot, J. M. F., 1859: Dipterorum aliquot nova genera. – Rev.Mag. Zool., ser. 2, 11: 305–315.

Boheman, C. H., 1852: Entomologiska Anteckningar under en resa i Södra Sverige 1851. – K.svenska Vetensk-Akad.Handl., 1851 (1852): 53–210.

Brauer, F., 1883: Zweiflügler des kaiserlichen Museum zu Wien. III. Systematische Studien auf Grundlage der Dipteren-Larven nebst einer Zusammenstellung von Beispielen aus der Literatur über Dieselben und Beschreibung neuer Formen. – Denkschr.Akad.Wiss., Wien, 47: 1–100.

Brindle, A., 1973: Taxonomic notes on the larvae of British Diptera. No. 28. The larva and pupa of *Hydromia stagnalis* (Haliday) (Empididae). – Entomologist, 106 (1326): 249–252.

Caspers, N., 1982: *Oedalea holmgreni* Zetterstedt, neu für Mitteleuropa. – Spixiana, 5: 171–174.

Chandler, P. J., 1972: Diptera. Middle-Thames Naturalist, 1972. 3 pp. (reprint).

– 1973: The flat-footed flies (Diptera, Aschiza – Platypezidae) known to occur in Kent, with a key to the genera and species so far recorded from the British Isles. – Trans.Kent Field Club, 5: 15–44.

– 1974: Additions and corrections to the British list of Platypezidae (Diptera), incorporating a revision of the Palaearctic species of *Callomyia* Meigen. – Proc.Brit.ent.nat.hist.Soc., 7: 1–32.

Chvála, M., 1975: The Tachydromiinae (Dipt. Empididae) of Fennoscandia and Denmark. – Fauna ent.scand., 3, 336 pp., Klampenborg.

– 1976: Swarming, mating and feeding habits in Empididae (Diptera), and their significance in evolution of the family. – Acta ent. bohemoslov., 73: 353–366.

– 1980: Swarming rituals in two *Empis* and one *Bicellaria* species (Diptera, Empididae). – Ibid., 77: 1–15.

– 1981a: Classification and phylogeny of Empididae, with a presumed origin of Dolichopodidae (Diptera). – Ent.scand.Suppl., 15: 225–236.

– 1981b: Revision of Central European species of the genus *Oedalea* (Diptera, Empididae). – Acta ent.bohemoslov., 78: 122–139.

– & K. G. V. Smith, 1982: *Anthalia* Zetterstedt, 1838 (Insecta, Diptera): Request for designation of type species. Z. N. (S.) 2380. – Bull.zool. Nom., 39 (3): 220–221.

Cockerell, T. D. A., 1917a: Arthropods in Burmese Amber. – Am.J.Sci. 4th ser., 44: 360–368.

– 1917b: Fossil insects. – Ann.ent.Soc.Am., 10: 1–22.

Coe, R. L., 1962: A further collection of Diptera from Jugoslavia, with localities and notes. – Bull.Mus.Hist.nat.Belgrade, 1962, Sér.B, Livre 18: 95–136.

Cole, J. H., 1964: A species of *Euthyneura* (Dipt., Empididae) new to Britain. - The Entomologist, 1964: 128.
Colless, D. H., 1963: An Australian species of *Microphorella* (Diptera: Empididae), with notes of the phylogenetic significance of the genus. - Proc.Linn.Soc.New South Wales, 88: 320–323.
Collin, J. E., 1926: Notes on the Empididae (Diptera) with additions and corrections to the British List. - Entomologist's mon.Mag., 62: 146–159, 185–190, 213–219, 231–237.
- 1928: New Zealand Empididae. B. M. (N. H.) London, 1928, 110 pp.
- 1933: Empididae, in: Diptera of Patagonia and South Chile, IV. B. M. (N. H.) London, 1933, 334 pp.
- 1941: Some Pipunculidae and Empididae from the Ussuri region on the far eastern border of the U.S.S.R. (Diptera). - Proc.R.ent. Soc.Lond. (B), 10: 218–248.
- 1960: Some Empididae from Palestine. - Ann.Mag.nat.Hist., Ser. 13, 2 (1959): 385–420.
- 1961: British Flies. VI: Empididae. 782 pp., Cambridge.
Coquillett, D. W., 1896: Revision of the North American Empidae - a family of two-winged insects. - Proc.U.S.natn.Mus., 18(1895): 387–440.
Downes, J. A., 1969: The swarming and mating flight of Diptera. - A.Rev.Ent., 14: 271–298.
- 1970: The feeding and mating behaviour of the specialized Empidinae (Diptera); observations on four species of *Rhamphomyia* in the high arctic and general discussion. - Can.Ent., 102: 769–791.
- & S. M. Smith, 1969: New or little known feeding habits in Empididae (Diptera). - Ibid., 101: 404–408.
Dyte, C. E., 1967: Some distinctions between the larvae and pupae of the Empididae and Dolichopodidae (Diptera). - Proc.R.ent.Soc. Lond. (A), 42: 119–128.
Emden, F. I. van & W. Hennig, 1956: Diptera, in: Tuxen, S. L. (ed.), Taxanomist's glossary of genitalia in insects. Munksgaard, Copenhagen, pp. 111–122.
Enderlein, G., 1936: 22. Ordnung: Zweiflügler, Diptera, in: Brohmer, P., P. Ehrmann & G. Ulmer, Die Tierwelt Mitteleuropas, 6, Insekten, III. Teil, Abt. 16, 259 pp., Leipzig.
Engel, E. O., 1940: Empidinae, *Microphorus*-Gruppe, in: Lindner, E., Die Fliegen der Palaearktischen Region, 28. Empididae, IV, 4, pp. 183–192, Stuttgart.
Fabricius, J. C., 1775: Systema entomologiae, 832 pp., Flensburgi et Lipsiae.
- 1794: Entomologia systematica emendata et aucta, 4, 472 pp., Hafniae.
- 1805: Systema antliatorum secundum ordines, genera, species, 373 + 30 pp., Brunsvigae.
Fallén, C. F., 1815: Empididae Sveciae, 34 pp., Lundae.
- 1816: Supplementum dipterorum Sveciae, 16 pp., Lundae.
Foster, G. N., 1971: Empididae (Diptera) caught in suction traps in South Northumberland, 1967–9. - Entomologist's mon.Mag., 106 (1970): 171–173.
Frey, R., 1913: Zur Kenntnis der Dipterenfauna Finlands. II. Empididae. - Acta Soc.Fauna Flora fenn., 37 (3): 1–89.
- 1950: Dipterfaunan vid Tana älv i Utsjoki sommaren 1949. Mit einem Anhang: Synonymische Bemerkungen und Beschreibungen einigen neuen Diptera brachycera aus Utsjoki in Finnish-Lappland. - Notul.ent., 30: 5–18.
- 1953: Studien über ostasiatische Dipteren. II. Hybotinae, Ocydromiinae, *Hormopeza* Zett. - Ibid., 33: 57–71.
- 1956: Ocydromiinae und Hybotinae, in: Lindner, E., Die Fliegen der Palaearktischen Region, 28. Empididae, IV, 4, pp. 584–617, Stuttgart.
Gistl, J. N. F. X. von, 1848: Naturgeschichte des Thierreichs für höhere Schulen, xvi + 216 (+4) pp., Stuttgart.
Gmelin, J. F., 1790: Caroli a Linné, Systema naturae per regna tria naturae, ed. 13, 1 (5): 2225–3020, Lipsiae.

Greenfeld, E. K., 1962: Origin of anthophily in insects. Izd.Leningrad.Univ., 186 pp., Leningrad (in Russian).
Griffith, E., 1832: The Animal Kingdom arranged in conformity with its organization, by the Baron Cuvier, with supplementary additions to each order. Vol. 15, 796 pp., London.
Griffiths, G. C. D., 1972: The phylogenetic classification of Diptera Cyclorrhapha with special reference to the structure of the male postabdomen. - Series Entomol., 8, 340 pp., The Hague.
- 1981: Book reviews. Manual of Nearctic Diptera. Volume 1. - Bull. ent.Soc.Canada, 13: 49-55.
Gruhl, K., 1955: Neue Beobachtungen an Schwarm- und Tanzgesellschaften der Dipteren (Dipt.). - Dt.ent.Z., (N. F.) 2: 332-353.
Grunin, K. J., 1953: Viviparity in coprobionts in the order Diptera. - Trudy Zool.Inst.AN USSR, 13: 387-389.
Haliday, A. H., 1833: Catalogue of Diptera occuring about Holywood in Downshire. - Entomologist's mon.Mag., 1: 147-180.
Harper, P. P., 1980: Phenology and distribution of aquatic dance flies (Diptera: Empididae) in a Laurentian watershed. - Am.Midl.Nat., 104: 110-117.
Harris, M., 1780: An exposition of English insects with curious observations and remarks wherein each insect is particularly described, its parts and properties considered, the different sexes distinguished, and the natural history faithfully related. Decads III-V: 73-166, London.
Hendel, F., 1928: Zweiflügler oder Diptera. II. Allgemeiner Teil, in: Dahl, F., Die Tierwelt Deutschlands. 135 pp., Gustav Fischer, Jena.
- 1937: Diptera, in: Kükenthal, W. & T. Krumbach, Handbuch der Zoologie, 4 (2), 2. Teil (1936-1938), pp. 1729-1998.
Hennig, W., 1950: Grundzüge einer Theorie der phylogenetischen Systematik, 370 pp., Berlin.
- 1952: Die Larvenformen der Dipteren. Vol. 3, 628 pp., Berlin.
- 1954: Flügelgeäder und System der Dipteren unter Berücksichtung der aus dem Mesozoikum beschriebenen Fossilien. - Beitr.Ent., 4: 245-388.
- 1970: Insektenfossilien aus der unteren Kreide II. Empididae (Diptera, Brachycera). - Stuttg.Beitr.Naturk., 214: 1-12.
- 1971: Insektenfossilien aus der unteren Kreide III. Empidiformia ("Microphorinae") aus der unteren Kreide und aus dem Baltischen Bernstein; ein Vertreter der Cyclorrhapha aus der unteren Kreide. - Ibid., 232: 1-28.
- 1972: Eine neue Art der Rhagionidengattung *Litoleptis* aus Chile, mit Bemerkungen über Fühlerbildung und Verwandschaftsbeziehungen einiger Brachycerenfamilien (Diptera: Brachycera). - Ibid., 242: 1-18.
- 1976: Das Hypopygium von *Lonchoptera lutea* Panzer und die phylogenetischen Verwandtschaftsbeziehungen der Cyclorrhapha (Diptera). - Ibid., 283: 1-63.
Hobby, B. M. & K. G. V. Smith, 1962: The larva of the viviparous fly *Ocydromia glabricula* (Fln.) (Dipt., Empididae). - Entomologist's mon.Mag., 98: 49-50.
Jones, C. G., 1940: Empididae: A, Hybotinae, Ocydromiinae, Clinocerinae and Hemerodromiinae. Ruwenzori Expedition 1934-5. II. No. 5 British Museum (Natural History), London.
Kato, A., 1971a: On the genus *Bicellaria* Macquart from Japan (Diptera, Empididae), - Kontyu, 39: 279-284.
- 1971b: A new species of the genus *Leptopeza* Macquart from Japan (Diptera, Empididae). - Ibid., 39: 284-287.
Kessel, E. L., 1955: The mating activities of balloon flies. - Syst. Zool., 4: 97-104.
- 1960: The systematic position of *Platycnema* Zetterstedt and *Melanderomyia*, new genus, together with the description of the genotype of the latter (Diptera: Platypezidae). - Wasmann J. Biol., 18: 87-101.

- & E. A. Maggioncalda, 1968: A revision of the genera of Platypezidae, with the descriptions of five new genera, and considerations of phylogeny, circumversion, and hypopygia (Diptera). - Ibid., 26: 33-106.
Knutson, L. V. & O. S. Flint, Jr., 1971: Pupae of Empididae in pupal cocoons of Rhyacophilidae and Glossosomatidae (Diptera - Trichoptera). - Proc.ent.Soc.Wash., 73: 314-320.
- 1979: Do dance flies feed on caddisflies? - further evidence (Diptera: Empididae; Trichoptera). - Ibid., 81: 32-33.
Kovalev, V. G., 1974: A new genus of Diptera of the family Empididae and its phylogenetic relationships. - Paleontol.Zh., 1974: 84-94 (in Russian).
- 1978: A new genus of flies of the family Empididae from the Upper Cretaceous resins of Taimyr. - Ibid., 1978: 72-78 (in Russian).
- 1979a: The main aspects in evolution of Diptera Brachycera in Mesozoic Era. Ecol. & morphol.principles of Diptera Systematics (Insecta). AN SSSR, Zool.Inst., Leningrad, pp. 35-37 (in Russian).
- 1979b: On two rare genera of the family Empididae (Diptera) in the fauna of the European part of the USSR. - Zool.Zh., 58 (8): 1242-1244 (in Russian).
Kröber, O., 1912: Die Thereviden der indo-australischen Region. (Dipt.). - Ent.Mitt., 1: 242-256.
- 1958: Nachträge zur Dipteren-Fauna Schleswig-Holsteins und Niedersachsen (1933-35). - Verh.Ver.naturw.Heimatforsch., 33: 39-96.
Krogerus, R., 1932: Über die Ökologie und Verbreitung der Arthropoden der Triebsandgebiete an den Küsten Finnlands. - Acta zool.fenn., 12: 1-308.
Krystoph, H., 1961: Vergleichend-morphologische Untersuchungen an den Mundteilen bei Empididae. - Beitr.Ent., 11: 824-872.
Lameere, A., 1906: Notes pour la classification des Diptères. - Mém.Soc.r.ent.Belg., 12: 105-140.
Larsson, S. G., 1978: Baltic Amber - a Palaeobiological Study. - Entomonograph, 1: 192 pp., Klampenborg.
Latreille, P. A., 1809: Genera Crustaceorum et Insectorum secundum ordinem naturalem in familias disposita, iconibus exemplisque plurimis explicita. 4: 1-399, Parisiis et Argentovati apud Amand Koenig, bibliopolam.
- 1825: Familles naturelles du règne animal, exposées succintement et dans un ordre analytique, avec l'indication de leurs genres. 570 pp., Paris.
Laurence, B. R., 1948: Observations on *Microphorus crassipes* Macquart (Dipt., Empididae). - Entomologist's mon.Mag., 84: 282-283.
- 1953: On the feeding habits of *Clinocera (Wiedemannia) bistigma* Curtis (Diptera: Empididae). - Proc.R.ent.Soc.Lond. (A), 28: 139-144.
Linley, J. R. & D. A. Carlson, 1978: A contact mating pheromone in the biting midge, *Culicoides melleus*. - Insect Physiol., 24: 423-427.
Linné, C., 1767: Systema naturae per regna tria naturae. Ed. 12 (rev.), Vol. 1, Pt. 2, pp. 533-1327, Holmiae.
Loew, H., 1840: Bemerkungen über die in der Posener Gegend einheimischen Arten mehrerer Zweiflügler-Gattungen. - Programm Posen, 1840: 1-40.
- 1850: *Meghyperus* und *Arthropeas*, zwei neue Dipterengattungen. - Stettin.ent.Ztg, 11: 302-308.
- 1857: Bidrag till kännedomen om Afrikas Diptera (cont.). - Öfver. K. Vetensk.Akad. Förh., 14: 337-383.
- 1859: Neue Beiträge zur Kenntnis der Dipteren. Sechster Beitrag. - K. Realschule zu Meseritz, Programm 1859: 1-50.
- 1864: Ueber die schlesischen Arten der Gattungen *Tachypeza* Meig. *(Tachypeza, Tachista, Dysaletria)* und *Microphorus* Macq. *(Trichina* und *Microphorus)*. - Z.Ent.Breslau, 14 (1860): 1-50.
- 1873: Beschreibung europäischer Dipteren. Systematische Beschreibung der bekannten

europäischen zweiflügeligen Insecten, von Johann Wilhelm Meigen. Vol. 3: Zehnter Theil oder vierter Supplementband, 320 pp., Halle.

Lundbeck, W., 1910: Empididae, in Diptera Danica, genera and species of flies hitherto found in Denmark. Vol. 3, 324 pp., Copenhagen.

- 1927: Platypezidae & Tachinidae, in Diptera Danica, genera and species of flies hitherto found in Denmark. Vol. 7, 571 pp., Copenhagen.

Lyneborg, L., 1965: The entomology of the Hansted Reservation, North Jutland, Denmark. 9. Diptera, Brachycera & Cyclorrhapha - Fluer. - Ent.Meddr, 30: 201-262.

Macquart, J., 1823: Monographie des insectes Diptères de la famille des Empides, observè dans le nord-ouest de la France. - Soc. d'Amateurs des Sci., de l'Agr.et des Arts, Lille, Rec.des Trav., 1819/1822: 137-165.

- 1827: Insectes Diptères du Nord de la France. Platypézines, Dolichopodes, Empides, Hybotides. 158 pp., Lille.

- 1834: Histoire naturelle des Insectes. Diptères. Vol. 1, 578 pp., Paris.

- 1836: Description d'un nouveau genre d'Insectes Diptères de la famille des Tanystomes. - Annls Soc.ent.Fr., 5: 517-520.

Meigen, J. W., 1800: Nouvelle classification des mouches à deux ailes (Diptera L.) d'après un plan tout nouveau. 40 pp., Paris.

- 1803: Versuch einer neuen Gattungseintheilung der europäischen zweiflügeligen Insekten. - Mag.f.Insektenkunde, 2: 259-281.

- 1804: Klassifikazion und Beschreibung der europäischen zweiflügeligen Insekten (Diptera Linn.). Erster Band, 1: XXVIII + 152, 2: VI + 153-314 pp., Braunschweig.

- 1820-1838: Systematische Beschreibung der bekannten europäischen zweiflügeligen Insekten. 2 (1820): X + 365 pp., 3 (1822): X + 416 pp., 4 (1824): XII + 428 pp., 6 (1830): IV + 401 pp., 7 (1838): XII + 434 pp., Hamm.

Melander, A. L., 1902: American Diptera. A monograph of the North American Empididae. I. - Trans.Am.ent.Soc., 28: 195-367.

- 1928: Diptera, fam. Empididae. - Genera Insect., 185 (1927), 434 pp., Bruxelles.

- 1965: Family Empididae, in: Stone, A. et al., A catalog of the Diptera of America North of Mexico, pp. 446-481, Washington.

Merritt, R. W., 1976: A review of the food habits of the insect fauna inhabiting cattle droppings in north central California. - Pan-Pacif.Ent., 52: 13-22.

Meunier, F., 1908: Monographie des Empidae de l'ambre de la Baltique et catalogue bibliographique complet sur les Diptères fossiles de cette résine. - Annls Sci.nat. (Zool.), 7: 81-135.

Mik, J., 1887: Ueber einige Empiden aus Kärnten. - Wien.ent.Ztg, 6: 99-103.

Müller, O. F., 1776: Zoologie Danicae prodromus, seu animalium Daniae et Norvegiae indigenarum characters, nomina, et synonyma imprimis popularium. XXXII + 282 pp., Hafniae.

Negrobov, O. P., 1978: Flies of the superfamily Empidoidea (Diptera) of the cretaceous resin of northern Siberia. - Paleontol.Zh., 1978: 81-90 (in Russian).

Oldroyd, H., 1977: The suborders of Diptera. - Proc.ent.Soc.Wash., 79: 3-10.

Pajunen, V. I., 1980: A note on the connexion between swarming and territorial behaviour in insects. - Annls Ent.Fenn., 46: 53-55.

Pokorny, E., 1887: (III.) Beitrag zur Dipterenfauna Tirols. - Verh. zool.-bot.Ges.Wien, 37: 381-420.

Poulton, E. B., 1913: Empidae and their prey in relation to courtship. - Entomologist's mon.Mag., 49: 177-180.

Ringdahl, O., 1951: Flugor från Lapplands, Jämtlands och Härjedalens fjälltrakter (Diptera Brachycera). - Opusc.ent., 16: 113-186.

Rondani, C., 1856: Dipterologie Italicae prodromus. Vol. 1. Genera Italica ordinis dipterorum

ordinatim disposita et distincta et in familias et stirpes aggregata. 228 pp., Parmae.
Roser, C. von, 1840: Erster Nachtrag zu dem in Jahre 1834 bekannt gemachten Verzeichnisse in Württemberg vorkommender zweiflügeliger Insekten. - KorespBl.Württ.landw.Ver., 1: 49–64.
Schiner, J. R., 1862: Fauna Austriaca. Die Fliegen (Diptera), 1. LXXX + 672 pp., Wien.
Scholz, H., 1851: Beiträge zur Kunde der schlesischen Zweiflügler. (Fortsetzung). - Z.Ent.Breslau, 5: 41–60.
Smith, K. G. V., 1968: The immature stages of *Rhamphomyia sulcata* Meigen (Diptera: Empididae) and their occurrence in large numbers in pasture soil. - Entomologist's mon.Mag., 104: 65–68.
- 1969: The Empididae of Southern Africa (Diptera). - Ann.Natal Mus., 19: 1–347.
- 1971: A replacement name for *Stuckenbergia* Smith (Dipt., Empididae). - Ibid., 20: 699.
- & M. Chvála, 1982: *Damalis* Fabricius, 1805 (Insecta, Diptera): Request for designation of type species, Z. N. (S.) 2369. - Bull. zool. Nom., 39 (1): 71–72.
Snodgrass, R. E., 1957: A revised interpretation of the external reproductive organs of male insects. - Smithson.misc.Collns, 135, no. 6, 60 pp.
Sommerman, K. M., 1962: Notes on two species of *Oreogeton*, predaceous on black fly larvae (Diptera: Empididae and Simuliidae). - Proc.ent.Soc.Wash., 64: 123–129.
Speight, M. C. D., 1969: The prothoracic morphology of Acalypterates (Diptera) and its use in systematics. - Trans.R.ent.Soc.Lond., 121: 325–421.
Stephens, J. F., 1829: A systematic catalogue of British insects: Being an attempt to arrange all the hitherto discovered indigenous insects in accordance with their natural affinities, vol. 2. 388 pp., London.
Strobl, G., 1893: Die Dipteren von Steiermark. I. - Mitt.naturw.Ver.Steierm., (1892) 29: 1–199.
- 1898: Die Dipteren von Steiermark. IV., Nachträge. - Ibid., (1897) 34: 192–298.
- 1899: Spanische Dipteren. - Wien.ent.Ztg, 18: 12–83.
- 1900: Tief's dipterologischer Nachlass aus Kärnten und Oesterr.-Schlesien. - Jahrb.nat.hist.Landesmus.Kärnten (Klagenfurt), Heft XXVI, 47: 171–246.
- 1902: Neue Beiträge zur Dipterenfauna der Balkanhalbinsel. - Glas. Zem.Mus.Bosni i Herzegov., Sarajevo, 14: 461–517 (in Serbian); Wiss.Mitt.Bosn.Hercegov., Sarajevo, (1904) 9: 519–581 (in German).
- 1909: Empididae, in: Czerny, L. & G. Strobl, Spanische Dipteren. III. Beitrag. - Verh.zool.-bot.Ges.Wien, 59: 121–301.
Teskey, H. J. & J. G. Chillcott, 1977: A revision of the Nearctic species of *Syndyas* Loew (Diptera: Empididae). - Can.Ent., 109: 1445–1455.
Trehen, P., 1971: Contribution à une étude d'intéret phylogénétique chez les Diptères Empididae: recherches morphologiques, écologiques et éthologiques chez les espèces à larves édaphiques. - These l'Univ.de Rennes, Ser. C, No. 44, 280 pp., Université de Rennes.
Tuomikoski, R., 1932: Zwei neue Empididen aus Finnland. - Notul.ent., 12: 46–50.
- 1935: Mitteilungen über die Empididen (Dipt.) Finnlands. I. Die Gattung *Trichina* Meig. - Annls Ent.Fenn., 1: 95–101.
- 1936a: Mitteilungen über die Empididen (Dipt.) Finnlands. II. Die Gattung *Bicellaria* Macq. - Ibid., 2: 74–85.
- 1936b: Mitteilungen über die Empididen (Dipt.) Finnlands. III. *Leptodromiella* n.gen. - Ibid., 2: 187–190.
- 1937: Mitteilungen über die Empididen (Dipt.) Finnlands. IV. Die Gattung *Ocydromia* Meig. - Ibid., 3: 17–20.
- 1938: Phänologische Beobachtungen über die Empididen (Dipt.) Süd- und Mittelfinnlands. - Ibid., 4: 213–247.
- 1939: Beobachtungen über das Schwärmen und die Kopulation einiger Empididen (Dipt.). - Ibid., 5: 1–30.

- 1952: Über die Nahrung der Empididen-Imagines (Dipt.) in Finland. - Ibid., 18: 170–181.
- 1955: Zur Kenntnis der paläarktischen Arten der Gattung *Bicellaria* Macq. (Dipt., Empididae). - Ibid., 21: 65–77.
- 1959: Mitteilungen über die Empididen (Dipt.) Finnlands. VI. *Trichinomyia* gen.n., eine neue Ocydromiinengattung. - Ibid., 25: 103–110.
- 1966: The Ocydromiinae group of subfamilies (Diptera, Empididae). - Ibid., 32: 282–294.

Tuxen, S. L., 1980: The phylogeny of Apterygota and on phylogeny in general. - Boll.Zool., 47 (Suppl.): 27–34.

Ulrich, H., 1971: Zur Skelett- und Muskelanatomie des Thorax der Dolichopodiden und Empididen (Diptera). - Veröff.zool.StSamml.Münch., 15: 1–44.
- 1972: Zur Anatomie des Empididen-Hypopygiums (Diptera). - Ibid., 16: 1-28.
- 1974: Das Hypopygium der Dolichopodiden (Diptera): Homologie und Grundplanmerkmale. - Bonn.zool.Monogr., 5: 1–60.
- 1975: Das Hypopygium von *Chelifera precabunda* Collin (Diptera, Empididae). - Bonn.zool.Beitr., 26: 264–279.

Usachev, D. A., 1968: New jurassic Asilomorpha (Diptera) in Karatau. - Ent.Obozr., 47: 617–628 (in Russian).

Vaillant, F., 1952: Un Empidide destructeur de Simulies. - Bull.Soc.zool.Fr., 76 (1951): 371-379.
- 1953: *Hemerodromia seguyi,* nouvel Empidide d'Algérie destructeur de Simulies. - Hydrobiologia, 5: 180–188.
- 1967: La répartition des *Wiedemannia* dans les cours d'eau et leur utilisation comme indicateurs de zone écologique (Diptera, Empididae). - Annls Limnologie, 3: 267–293.

Verrall, G. H., 1901: Platypezidae, Pipunculidae and Syrphidae, in: British Flies, 8, 812 pp., London.
- 1909: Stratiomyidae and succeeding families of the Diptera Brachycera of Great Britain, in: British Flies, 5, 780 pp., London.

Walker, F., 1837: Notes on Diptera. - Entomologist's mon.Mag., 4: 226–230.
- 1849: List of the specimens of dipterous insects in the collection of the British Museum, vol. 3, pp. 485–687, London.
- 1852: Diptera, in: Saunders, W. W., ed., Insecta Saundersiana, vol. 1, pp. 157-252, 253–414, London.
- 1860: Catalogue of the dipterous insects collected in Amboyna by Mr. A. R. Wallace, with descriptions of new species. - J.Proc.Linn.Soc. (Zoology), 5: 144–168.

Wiedemann, C. R. W., 1828 & 1830: Aussereuropäische zweiflügelige Insekten, I. Theil (1828) XXII + 608 pp., II. Theil (1830) XII + 684 pp., Hamm.

Wilder, D. D., 1974: A revision of the genus *Syneches* Walker (Diptera: Empididae) for North America and the Antilles. - Contrib.Am.ent.Inst., 10 (5): 1-30.

Zetterstedt, J. W., 1838: Insecta Lapponica descripta. Diptera, pp. 477–868, Lipsiae.
- 1842–1859: Diptera Scandinaviae. Disposita et descripta, 1 (1842): 1–440, 8 (1849): 2935–3366, 11 (1852): 4091–4545, 12 (1855): 4547–4942, 13 (1859): 4943–6190, Lund.

Index

Synonyms are given in italics. The number in bold refers to the main treatment of the taxon.

Author's address:

Dr. Milan Chvála
Department of Systematic Zoology
Charles University
Viničná 7
CS-128 44 Praha 2
Czechoslovakia

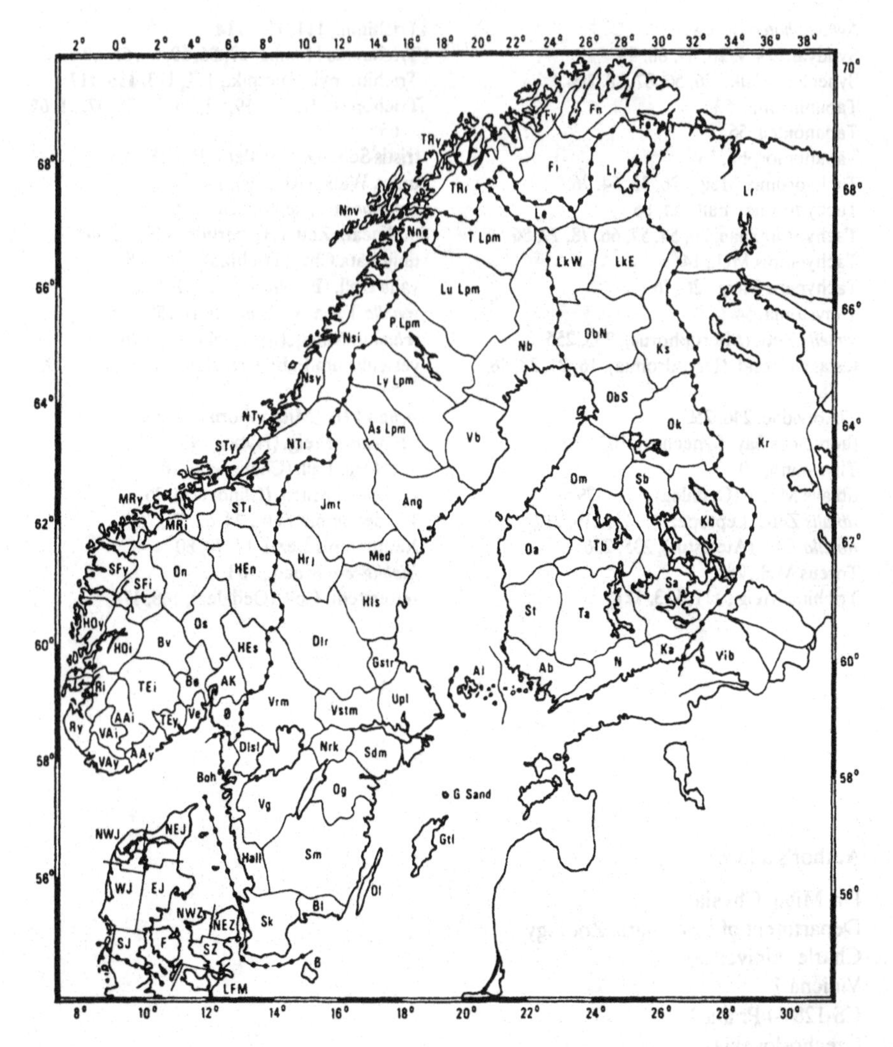

List of abbreviations for the provinces used throughout the text, on the map and in the following tables.

DENMARK

SJ	South Jutland	LFM	Lolland, Falster, Møn
EJ	East Jutland	SZ	South Zealand
WJ	West Jutland	NWZ	North West Zealand
NWJ	North West Jutland	NEZ	North East Zealand
NEJ	North East Jutland	B	Bornholm
F	Funen		

SWEDEN

Sk.	Skåne	Vrm.	Värmland
Bl.	Blekinge	Dlr.	Dalarna
Hall.	Halland	Gstr.	Gästrikland
Sm.	Småland	Hls.	Hälsingland
Öl.	Öland	Med.	Medelpad
Gtl.	Gotland	Hrj.	Härjedalen
G. Sand.	Gotska Sandön	Jmt.	Jämtland
Ög.	Östergötland	Ång.	Ångermanland
Vg.	Västergötland	Vb.	Västerbotten
Boh.	Bohuslän	Nb.	Norrbotten
Dlsl.	Dalsland	Ås. Lpm.	Åsele Lappmark
Nrk.	Närke	Ly. Lpm.	Lycksele Lappmark
Sdm.	Södermanland	P. Lpm.	Pite Lappmark
Upl.	Uppland	Lu. Lpm.	Lule Lappmark
Vstm.	Västmanland	T. Lpm.	Torne Lappmark

NORWAY

Ø	Østfold	HO	Hordaland
AK	Akershus	SF	Sogn og Fjordane
HE	Hedmark	MR	Møre og Romsdal
O	Opland	ST	Sør-Trøndelag
B	Buskerud	NT	Nord-Trøndelag
VE	Vestfold	Ns	southern Nordland
TE	Telemark	Nn	northern Nordland
AA	Aust-Agder	TR	Troms
VA	Vest-Agder	F	Finnmark
R	Rogaland		

n northern   s southern   ø eastern   v western   y outer   i inner

FINLAND

Al	Alandia	Kb	Karelia borealis
Ab	Regio aboensis	Om	Ostrobottnia media
N	Nylandia	Ok	Ostrobottnia kajanensis
Ka	Karelia australis	ObS	Ostrobottnia borealis, S part
St	Satakunta	ObN	Ostrobottnia borealis, N part
Ta	Tavastia australis	Ks	Kuusamo
Sa	Savonia australis	LkW	Lapponia kemensis, W part
Oa	Ostrobottnia australis	LkE	Lapponia kemensis, E part
Tb	Tavastia borealis	Li	Lapponia inarensis
Sb	Savonia borealis	Le	Lapponia enontekiensis

USSR

Vib	Regio Viburgensis	Kr	Karelia rossica	Lr	Lapponia rossica

Printed in the United States
By Bookmasters